T0297738

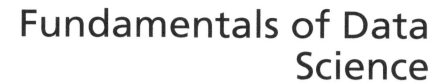

Fundamentals of Data Science

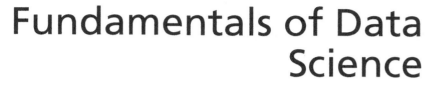

Fundamentals of Data Science
Theory and Practice

Jugal K. Kalita
Dhruba K. Bhattacharyya
Swarup Roy

ACADEMIC PRESS
An imprint of Elsevier

Academic Press is an imprint of Elsevier
125 London Wall, London EC2Y 5AS, United Kingdom
525 B Street, Suite 1650, San Diego, CA 92101, United States
50 Hampshire Street, 5th Floor, Cambridge, MA 02139, United States
The Boulevard, Langford Lane, Kidlington, Oxford OX5 1GB, United Kingdom

Notices

Knowledge and best practice in this field are constantly changing. As new research and experience broaden our understanding, changes in research methods, professional practices, or medical treatment may become necessary.

Practitioners and researchers must always rely on their own experience and knowledge in evaluating and using any information, methods, compounds, or experiments described herein. In using such information or methods they should be mindful of their own safety and the safety of others, including parties for whom they have a professional responsibility.

To the fullest extent of the law, neither the Publisher nor the authors, contributors, or editors, assume any liability for any injury and/or damage to persons or property as a matter of products liability, negligence or otherwise, or from any use or operation of any methods, products, instructions, or ideas contained in the material herein.

ISBN: 978-0-323-91778-0

For information on all Academic Press publications
visit our website at https://www.elsevier.com/books-and-journals

Publisher: Mara Conner
Acquisitions Editor: Chris Katsaropoulos
Editorial Project Manager: Rafael Guilherme Trombaco
Production Project Manager: Erragounta Saibabu Rao
Cover Designer: Christian Bilbow

Typeset by VTeX

Working together
to grow libraries in
developing countries

www.elsevier.com • www.bookaid.org

Dedicated to

To my parents Nirala and Benudhar Kalita
Jugal K. Kalita

To my mother Runu Bhattacharyya
Dhruba K. Bhattacharyya

To His Holiness, Sri Sri Babamoni, and my Loving Family
Swarup Roy

Contents

Preface   xix

Acknowledgment   xxi

Foreword   xxiii

Foreword   xxv

1. Introduction   1

    1.1. Data, information, and knowledge   2

    1.2. Data Science: the art of data exploration   3

       1.2.1. Brief history   3

       1.2.2. General pipeline   4

       1.2.3. Multidisciplinary science   6

    1.3. What is not Data Science?   7

    1.4. Data Science tasks   7

       1.4.1. Predictive Data Science   8

       1.4.2. Descriptive Data Science   8

       1.4.3. Diagnostic Data Science   9

       1.4.4. Prescriptive Data Science   9

    1.5. Data Science objectives   9

       1.5.1. Hidden knowledge discovery   9

       1.5.2. Prediction of likely outcomes   10

       1.5.3. Grouping   10

       1.5.4. Actionable information   10

    1.6. Applications of Data Science   10

    1.7. How to read the book?   12

    References   13

2. Data, sources, and generation   15

    2.1. Introduction   15

    2.2. Data attributes   15

        2.2.1. Qualitative   15

        2.2.2. Quantitative   17

    2.3. Data-storage formats   19

        2.3.1. Structured data   19

        2.3.2. Unstructured data   20

        2.3.3. Semistructured data   20

    2.4. Data sources   21

        2.4.1. Primary sources   21

        2.4.2. Secondary sources   21

        2.4.3. Popular data sources   22

        2.4.4. Homogeneous vs. heterogeneous data sources   23

    2.5. Data generation   24

        2.5.1. Types of synthetic data   25

        2.5.2. Data-generation steps   25

        2.5.3. Generation methods   26

        2.5.4. Tools for data generation   27

    2.6. Summary   29

    References   30

3. Data preparation   31

    3.1. Introduction   31

    3.2. Data cleaning   32

        3.2.1. Handling missing values   32

        3.2.2. Duplicate-data detection   35

    3.3. Data reduction   36

        3.3.1. Parametric data reduction   36

3.3.2. Sampling   36

3.3.3. Dimensionality reduction   37

3.4. Data transformation   37

3.4.1. Discretization   38

3.5. Data normalization   41

3.5.1. Min–max normalization   41

3.5.2. Z-score normalization   42

3.5.3. Decimal-scaling normalization   42

3.5.4. Quantile normalization   42

3.5.5. Logarithmic normalization   43

3.6. Data integration   43

3.6.1. Consolidation   44

3.6.2. Federation   44

3.7. Summary   44

References   44

4. Machine learning   47

4.1. Introduction   47

4.2. Machine Learning paradigms   48

4.2.1. Supervised learning   49

4.2.2. Unsupervised learning   52

4.2.3. Semisupervised learning   53

4.3. Inductive bias   53

4.4. Evaluating a classifier   55

4.4.1. Evaluation steps   56

4.4.2. Handling unbalanced classes   57

4.4.3. Model generalization   58

4.4.4. Evaluation metrics   61

4.4.5. Hypothesis testing   66

4.5. Summary   67

References   68

5. Regression   69

5.1. Introduction   69

5.2. Regression   70

    5.2.1. Linear least-squares regression   70

5.3. Evaluating linear regression   74

    5.3.1. Coefficient of determination R^2 74

    5.3.2. Standard error of regression and F-statistic   75

    5.3.3. Is the model statistically significant?   76

5.4. Multidimensional linear regression   76

5.5. Polynomial regression   78

5.6. Overfitting in regression   81

5.7. Reducing overfitting in regression: regularization   81

    5.7.1. Ridge regression   82

    5.7.2. Lasso regression   84

    5.7.3. Elastic-net regression   85

5.8. Other approaches to regression   86

5.9. Summary   88

References   88

6. Classification   91

6.1. Introduction   91

6.2. Nearest-neighbor classifiers   93

    6.2.1. Storing classified examples   94

    6.2.2. Distance measure   95

    6.2.3. Voting to classify   96

6.3. Decision trees   97

    6.3.1. Building a decision tree   100

6.3.2. Entropy for query construction   101

6.3.3. Reducing overfitting in decision trees   104

6.3.4. Handling variety in attribute types   106

6.3.5. Missing values in decision tree construction   107

6.3.6. Inductive bias of decision trees   107

6.4. Support-Vector Machines (SVM)   108

6.4.1. Characterizing a linear separator   108

6.4.2. Formulating the separating hyperplane   110

6.4.3. Maximum margin classifier   110

6.4.4. Formulating a maximum-margin classifier   111

6.4.5. Solving the maximum-margin optimization problem   115

6.4.6. SVM in multiclass classification   116

6.5. Incremental classification   117

6.5.1. Incremental decision trees   117

6.6. Summary   118

References   119

7. Artificial neural networks   121

7.1. Introduction   121

7.2. From biological to artificial neuron   121

7.2.1. Simple mathematical neuron   122

7.2.2. Perceptron model   123

7.2.3. Perceptron learning   124

7.2.4. Updating perception weights   125

7.2.5. Limitations of the perceptron   125

7.3. Multilayer perceptron   126

7.4. Learning by backpropagation   127

7.4.1. Loss propagation   128

7.4.2. Gradient descent for loss optimization   130

7.4.3. Epoch of training   134

7.4.4. Training by batch   134

7.5. Loss functions   135

7.5.1. Mean-squared loss   135

7.5.2. Crossentropy loss   136

7.6. Activation functions   138

7.6.1. Binary step function   139

7.6.2. Sigmoid activation   139

7.6.3. *tanh* activation   140

7.6.4. ReLU activation   141

7.7. Deep neural networks   141

7.7.1. Convolutional neural networks   145

7.7.2. Encoder–decoder architectures   150

7.7.3. Autoencoders   152

7.7.4. Transformer   157

7.8. Summary   159

References   159

8. Feature selection   161

8.1. Introduction   161

8.1.1. Feature extraction vs. feature selection   162

8.2. Steps in feature selection   163

8.2.1. Generation of feature subsets   164

8.2.2. Feature-subset evaluation   166

8.2.3. Feature-selection methods   166

8.3. Principal-component analysis for feature reduction   174

8.3.1. Summary   177

References   178

9. Cluster analysis   181

   9.1. Introduction   181

   9.2. What is cluster analysis?   181

   9.3. Proximity measures   183

      9.3.1. Standard measures   185

      9.3.2. Statistical measures   185

      9.3.3. Divergence measures   186

      9.3.4. Kernel similarity measures   187

   9.4. Exclusive clustering techniques   188

      9.4.1. Partitional clustering   190

      9.4.2. Hierarchical clustering   193

      9.4.3. Density-based clustering   197

   9.5. High-dimensional data clustering   202

      9.5.1. Dimensionality reduction and feature selection   203

      9.5.2. Projected clustering   203

      9.5.3. Subspace clustering   204

   9.6. Biclustering   206

      9.6.1. Biclustering techniques   207

   9.7. Cluster-validity measures   208

      9.7.1. External evaluation   209

      9.7.2. Internal evaluation   210

      9.7.3. Relative evaluation   211

   9.8. Summary   212

   References   212

10. Ensemble learning   215

   10.1. Introduction   215

      10.1.1. What is ensemble learning?   215

      10.1.2. Building an ensemble   217

        10.1.3. Categories of ensemble learning   217

    10.2. Ensemble-learning framework   218

        10.2.1. Base learners   220

        10.2.2. Combination learners   220

    10.3. Supervised ensemble learning   224

        10.3.1. Requirements for base-classifier selection   224

        10.3.2. Ensemble methods   225

    10.4. Unsupervised ensemble learning   227

    10.5. Semisupervised ensemble learning   229

    10.6. Issues and challenges   231

    10.7. Summary   231

    References   231

11. Association-rule mining   233

    11.1. Introduction   233

    11.2. Association analysis: basic concepts   233

        11.2.1. Market-basket analysis   234

        11.2.2. Sources of market-basket databases   235

        11.2.3. Interestingness measures   236

        11.2.4. Association-rule mining process   237

    11.3. Frequent itemset-mining algorithms   238

        11.3.1. Apriori algorithm   239

        11.3.2. Candidate generation   240

        11.3.3. Cost of the Apriori algorithm   241

        11.3.4. FP-growth   242

    11.4. Association mining in quantitative data   245

        11.4.1. Partitioning approach   246

        11.4.2. Clustering approach   246

	11.4.3. Information-theoretic approach	248
	11.5. Correlation mining	250
	11.6. Distributed and parallel association mining	254
	11.6.1. Count distribution	254
	11.6.2. Data distribution	255
	11.6.3. Parallel-rule generation	255
	11.7. Summary	256
	Acknowledgment	257
	References	257
12.	Big Data analysis	259
	12.1. Introduction	259
	12.2. Characteristics of Big Data	260
	12.3. Types of Big Data	261
	12.4. Big Data analysis problems	262
	12.5. Big Data analytics techniques	263
	12.6. Big Data analytics platforms	264
	12.7. Big Data analytics architecture	266
	12.7.1. MapReduce architecture	266
	12.7.2. Fault-tolerant graph architecture	267
	12.7.3. Streaming-graph architecture	269
	12.8. Tools and systems for Big Data analytics	269
	12.8.1. Bioinformatics tools	269
	12.8.2. Computer-vision tools	271
	12.8.3. Natural language processing tools	272
	12.8.4. Network-security tools	272
	12.9. Active challenges	273
	12.10. Summary	273
	References	274

13. Data Science in practice   277

    13.1. Need of Data Science in the real world   277

    13.2. Hands-on Data Science with Python   278

        13.2.1. Necessary Python libraries   279

        13.2.2. Loading the dataset   280

        13.2.3. A quick look at the dataset   280

        13.2.4. Checking dataset header   281

        13.2.5. Dimensions of the dataset   281

    13.3. Dataset preprocessing   281

        13.3.1. Detecting nonnumeric columns   282

        13.3.2. Encoding nonnumeric columns   282

        13.3.3. Detecting missing values   283

        13.3.4. Checking the class distribution   283

        13.3.5. Separating independent and dependent variables   283

    13.4. Feature selection and normalization   283

        13.4.1. Correlation among features   284

        13.4.2. Normalizing the columns   285

        13.4.3. Viewing normalized columns   285

    13.5. Classification   286

        13.5.1. Splitting the dataset   286

        13.5.2. Logistic regression   287

        13.5.3. Support-vector machine   287

        13.5.4. Artificial neural network (ANN)   288

        13.5.5. Predictions on test data with a high-performing
                ANN model   288

        13.5.6. Evaluating the model   288

        13.5.7. Performance measurement   289

        13.5.8. Curve plotting   289

    13.6. Clustering   290

13.6.1. K-means—using the elbow method 290

13.6.2. Fitting the data 291

13.6.3. Validation with labels 291

13.6.4. Agglomerative clustering—using dendrograms to find optimal clusters 292

13.6.5. Finding optimal clusters 292

13.6.6. Fitting the data 293

13.6.7. Validation with labels 293

13.7. Summary 293

References 294

14. Conclusion 295

Index 297

Preface

Data science is an evolving area of study that is extensively used in solving real-life problems that are empirical in nature. It is not just about machine learning, statistics, or databases. Instead, it is a comprehensive study of topics that help extract novel knowledge from data, starting with preparing the data, applying suitable intelligent learning models, and interpreting the outcome. The models applied are not "one-size-fits-all" and vary with the nature of the data and the applications under consideration. This book presents basic as well as advanced concepts in Data Science, along with real-life applications. We believe this book will provide students, researchers, and professionals at multiple levels with a good understanding of Data Science, Machine Learning, Big Data Mining, and analytics.

The overwhelming and growing need for data scientists in the industry has created the demand for extensive, easy-to-understand study materials that cover basic concepts to state-of-the-art applications of the concepts of Data Science in different domains. This book includes discussions of theoretical and practical approaches in Data Science, aiming to produce a solid understanding of the field with the ultimate aim of novel knowledge discovery. It contains an indepth discussion on some essential topics for Data Science projects. It starts with an introduction to data types, sources, and generation pipelines. It presents a detailed discussion of various data-preparation steps and techniques. Prepared data are then analyzed through predictive and descriptive data-analysis techniques. The book includes a systematic presentation of many predictive and descriptive learning algorithms, including recent developments that have successfully handled large datasets with high accuracy. It presents a detailed discussion of Machine-Learning techniques and evaluation methods. Several chapters cover regression analysis, classification techniques, artificial neural networks, and deep-learning models. The cluster-analysis chapter starts with various proximity measures, followed by an introduction to several clustering techniques. The chapter also covers biclustering techniques with cluster-validation measures. The chapter on association-rule mining introduces the basic idea of frequent itemset mining, and a new topic on qualitative and correlation association mining. A unique feature of this book is a chapter on Big Data analysis and Ensemble Learning as emphasized topics. A dedicated chapter on how Data Science methods can be implemented using popular Python libraries is also included in the penultimate chapter.

This book will not only be useful for beginners, it will also become a handbook for those already working in Data Science, Data Mining, and related areas.

The book is organized into 13 chapters. Chapter 1 introduces the basic concepts of Data Science. Chapter 2 deals with data types, storage formats, data sources, and data-generation techniques. A detailed discussion of data preparation or preprocessing techniques is presented in Chapter 3. The basic concept of Machine Learning is presented in Chapter 4. Chapter 5 is dedicated to Regression Analysis. Chapter 6 is on the topic of Classification. Artificial Neural Networks (ANN) and Deep Learning are currently popular and effective techniques, and are discussed in Chapter 7. Classification requires a "best set" of features to work optimally. Feature-Selection techniques presented in Chapter 8 discuss how to find the "optimal" feature set. Chapter 9 deals with Cluster Analysis. Ensemble Learning is a unique topic introduced in Chapter 10. Association-rule mining, another data-analytics technique, is introduced in Chapter 11. Big Data analysis is another unique topic that is presented in Chapter 12. Finally, the book ends with a significant chapter on applications of Data Science in real life and a brief hands-on tutorial on Data Science using Python.

Acknowledgment

Writing a book is not an easy task, and it is impossible without the help and support of many individuals. With heartfelt gratitude, we express our sincere appreciation to all those who were directly and indirectly involved and helped in the writing of the book.

We are grateful to Elsevier, Inc, for considering the project worthy of publication. Thanks to Chris Katsaropoulos, Rafael G. Trombaco, and their team for translating the dream into reality through proposal finalization, progress tracking, and final production. It would never have been possible to complete the project without their support and help. We are grateful to all the reviewers for their valuable suggestions.

We acknowledge the valuable help and support from Mr. Binon Teji (JRF, NETRA Lab-Sikkim University), Ms. Upasana Sharma, and Mr. Parthajit Borah (PhD students of Tezpur University) during the preparation of some chapters. We thank Prof Dinabandhu Bhandari (Heritage Institute of Technology, Kolkata), Ms. Softya Sebastian (JRF, NETRA Lab-Sikkim University), and Dr. Keshab Nath (IIIT-Kottyam) for their wonderful support.

We are grateful to the University of Colorado at CS, USA, Tezpur University, and Sikkim University for providing a conducive environment and infrastructure facilities on the University premises.

While writing the book, we referred to several related works in this domain. We acknowledge all the authors whose work has helped prepare this book.

Last but not least, we are grateful to the Almighty for everything. To our families and loved ones, thank you for your unwavering support, encouragement, and understanding throughout the writing process. Your love and encouragement have kept us motivated and focused on this project.

Readers are the most important part of any book. We extend our deepest gratitude to the readers of this book. We hope that this work will provide you with a comprehensive understanding of the theoretical and practical aspects of Data Science and that it will serve as a valuable resource for your professional development.

Jugal K. Kalita
Dhruba K. Bhattacharyya
Swarup Roy

Foreword

Data (-driven) science research is concerned with tasks such as acquisition, storage, retrieval, processing, reasoning, mining, and finally the conversion of data into knowledge. It involves a huge challenge. It employs techniques/theories drawn from several fields such as mathematics, statistics, information science, and computer science, on top of application-domain knowledge. The subject is highly multidisciplinary. Today in the digital world, this discipline involving AI and machine-learning techniques has high potential from an R & D perspective, both in academia and industry worldwide. The demand and popularity of the subject makes it sound sometimes like a buzzword too. Although the scenario is extremely encouraging with respect to job opportunity and initiative, and its necessity has been felt strongly for more than a decade in big-data analytics, there is still a scarcity of deep analytics globally.

The present volume "Fundamentals of Data Science: Theory and Practice", coauthored by Professors Jugal Kalita, Dhruba Bhattacharyya, and Swarup Roy, is timely in that context that provides a user-friendly and application-oriented document to help understand the basics of the subject. After introduction of data science, the book starts with chapters like data generation and preprocessing of raw data. These are followed by techniques of machine learning, regression, predictive and descriptive learning, ensemble learning, classification, clustering, artificial neural networks and deep learning, feature selection, and association-rule mining. Finally, two chapters on big-data analytics, and data science in practice using Python are included.

The authors of the volume deserve congratulations for their long research contributions in AI, machine learning, and related areas, and bringing out the book of enormous archival value.

Prof. Sankar K. Pal   July 2023
FNA, FASc, FTWAS, FIEEE, FIAPR
National Science Chair, Govt. of India,
President, Indian Statistical Institute

Foreword

Data science plays a crucial role in various fields by utilizing statistical and computational methods to extract insights and knowledge from massive amounts of data using various scientific approaches. It helps in making informed decisions, predicting future trends, improving business operations, and enhancing customer experience. Data science is now more important than ever as most data nowadays are either semistructured or unstructured and they even take the form of networks, photos, audio, and videos. Therefore, sophisticated computational methods that can handle vast volumes of such heterogeneous data are needed.

Data science is crucial because of the numerous applications it may be used for, ranging from simple activities, such as asking Siri or Alexa for recommendations, to more sophisticated ones, such as operating a self-driving automobile. Data science is used in industry to assist organizations in comprehending the patterns, including client information, business growth rates, data volume, etc. In the healthcare sector, data on patient demographics, treatment plans, medical exams, insurance, etc., can be processed, analyzed, assimilated, and managed.

Generative artificial intelligence is a key component and will likely be the most powerful tool that data scientists will have to work with in the coming years. This book, comprising of 14 chapters (including introductory and concluding chapters), presents topics that are necessary for a data scientist to extract useful information, given voluminous raw data. While some books briefly touch upon data preprocessing, this book devotes an entire chapter to this critical phase. The authors have delved into data cleaning, reduction, transformation, integration, and normalization, addressing the often-overlooked intricacies that profoundly impact the quality of analysis and results. The chapter on Data Sources and their generation process is one of the unique chapters and breaks new ground by offering a comprehensive exploration of data-generation techniques, including the often-overlooked synthetic data creation. It provides a thorough understanding of data attributes, sources, and storage formats, setting the stage for robust data preparation. This book also goes beyond the fundamentals by exploring advanced subjects such as deep learning, ensemble learning, big-data analytics, autoencoders, and transformer learning models.

The authors have also illustrated recent developments on large datasets, accurately focusing on cutting-edge techniques to tackle real-world challenges. "Data Science in Practice" provides readers with practical insights and tools essential for the modern data scientist. This chapter offers a hands-on guide to data science using Python. A book like these matters because it teaches us along with our students, how to do something that is both challenging and useful in this modern world. I am sure that readers will enjoy the contents and thank you Jugal, Dhruba, and Swarup for this wonderful work. All the best!

Prof. Ajith Abraham   August 15, 2023
Pro Vice Chancellor, Bennett University, India.
Founding Director-Machine Intelligence Research (MIR) Labs, USA (2008–2022).
Editor in Chief-Engineering Applications of Artificial Intelligence (EAAI), Elsevier (2016–2021).
http://www.softcomputing.net

1

Introduction

"The secret of business is to know something that nobody else knows."
— **Aristotle Onassis**

Consumer satisfaction is a fundamental performance indicator and a key element of an enterprise's success. The success of any enterprise relies on its understanding of customer expectations and needs, buying behaviors, and levels of satisfaction. Modern giant business houses analyze customer expectations and perceptions of the quality and value of products to make effective decisions regarding product launch and update, servicing, and marketing.

Due to the availability of fast internet technologies and low-cost storage devices, it has become convenient to capture voluminous amounts of consumer opinions and records of consumer activities. However, discovering meaningful consumer feedback from a sea of heterogeneous sources of reviews and activity records is just like finding a needle in a haystack. Data Science appears to be the savior in isolating novel, unknown, and meaningful information that helps proper decision making.

Data Science is the study of methods for programming computers to process, analyze, and summarize data from various perspectives to gain revealing and impactful insights and solve a vast array of problems. It is able to answer questions that are difficult to address through simple database queries and reporting techniques. Data Science aims to address many of the same research questions as statistics and psychology, but with differences in emphasis. Data Science is primarily concerned with the development, accuracy, and effectiveness of the resulting computer systems. Statistics seek to understand the phenomena that generate the data, often with the goal of testing different hypotheses about the phenomena. Psychological studies aspire to understand the mechanisms underlying the behaviors exhibited by people such as concept learning, skill acquisition, and strategy change.

Google Maps is a brilliant product developed by Google, Inc., using Data Science to facilitate easy navigation. But how does it work? It collects location data continuously from different reliable heterogeneous sources, including GPS locations via mobile phones of millions of users who keep their location services on. It captures location, velocity, and itinerary-related data automatically. Efficient Data Science algorithms are applied to the collected data to predict traffic jams and road hazards, the shortest routes, and the time to reach the destination. Massive quantities of collected past, current, and near current traffic data help Google predict real-time traffic patterns.

Fundamentals of Data Science. https://doi.org/10.1016/B978-0-32-391778-0.00008-9

1

1.1 Data, information, and knowledge

To introduce the arena of Data Science, it is of utmost importance to understand the data-processing stack. Data Science-related processing starts with a collection of raw data. Any facts about events that are unprocessed and unorganized are called *data*. Generally, data are received raw and hardly convey any meaning. Data, in their original form, are useless until processed further to extract hidden meaning. Data can be (i) operational or transactional data such as customer orders, inventory levels, and financial transactions, (ii) nonoperational data, such as market-research data, customer demographics, and financial forecasting, (iii) heterogeneous data of different structures, types, formats such as MR images and clinical observations, and (iv) metadata, i.e., data about the data, such as logical database designs or data dictionary definitions.

Information is the outcome of processing raw data in a meaningful way to obtain summaries of interest. To extract information from data, one has to categorize, contextualize, and condense data. For example, information may indicate a trend in sales for a given period of time, or it may represent a buying pattern for customers in a certain place during a season. With rapid developments in computer and communication technologies, the transformation of data into information has become easier. In a true sense, Data Science digs into the raw data to explore hidden patterns and novel insights from the data.

Knowledge represents the human understanding of a subject matter, gained through systematic analysis and experience. Knowledge results from an integration of human perspectives and processes to derive meaningful conclusions. Some researchers [5] define knowledge with reference to a subject matter from three perspectives, (i) understanding (know-why), (ii) cognition or recognition (know-what), and (iii) capacity to act (know-how). Knowledge in humans can be stored only in brains, not in any other media. The brain has the ability to interconnect it all together. Unlike human beings, computers are not capable of understanding what they process, and they cannot make independent decisions. Hence, computers are not artificial brains! While building knowledge, our brain is dependent on two sources, i.e., data and information. To understand the relationship between data and information, consider an example. If you take a photograph of your house, the raw image is an example of data. However, details of how the house looks in terms of attributes such as the number of stories, the colors of the walls, and its apparent size, constitute information. If you send your photograph via email or message to your friend, you are actually not sending your house or its description to your friend. From the photograph, it is up to your friend, how he/she perceives its appearance or looks. If it so happens that the image is corrupted or lost, still your original house will be retained. Hence, even if the information is destroyed, the data source remains.

The key concepts of data, information, and knowledge are often illustrated as a pyramid, where data are the starting point placed at the base of the pyramid (see Fig. 1.1), and it ends in knowledge generation. If we collect knowledge from related concepts, domains, and processes further, it gives rise to wisdom. We skip discussions on wisdom as the concept is highly abstract and controversial and difficult to describe. Usually, the sizes of repositories to store data, information, and knowledge become smaller as we move upward

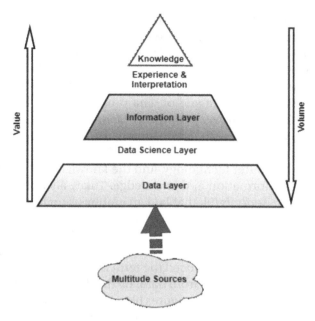

FIGURE 1.1 Data, Information, and Knowledge pyramid, and intermediate conversion layers. The directions of the arrowheads indicate increase in size and importance.

in the pyramid, where data in their original form have lower importance than information and knowledge. It is worth mentioning that quality raw data lead to more significant information and knowledge generation. Hence, good-quality data collection is a stepping stone for effective information and knowledge mining.

1.2 Data Science: the art of data exploration

Data Science is a multifaceted and multidisciplinary domain dedicated to extracting novel and relevant patterns hidden inside data. It encompasses mathematical and statistical models, efficient algorithms, high-performance computing systems, and systematic processes to dig inside structured or unstructured data to explore and extract nontrivial and actionable knowledge, ultimately being useful in the real world and having an impact.

The success of Data Science depends on many factors. The bulk of efforts have concentrated on developing effective exploratory algorithms. It usually involves mathematical theories and expensive computations to apply the theory to large-scale raw data.

1.2.1 Brief history

The dawn of the 21st century is known as the *Age of Data*. Data have become the new fuel for almost every organization as references to data have infiltrated the vernacular of various communities, both in industry and academia. Many data-driven applications have become amazingly successful, assisted by research in Data Science. Although *Data Science* has become a buzzword recently, its roots are more than half a century old. In 1962, John Wilder Tukey, a famous American mathematician published an article entitled *The Future Of Data Analysis* [8] that sought to establish a science focused on learning from data. After six years, another pioneer named Peter Naur, a Danish computer scientist introduced the term *Datalogy* as the science of data and of data processes [6], followed by the publication of a book in 1974, *Concise Survey of Computer Methods* [7], that defined the term *Data Science* as the science of dealing with data. Later, in 1977, The International Association for Statistical Computing (IASC) was founded with a plan for linking traditional statistical methodology, modern computer technology, and the knowledge of domain experts in order to convert data into information and knowledge. Tukey also published a major work entitled, *Exploratory Data Analysis* [9], that laid an emphasis on hypothesis testing during data analysis, giving rise to the term *data-driven discovery*. Following this, the first Knowledge Discovery in Databases (KDD) workshop was organized in 1989, becoming the annual ACM SIGKDD Conference on Knowledge Discovery and Data Mining (KDD).[1]

Later, in 1996, Fayyad et al. [2] introduced the term *Data Mining*, the application of specific algorithms for extracting patterns from data. By the dawn of the 2000s, many journals started recognizing the field and notable figures like William S. Cleveland, John Chambers, and Leo Breiman expanded boundaries of statistical modeling, envisioning a new epoch in statistics focused on Data Science [1].

The term *Data Scientist* was first introduced in 2008 by Dhanurjay Patil and Jeff Hammerbacher of LinkedIn and Facebook [10].

1.2.2 General pipeline

Data Science espouses a series of systematic steps for converting data into information in the form of patterns or decisions. Data Science has evolved by borrowing concepts from statistics, machine learning, artificial intelligence, and database systems to support the automatic discovery of interesting patterns in large data sets. A Data Science pipeline is made of the following four major phases. An illustrative representation of a typical Data Science workflow [4] is depicted in Fig. 1.2.

1.2.2.1 Data collection and integration

Data are initially collected, and integrated if collection involves multiple sources. For any successful Data Science and -analysis activity, data collection is one of the most important steps. The quality of collected data carries great weight. If the collected samples are not sufficient to describe the overall system or process under study, downstream activities are likely to become useless despite employing sophisticated computing methods. The quality of the outcome is highly dependent on the quality of data collection.

It has been observed that dependence on a single source of data is always precarious. Integration of multifaceted and multimodal data may offer better results than working

---

[1] https://www.kdnuggets.com/gpspubs/sigkdd-explorations-kdd-10-years.html.

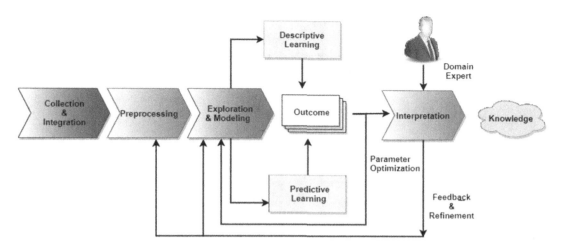

FIGURE 1.2 Major steps in Data Science pipeline for decision making and analysis.

with a single source of information. In fact, information from one source may complement those from other sources when one source of data is not sufficient to understand a system or process well. However, the integration of multisource data itself is a challenging task and needs due attention. Integration should be deliberate rather than random mixing to deliver better results.

1.2.2.2 Data preparation
Raw data collected from input sources are not always suitable for downstream exploration. The presence of noise and missing values, and the prevalence of nonuniform data structures and standards may negatively affect final decision making. Hence, it is of utmost importance to prepare the raw data before downstream processing. Preprocessing also filters uninformative or possibly misleading values such as outliers.

1.2.2.3 Learning-model construction
Different machine learning models are suitable for learning different types of data patterns. Iterative learning via refinement is often more successful in understanding data distributions. A plethora of models are available to a data scientist and choices must be made judiciously. Models are usually used to explain the data or extract relevant patterns to describe the data or predict associations.

1.2.2.4 Knowledge interpretation and presentation
Finally, results need to be interpreted and explained by domain experts. Each step of analysis may trigger corrections or refinements that are applied to the preceding steps.

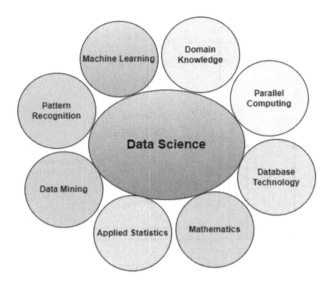

FIGURE 1.3 Data Science joins hands with a variety of other disciplines.

1.2.3 Multidisciplinary science

Data Science is a multidisciplinary domain of study to extract, identify, and analyze novel knowledge from raw data by applying computing and statistical tools, together with domain experts for the interpretation of outcomes. It involves mathematical and statistical tools for effective data analysis and modeling, pattern recognition and machine learning to assist in decision making, data and text mining for hidden pattern extraction, and database technologies for effective large data storage and management (Fig. 1.3). Due to the complex nature of the data elements and their relationships, most often, understanding the data itself is challenging. Before understanding the underlying distribution of data elements, it may not be very fruitful to apply any statistical or computational tools for knowledge extraction. Visualization may play a large role in deciphering the interrelationships among the data elements, thereby helping decide the appropriate computational models or tool for subsequent data analysis. The presence of noise in the data may be discovered and eliminated by looking into distribution plots. However, it is a well-known fact that visualizing multidimensional data itself is challenging and needs special attention. With the availability of low-cost data-generation devices and fast communication technologies, "Big Data"or vast amount of data have become ubiquitous. Dealing with Big Data for Data Science needs high-performance computing platforms. The science and engineering of parallel and distributed computing are important discipline that need to be integrated into the Data Science ecosystem. Recently, it has become convenient to integrate parallel computing due to the wide availability of relatively inexpensive Graphical Processing Units (GPU). Last but not least, knowledge of and expertise in the domain in which Data Science approaches are applied play major roles during problem formulation and interpretation of solutions.

1.3 What is not Data Science?

In recent years, the term Data Science has become a buzzword in the world of business and intelligent computing. As usual, high demand and popularity invite misinterpretation and hype. It is important to be aware of the terms that are used as well as misused in the context of Data Science.

Machine Learning is not a branch of Data Science. It provides the technology or the tools to facilitate smart decision making using the software. Data Science uses Machine Learning as a tool for autonomous pattern analysis and decision making.

There is a prevalent fallacy that techniques of Data Science are applicable to only a very large amounts of data or so-called Big Data. This is not true, as even smaller amounts of data can also be analyzed usefully. The quality of the data in hand and the completeness of the data are always important. It is true that a Machine Learning system is likely to be able to extract more accurate knowledge when large amounts of relevant data are used to draw intuitions about the underlying patterns.

It is true that statistical techniques play a great role in effective data analysis. Statistics complements and enhances Data Science [3] for efficient and effective analysis of large collections of data. Statistics use mathematical models to infer and analyze data patterns by studying data distributions of collected samples from the past. However, Data Science cannot be considered as completely dependent on statistics alone. Statistics is used mostly to describe past data, whereas Data Science performs predictive learning for actionable decision making. A number of nonparametric Data Science learning models help understand the data very well without knowing the underlying data distributions.

Last but not least, people often give more importance to scripting languages, such as Python or R, and tools available for ready use rather than understanding the theory of Data Science. Of course, knowledge of tools greatly helps in developing intelligent systems quickly and effectively. Without understanding the models and formalisms, quite often many users concentrate on using the readily and freely available preimplemented models. Knowing only prescribed or programmed tools is not sufficient to have a good overall understanding of Data Science so that existing tools can be adapted and used efficiently to solve complex problems. Proficiency in data-analysis tools without deeper knowledge of data analysis does not make for a good data scientist.

1.4 Data Science tasks

Data Science-related activities are broadly classified into predictive and descriptive tasks. The former deals with novel inferences based on acquired knowledge and the latter describes the inherent patterns hidden inside data. With the rise in business-analysis applications, the span of Data Science tasks has extended further into two related tasks, namely diagnostic and prescriptive. Somewhat simplistic, but differentiating, views of the four tasks can be obtained by asking four different questions: "What is likely to happen?" (Predictive), "What is happening?" (Descriptive), "Why is it happening?" (Diagnostic) and "What do I need to do?" (Prescriptive), respectively.[2]

1.4.1 Predictive Data Science

Predictive tasks apply supervised Machine Learning to predict the future by learning from past experiences. Examples of predictive analysis are classification, regression, and deviation detection. Some predictive techniques are presented below. **Classification** attempts to assign a given instance into one of several prespecified classes based on behaviors and correlations of collected and labeled samples with the target class. A classifier is designed based on patterns in existing data samples (training data). The trained model is then used for inferring the class of unknown samples. The overall objective of any good classification technique is to learn from the training samples and to build accurate descriptions for each class. For example, spam filtering separates incoming emails into safe and suspicious emails based on the signatures or attributes of the email. Similar to classification, **prediction** techniques infer the future state based on experiences from the past. The prime difference between classification and prediction models is in the type of outcome they produce. Classification assigns each sample to one of several prespecified classes. In contrast, prediction outcomes are continuous valued as prediction scores. The creation of predictive models is similar to classification. The prediction of the next day's or next week's weather or temperatures is a classic example of a prediction task based on observations of the patterns of weather for the last several years in addition to current conditions. **Time-series data analysis** predicts future trends in time-series data to find regularities, similar sequences or subsequences, sequential patterns, periodicities, trends, and deviations. For example, predicting trends in the stock values for a company based on its stock history, business situation, competitor performance, and current market.

1.4.2 Descriptive Data Science

Descriptive analysis is also termed exploratory data analysis. Unlike predictive models, descriptive models avoid inference, but analyze historical or past data at hand and present the data in a more interpretable way such that it may better convey the underlying configuration of the data elements. Leveraging powerful visualization techniques usually enhances descriptive analysis for better interpretation of data. A descriptive task may act as the starting point to prepare data for further downstream analysis. Clustering, association, and sequence mining are some descriptive analysis techniques.

Clustering is an unsupervised technique that groups data into meaningful clusters such that data within a cluster possess high similarity, but low similarity with data in other clusters. A similarity or distance measure is needed to decide the inclusion of a data sample in a particular cluster. The basic difference between classification and clustering is that classification assumes prior knowledge of class labels assigned to data examples, whereas clustering does not make such an assumption. For example, clustering can be used to separate credit-card holders into groups, such that each group can be treated with a different strategy. **Association-rule mining** finds interesting associations among data elements. The market-basket analysis is the initial application where association rule mining was extensively used. Such market-basket analysis deals with studying the buying behaviors of customers. It can be used by business houses in creating better marketing strategies, logistics plans, inventory management, and business promotions. **Summarization** is an effective data-abstraction technique for the concise and comprehensive representation of raw data. With summarization, data may become better interpretable because irrelevant details are removed. Text summarization is a popular technique for generating a short, coherent description of a large text document.

1.4.3 Diagnostic Data Science

The diagnostic analysis builds on the outcomes of descriptive analysis to investigate the root causes of a problem. It includes processes such as data discovery, data mining, drilling down, and computing data correlations to identify potential sources of data anomalies. Probability theory, regression analysis, and time-series analysis are key tools for diagnostic analysis. For example, Amazon can drill down the sales and profit numbers to various product categories to identify why sales have gone down drastically in a certain span of time in certain markets.

1.4.4 Prescriptive Data Science

This is the most recent addition to analysis tasks. A prescriptive model suggests guidelines or a follow-up course of action to achieve a goal. It uses an understanding of what has happened, why it has happened, and what might happen to suggest the best possible alternatives. Unlike predictive analysis, prescriptive analysis does not need to be perfect but suggests the "best" possible way toward the future. Google Maps is an excellent prescriptive model that suggests the best route from the current location to the destination on the fly, considering traffic conditions and the shortest route.

1.5 Data Science objectives

Data Science aims to achieve four basic objectives: (i) to extract interesting patterns in data without human intervention, (ii) to predict the most likely future outcomes based on past data, (iii) to create actionable knowledge to achieve certain goals, and (iv) to focus on how to handle voluminous data in achieving the previous three objectives.

1.5.1 Hidden knowledge discovery

To discover hidden knowledge, Data Science builds models. A model is built using an algorithm that operates on a data set. It is a computational structure that learned from collected and pre-processed data and used to solve data analytics tasks with possibly previously unseen data. To facilitate the automatic extraction of hidden patterns, the data scientist executes appropriate data-mining models. Data Science models are typically used to explore the types of data for which they are built. Many types of models can be adapted to new data.

1.5.2 Prediction of likely outcomes

One way Data Science can help predict likely outcomes is by generating rules, which are in antecedent-consequent form. Data scientists use statistical and machine-learning models to analyze data and identify patterns, trends, and relationships that can be used to predict future outcomes. The goal is to use data to gain insights and make accurate predictions that can help businesses and organizations make better decisions. The ability to make accurate predictions is critical for businesses and organizations to stay competitive and make informed decisions. Data Science plays a vital role in enabling organizations to leverage their data to gain insights and make better predictions about future outcomes. Identifying potential fraudulent activities by analyzing patterns and anomalies in financial transactions or predicting future sales based on historical sales data, customer demographics, and other factors are some examples of prediction outcomes.

1.5.3 Grouping

Another major objective of Data Science is to obtain natural groupings in data to extract interesting patterns. An example is finding a group of employees in an organization that is likely to benefit from certain types of investments based on income, age, years of work, temperament, and desired investment goals.

1.5.4 Actionable information

Data Science can handle voluminous data, popularly termed Big Data (as alluded to earlier), and can extract relevant information, which can be of direct help in the decision-making process. For example, a bank may use a predictive model to identify groups of clients with a net worth above a predefined threshold such that they are likely to be receptive to investing large amounts of money in certain high-risk, high-reward business initiatives being proposed.

1.6 Applications of Data Science

Thanks to low-cost high-performance computing devices, superfast internet technologies, and ample cloud storage, it is now possible for business houses to rely on intelligent decision making based on large amounts of data. There is hardly any reputable organization in the world that is not leveraging Data Science for smart decision making. In a nutshell, Data Science is concerned with two broad endeavors. The first is the effort to make devices or systems smart and the other involves deciphering data generated by natural or engineered systems and learning from them. A few of the many potential application domains are introduced next.

- **Healthcare:** Recent advances in Data Science are a blessing to healthcare and allied sectors. Data Science has positively and extensively impacted upon these application areas, in turn helping mankind significantly. Health informatics and smart biomedical devices are pushing medical and health sciences to the next level. Precision medicine is likely to be a game changer in extending the human life span. Computer-vision and biomedical technologies are making quick and accurate disease diagnosis ubiquitous. Even a handheld smartphone may be able to continuously monitor the health metrics of a person and generate smart early alarms when things go wrong.

- **Computational Biology:** With the availability of high-throughput omics data, it is now possible to understand the genetic causes of many terminal diseases. Computational biology and bioinformatics mine massive amounts of genomic and proteomic data to better understand the causes of diseases. Once the causes are identified with precision, it becomes possible to develop appropriate drug molecules for treatment. On average, traditional drug development requires more than 14 years of effort, which can now be reduced drastically due to effective Data Science techniques. Precision medicine is the future of drug technology, customizing drugs for the individual and avoiding the one-size-fits-all approach that does not always work.

- **Business:** The rise of Data Science was originally intended to benefit business sectors. With the need for business intelligence, the use of data analytics has gained momentum. Almost every business venture is investing significant resources in smart business decision making with Data Science. Analyzing customer purchasing behavior is a great challenge, and important for improving revenue and profit. Integration of heterogeneous data is an effective way to promote sales of products. Data-analysis experts apply statistical and visualization tools to understand the moods, desires, and wants of customers and suggest effective ways for business and product promotion and plans for progress and expansion.

- **Smart Devices:** A mobile device is no longer just a communication device, but rather a miniature multipurpose smart tool that enhances the lifestyle of the common man or woman. Apps installed in smart devices like smartphones or smart tablets can be used to understand a person well in regard to their choices, preferences, likes, and dislikes. Technology is becoming so personalized that installed apps in such devices will be able to predict a user's actions in advance and make appropriate recommendations. In the near future, smart devices will be able to monitor user health status, recommend doctors, book appointments, place orders for medicine, and remind users of medicine schedules. The Internet-of-Things (IoT) and speaking smart devices make our life easier. For example, a home automation system can monitor and control an entire home remotely with the help of a smartphone, starting from kitchen to home security.

- **Transportation:** Another important application area is smart transportation. Data Science is actively involved in making it possible to take impressive steps toward safe and secure driving. The driverless car will be a big leap into the future in the automobile sector. The ability to analyze fuel-consumption patterns and driving styles and monitor vehicle status makes it possible to create optimized individual driving experiences, spurring new designs for cars by manufacturers. A transportation company like Uber uses Data Science for price optimization and offers better riding experiences to customers.

In addition to the above, the list of applications is huge and is growing every day. There are other industry sectors like banking, finance, manufacturing, e-commerce, internet, gaming, and education that also use Data Science extensively.

1.7 How to read the book?

To read this book, follow the Table of Contents to gain a comprehensive understanding of the key concepts in Data Science. Start with the Introduction, which provides an overview of Data Science and its applications. This chapter will give you an understanding of the scope of the book and its objectives. Next, move on to Chapter 2, titled Data, Sources, and Generation. Here, you will learn about the different types of data and sources, as well as methods for generating data. This chapter lays the groundwork for the rest of the book and is essential for understanding the data used in the subsequent chapters. Chapter 3, Data Preparation, discusses the process of cleaning, transforming, and preprocessing data before analysis. This chapter covers important topics like data normalization and missing-data imputation. Chapter 4, Introduction to Machine Learning, provides an overview of the key concepts in machine learning, including supervised and unsupervised, and semisupervised learning, as well as different model-evaluation metrics. Chapter 5, Regression, focuses on regression analysis and covers techniques for building and evaluating regression models. Chapter 6, Classification, covers the key concepts in classification, including several classification algorithms and techniques for evaluating classification models. Chapter 7, Artificial Neural Networks, introduces the basics of artificial neural networks, including feedforward and convolutional neural networks. Chapter 8, Feature Selection and Extraction, covers techniques for selecting and extracting the most relevant features from a dataset. Chapter 9, Cluster Analysis, focuses on unsupervised learning and covers key concepts in cluster analysis. Chapter 10, Ensemble Learning, covers techniques for combining multiple learning algorithms to improve predictive performance. Chapter 11, Association Rule Mining, covers techniques for identifying patterns and relationships in data. Chapter 12, Big-Data Analysis, focuses on the unique challenges and opportunities presented by Big Data and covers techniques for analyzing large datasets. Chapter 13, Data Science in Practice, provides an overview of Data Science in real-life applications and a demonstration of experimenting with Data Science using Python. Finally, the Conclusion chapter summarizes the key concepts covered in the book.

To get the most out of this book, it is recommended to read each chapter in order and to practice the techniques covered in each chapter using real-world datasets. However, depending on acedemic and professional need, a reader may jump to a chapter of choice.

The text of each chapter is written in a manner suitable for a motivated learner to follow in isolation.

In the next chapter we begin with the journey of Data Science on how data can be collected, generated, and stored.

References

[1] Longbing Cao, Data science: a comprehensive overview, ACM Computing Surveys (CSUR) 50 (3) (2017) 1–42.

[2] Usama Fayyad, Gregory Piatetsky-Shapiro, Padhraic Smyth, From data mining to knowledge discovery in databases, AI Magazine 17 (3) (1996) 37–37.

[3] Hossein Hassani, Christina Beneki, Emmanuel Sirimal Silva, Nicolas Vandeput, Dag Øivind Madsen, The science of statistics versus data science: what is the future?, Technological Forecasting & Social Change 173 (2021) 121111.

[4] Jayanta Kumar Das, Giuseppe Tradigo, Pierangelo Veltri, Pietro H. Guzzi, Swarup Roy, Data science in unveiling COVID-19 pathogenesis and diagnosis: evolutionary origin to drug repurposing, Briefings in Bioinformatics 22 (2) (2021) 855–872.

[5] Anthony Liew, Understanding data, information, knowledge and their inter-relationships, Journal of Knowledge Management Practice 8 (2) (2007) 1–16.

[6] Peter Naur, 'Datalogy', the science of data and data processes, in: IFIP Congress (2), 1968, pp. 1383–1387.

[7] Peter Naur, Concise Survey of Computer Methods, Petrocelli Books, 1974.

[8] John W. Tukey, The future of data analysis, The Annals of Mathematical Statistics 33 (1) (1962) 1–67.

[9] John W. Tukey, Exploratory Data Analysis, Pearson, Washington, DC, 1977 [Google Scholar].

[10] Bradley Voytek, Social media, open science, and data science are inextricably linked, Neuron 96 (6) (2017) 1219–1222.

Data, sources, and generation

2.1 Introduction

The modern world generates enormous amounts of data almost every day. According to one source,[1] *2.5 quintillion bytes* of data are being created every day. It is expected that by the end of 2023, more than *100 zettabytes* of data will need to be stored and handled across the world.[2] Such massive amounts of data are generated primarily due to widespread access to the World Wide Web and social-media platforms. Facebook generates 4 petabytes of data per day. A single person can create up to 1.7 MB of data every second. Google processes over 40 000 searches every second, or 3.5 billion searches daily. Through video platforms such as Zoom, 3.3 trillion minutes are spent in meetings yearly. One hour of a meeting can create data ranging from 800 MB to 2.47 GB if recorded in video.

It is of utmost importance to dig into such large volumes of data to extract hidden novel and nontrivial patterns in the form of knowledge. Often, this process of knowledge discovery is termed *Knowledge Discovery in Databases (KDD)*, and the intermediate step toward KDD that requires extraction of hidden patterns is popularly called *Data Mining*. Mining such a large volume of data necessitates efficient storage and processing. Many types, shapes, and formats of data must be ultimately stored in binary form to be processed by computing devices.

Below, we discuss the basic data types or attributes used for storage and processing.

2.2 Data attributes

The study and use of programming languages or database-management systems require dealing with data types or data attributes. A data attribute defines a data element using a single descriptor with specific characteristics. In statistics, it is also called a *measurement scale*. The way of handling each attribute varies with its nature. Hence, it is a necessary as well as important to understand attribute types (Fig. 2.1) before processing. Below, we discuss major attribute types or variables with examples to facilitate understanding.

2.2.1 Qualitative

As the name suggests, qualitative attributes are data types that cannot be measured using standard measuring units; rather, they can be observed, compared, and classified.

---

[1] https://seedscientific.com/how-much-data-is-created-every-day.

[2] https://bernardmarr.com/how-much-data-is-there-in-the-world/, Access Date: 19th July 2022.

Fundamentals of Data Science. https://doi.org/10.1016/B978-0-32-391778-0.00009-0

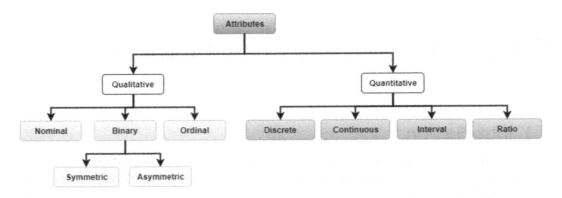

FIGURE 2.1 Attribute types in databases.

2.2.1.1 Nominal

A nominal attribute represents a specific symbol or the name of an entity; it is also called a *categorical attribute*. It is a qualitative attribute that cannot be ordered or ranked. No meaningful relationship can be established among the values. Only the logical or "is equal" operator can be applied among the values. Categorical values cannot be dealt with directly for various Data Science tasks. For example, neural networks are not built to handle categorical data straightforwardly. A few examples of nominal attributes are given in Table 2.1.

Table 2.1 Examples of nominal attributes and values.

Attribute	Values
Color	Red, Blue, Green
City	Moscow, Kolkata, New York
Name	John, Harry, Ron

2.2.1.2 Binary

A binary attribute holds discrete values with exactly two mutually exclusive states, either 1 (true) or 0 (false). A person is either infected by a virus or not infected. Both states never occur at the same time.

A binary attribute is said to be a *symmetric binary attribute* if both states possess equal importance. In such a case, it hardly matters which attribute value, such as 0 or 1, to assign. On the other hand, an *asymmetric binary attribute* does not give equal importance to both states. Table 2.2 gives a few examples of binary attributes.

2.2.1.3 Ordinal

Attribute values that can be ordered or ranked are called ordinal attributes. It is worth mentioning that magnitude-wise, the difference between ordered values may not be known or

Table 2.2 Examples of binary attribute types.

Attribute	Values	Type
Gender	Male, Female	Symmetric
Status	Present, Absent	Asymmetric

may not even be significant. The ordering exhibits the importance of the attribute value but does not indicate how important. For example, the allowable values for an attribute may be Good, Better, and Best. One may rank the values based on their importance without being clear about how Good differs from Better or Best. A few more examples are listed in Table 2.3.

Table 2.3 Examples of ordinal attribute and values.

Attribute	Values
Grade	A+, A, B+, B, C+, C
Position	Asst. Prof., Assoc. Prof., Prof.
Qualification	UG, PG, PhD

2.2.2 Quantitative

Unlike Qualitative attributes, Quantitative attributes are measurable quantities. The values are quantified using straightforward numbers. Hence, they are also called *numerical attributes*. Statistical tools are easily applicable to such types of attributes for analysis. They can also be represented by data-visualization tools like pie charts, line and bar graphs, scatter plots, etc. The values may be countable or noncountable. Accordingly, they are of the two following types: Table 2.4 shows some quantitative attributes.

2.2.2.1 Discrete
Countable quantitative attributes are also called discrete attributes. Usually, if the plural of the attribute name may be prefixed with "the number of," it can be considered a discrete value attribute. Examples of discrete attributes includes the number of persons, cars, buildings, population of a city, etc. They are denoted with whole numbers or integer values and can't have fraction.

2.2.2.2 Continuous
Continuous values are noncountable and measurable quantities with infinite possible states. They are represented with fractional or floating point values. Attributes such as Weight, Height, Price, Blood Pressure, Temperature, etc., hold continuous values.

Whether discrete or continuous type, quantitative attributes can be of the following two types.

2.2.2.3 Interval

Like ordinal attributes, quantitative attributes (both discrete and continuous) can be ordered. The distinction comes from whether differences or intervals among the ordered values are known or not. Interval values do not have any correct references or true zero points. The zero point is the value at which no quantity or amount of the attribute is measured. In other words, it is the point on the scale that represents the absence of the attribute. Values can be added and subtracted from the list of attribute values, but individual values cannot be multiplied or divided. For example, temperature intervals (Table 2.4), such as 10°C to 20°C, represents a range of values within which temperature can vary. The temperatures for two consecutive days may differ by a certain degree, but we cannot say there is "no temperature" (zero-point).

2.2.2.4 Ratio

Ratio is another quantitative attribute representing a relative or comparative quantity relating to two or more values. Unlike intervals, ratios have fixed zero points. A fixed zero point is a reference point on a fixed scale that does not vary based on the measured attribute. In other words, it is a point on the scale representing a constant value, regardless of the quantity or amount of the attribute being measured. Ratio-scale attributes can also be ordered, but with a perfect zero. However, no negative value is allowed in ratio-scale attributes. If a value is ratio scaled, it implies a multiple (or ratio) of another value. Parametric measures such as mean, median, mode, and quantile range are considered ratio-scale attributes. Family Dependents are considered Ratio data because they possess all the properties of interval data and, in addition, have a meaningful zero point. As given in Table 2.4, Family Dependents are considered Ratio data because they possess all the properties of interval data and, in addition, have a meaningful zero point. In this case, a Family Dependents value of 0 indicates the absence of family dependents, which represents a true absence and not just an arbitrary point on a scale. We can say that having 3 dependents is 50% more than having 2 dependents, or having 2 dependents is twice as many as having 1 dependent. Additionally, we can perform meaningful ordering of the values and apply mathematical operations like addition and multiplication on Family Dependents values, making them Ratio data.

Table 2.4 Quantitative attributes.

Attribute	Values	Type
Population	5000, 7680, 2500	Discrete
Price	3400.55, 6600.9	Continuous
Temperature	10°C–20°C, 20°C–30°C	Interval
Family Dependents	0, 2, 3	Ratio

How can data be stored effectively? This depends on the nature and source of data generation. Appropriate treatment is necessary to store the data in any data repository and retrieve data from the repository. Accordingly, appropriate mechanisms of data retrieval and mining need to be devised for the same. Available data-storage formats are discussed next.

2.3 Data-storage formats

A data structure is a specific way to organize and store data in a computer so that the data may be accessed and changed effectively. It is a grouping of data values of various types that preserves the logical dependencies among the data elements and the functions or operations that apply to the stored data for retrieval, access, and update.

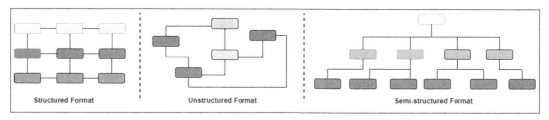

FIGURE 2.2 Three data formats. A node represents an entity or entity set, and an edge shows logical dependency. The color of a node indicates the characteristics of the entity. Similar color depicts the same entity or the same label for data representation.

The nature of the data dictates how data must be stored in a repository or database. Data may be stored in the following formats.

2.3.1 Structured data

Structured data can be either quantitative or qualitative. Structured data refers to data that have been organized into a structured format, such as a table or a spreadsheet, with a clear schema or data model that defines the relationships between different data elements. Structured data can be easily analyzed since such data follow a predefined data model. A data model, such as relational, object-oriented, or hierarchical, defines how data can be stored, processed, and accessed. For instance, in the relational data model, data are usually stored in row–column or tabular format. Excel files and relational databases are common examples of structured data. Structured data can easily be arranged in a specific order for effective searching. Every data field is distinct and can be accessed independently or in conjunction with information from other fields. In the illustrated structured format shown in Fig. 2.2, each row depicts a collection of attributes or field values that belong to a particular record. A node contains a specific value for a specific record. Nodes are connected since the nodes belong to a single record with related attributes. As it is feasible to aggregate data from many areas in the database swiftly, structured data is incredibly powerful. It is one of the most common forms of data storage. Popular database-management systems (DBMS) follow a structured storage format for quick and efficient data access.

2.3.2 Unstructured data

Unlike structured data, unstructured data do not follow any specific format or adhere to a predefined model. Data or entities are associated arbitrarily due to logical association or dependency among them. As shown in Fig. 2.2, such data can be viewed as a graph or network of data attributes (nodes). The lack of uniform structure leads to anomalies and ambiguities that make it more challenging to process such data using conventional programming platforms. Unstructured data are usually qualitative, including text, video, audio, images, reports, emails, social-media posts, graph data, and more. NoSQL databases nowadays provide a popular platform to store unstructured data. Storage and processing of unstructured data are growing needs as most of the current data-generation sources do not adhere to any rigid framework. In comparison to structured data, unstructured data formats are highly scalable, leading to the need for large or Big Data storage and processing.

2.3.3 Semistructured data

Another form of data arrangement is between structured and unstructured formats. Semistructured data represent a form of unstructured data that is loosely or partially structured. Although such data do not conform to specific data models as such, the containing data can be categorized with the help of tags or other markers incorporating certain semantic elements. As shown in Fig. 2.3, attribute categorization may enforce a hierarchical arrangement composed of records and fields within the data. This format is also known as self-describing. A majority of the large-data sources produce semistructured data. Unlike the other two formats, it is relatively less complex and easy to analyze. Semistructured data analysis has become popular from a Big-Data perspective. Popular data-storage and -transaction formats such as JSON and XML are built for communication with semistructured data. To understand this better, we consider email as an example of semistructured data arrangement. Data components may be categorized as Sender, Recipient, Subject, Date, etc., and stored hierarchically.

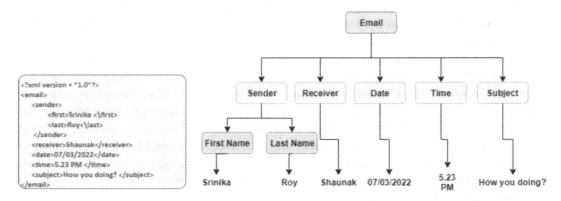

FIGURE 2.3 Snippet of XML code and semistructured arrangement of an email entity.

2.4 Data sources

The location from where the data are sourced is called the data source. The database, which may be kept on a local hard drive or a distant server, is the main data source for database-management systems. A computer program may use a file, a data sheet, a spreadsheet, an XML file, or even hard-coded data into the program itself as its data source. Sources are of two types, considering how data are collected or generated. A simple illustration (Fig. 2.4) may be helpful to show how they are related to each other.

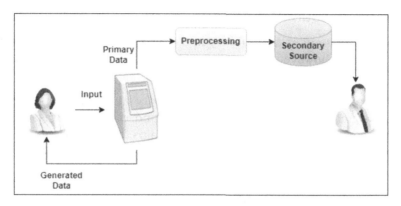

FIGURE 2.4 Transition from Primary to Secondary Data.

2.4.1 Primary sources

First-hand data are generated directly by data-generation sources and stored without major alteration or preprocessing. They can also be called original or raw data. Original and first-hand collection increases the reliability and authenticity of data. However, reliability further depends on how the data are produced and how reliable the source is. Primary-data sources help retrieve hidden and novel facts. Generating primary data is an expensive and time-consuming activity. Once analyzed, primary data can be further processed to produce secondary data.

For instance, data collected directly by a researcher through a social survey or directly from IoT devices are primary data. Similarly, raw nucleotide sequences generated and recorded using a Next-Generation Sequencer are also primary data.

2.4.2 Secondary sources

Secondary data sources are repositories where the data have been derived from primary sources and are stored after curation. Usually, scientists generate the data and make them available for public use. The data may be passed through several rounds of experiments and statistical analysis. Although not generated directly, a secondary source ensures reliability and validity if it has tested the data. Secondary data are good for quick and cost-effective validation of a new system before deploying it or testing with original data. The suitability of available secondary data depends on the experimental purpose at hand. For instance, someone may want to validate her newly designed IoT intrusion detection system (IDS). Assume further that various computer-network attack examples are available but no examples of IoT attacks. In such a case, validating the IoT IDS with such data may not be appropriate. Either she needs to generate primary data or use further preprocessed available secondary data to fit.

2.4.3 Popular data sources

Many secondary data sources are publicly and privately available for Data Science research. The list is large; however, this section highlights a few popular and significant data repositories below.

UCI ML Repository: The machine-learning repository at the University of California, Irvine (UCI), is a great place to find open-source and cost-free datasets for machine-learning experiments, covering a wide range of domains from biology to particle physics. It houses 622 datasets for machine-learning research. The datasets contain attributes that are of Categorical, Integer, and Real types. The datasets have been curated and cleaned. The repository includes only tabular data, primarily for classification tasks. This is limiting for anyone interested in natural language, computer vision, and other types of data. Data sizes are small and not suitable for Big-Data analysis.

Kaggle: This platform, recently acquired by Google, publishes datasets for machine-learning experiments. It also offers GPU-integrated notebooks to solve Data Science challenges. At present, it houses approximately 77k datasets. Kaggle fosters advancements in machine learning through open community competitions. The datasets on Kaggle are divided into three categories: general-public datasets, private-competition datasets, and public-competition datasets. Except for the private datasets, all others are free to download. A majority of datasets are based on real-life use cases.

Awesome Public Datasets: This is a public repository of datasets covering 30 topics and tasks in diverse domains hosted in GitHub. The large datasets can be used to perform Big-Data-related tasks. The quality and uniformity of the datasets are high.

NCBI: This is an open-access large collection of databases for exploratory molecular-biology data analysis provided by the National Center for Biotechnology Information (NCBI) in the USA. Databases are grouped into six categories: Literature, Health, Genomes, Genes, Proteins, and Chemicals. It is a rich source of sequences, annotations, metadata, and other data related to genes and genomes.

SNAP: The Stanford Network Analysis Platform (SNAP) [5] offers large network datasets. It covers large graphs derived from 80+ categories of domains, such as social networks, citation networks, web graphs, online communities, and online reviews. Libraries implementing more than 140 graph algorithms are available for download in SNAP. These algorithms can effectively process the characteristics and metadata of nodes and edges, manipulate massive graphs, compute structural properties, produce regular and random graphs, and generate nodes and edges.

MNIST: The Modified National Institute of Standards and Technology (MNIST) dataset is a large collection of hand-written digits, popular among deep-learning beginners and the image-processing community. The MNIST database contains 60 000 training images and 10 000 testing images. Extended MNIST (EMNIST) [2] is a more refined database with 28×28 pixel-sized images. EMNIST is a dataset with balanced and unbalanced classes. MNIST is a very popular dataset for machine-learning experiments.

VoxCeleb Speech Corpus: This is a large collection of audiovisual data for speaker recognition [6]. It includes more than one million real-world utterances from more than 6000 speakers. The dataset is available in two parts, VoxCeleb1 and VoxCeleb2. In VoxCeleb1, there are over 100 000 utterances for 1251 celebrities; in VoxCeleb2, there are over 1 million utterances for more than 6000 celebrities taken from YouTube videos. Different development and test sets are segregated with different speakers.

Google has developed a dedicated dataset search engine, **Google Dataset Search**, that helps researchers search and download publicly available databases. Google claims that its Dataset Search engine has indexed about 25 million datasets, and one may obtain useful information about them. To locate these datasets in the search results, Google Dataset Search leverages *schema.org* and other metadata standards.

2.4.4 Homogeneous vs. heterogeneous data sources

It is important to consult multiple reliable data sources for effective data analytics and smart decision making. However, reliability is difficult to ensure for data obtained from a single arbitrary source. In addition, data obtained from a single source may be incomplete, biased, and/or unbalanced. The alternative is to amalgamate similar data from multiple sources and make decisions based on majority agreement.

Considering multiple datasets, if they are uniform and similar to one another with reference to data sources, data types, and formats, the sources are called *homogeneous data sources.* Homogeneous data sources can be combined easily without much additional processing. For example, Walmart may want to analyze countrywide sales patterns. It needs to integrate its transaction databases across different stores located in different places. Since various stores use a common database structure, it is easy to integrate large collection transactions for analysis.

On the contrary, heterogeneous data sources are sources with high variability of data types and formats. It is difficult to integrate heterogeneous data sources without preprocessing necessary to transform the datasets to a uniform format. This may sometimes lead to a loss of information while integrating. For instance, devices on the Internet of Things (IoT) frequently produce heterogeneous data. The heterogeneity of IoT data sources is one of the biggest challenges for data processing. This is due to the variety of data acquisition devices and the lack of common data-acquisition, -structuring, and -storage protocols. Four different types of data heterogeneity have been identified [4].

- **Syntactic heterogeneity:** This appears when data sources are created using different programming languages with varying structures.
- **Conceptual heterogeneity:** This is also known as semantic or logical heterogeneity. It represents the differences that arise while modeling the data in various sources within the same domain of interest.
- **Terminological heterogeneity:** Different data sources referring to the same entities with different names lead to such heterogeneity.
- **Semiotic heterogeneity:** This is also known as pragmatic heterogeneity due to varying interpretations of the same entity by different users.

2.5 Data generation

Researchers and data scientists frequently encounter circumstances where they either lack suitable real-world (primary or secondary) data or may be unable to use such data due to closed access, privacy issues, and/or litigation potential. The alternative solution in such cases is to use artificially generated datasets produced using *synthetic-data generation.* Synthetic-data generation is the process of creating artificial data as a replacement for real-world data, either manually or automatically, using computer simulations or algorithms. If the original data are unavailable, "fake" data may be created entirely from scratch or seeded with a real dataset. A realistic and statistically well-distributed synthetic dataset that works well with the target Data Science algorithms is desired while replacing real data to make the algorithms learn and behave as close to the real system as possible.

Concepts such as data augmentation and randomization are related to creating synthetic data. However, conceptually, there is a difference between them. *Data augmentation* involves adding slightly altered copies of already-existing dataset samples. Using data augmentation, a dataset can be expanded with nearly identical samples but with slight differences in data characteristics. This is useful when adequate data samples are scarce. For example, adding new face images into a face database by altering the orientation, brightness, and shape of the faces augments the face database. In contrast, *data randomization* does not produce new data elements; it simply changes slightly the items already present in the dataset. With randomization, data attributes or features may be altered in some samples in the dataset. Randomization helps protect sensitive data by making it harder for unauthorized persons to identify or infer personal information from the data. By randomizing the order of the data, any patterns or relationships that could be used to identify individuals are disrupted. For example, a clinical dataset of cancer patients may be randomized by altering a few selected features or attributes of the real data in order to hide the patient's personal details.

On the other hand, synthetic data allow us to create entirely new data elements similar to real data. However, such data may not accurately represent the real data in its entirety. In essence, synthetic data duplicate some patterns that may already exist in reality, introducing such properties without explicitly representing them.

The advantages of using synthetic data come from cost and time effectiveness, better data-labeling accuracy because the labels are already established, scalability (it is simple to produce enormous amounts of simulated data), and controlled data distribution and variability. Data samples can be generated for exceptional scenarios where such events do not occur regularly in the real world, however high the possibility of occurrence in the near future. Usually, real-data samples are mimicked to generate additional data samples with similar distribution and nature. An appropriate mathematical or statistical model is built by observing the real data for this purpose.

2.5.1 Types of synthetic data

The intention behind synthetic-data generation is to provide alternatives to real data when real data are unavailable or have privacy, intellectual property, or other similar issues. Based on the amount of artificiality or syntheticity in the generated data, they can be classified into two types.

- **Fully Synthetic:** Data that are fully synthetic and have no direct relationship to actual data. Even though all the necessary characteristics of real data are present, it is likely that there is no real individual or example that corresponds to a generated data example. When creating fully synthetic data, the data generator often finds the density function of the features in the real data and estimates the appropriate parameters for data generation. The bootstrap method and multiple approaches to imputation are common ways of producing fully synthetic data.
- **Partially Synthetic:** All information from the real data is retained in partially synthetic data, with the exception of sensitive information. Only selected sensitive feature values are replaced with synthetic values in such data. Most real values are likely to be retained in the carefully curated synthetic-data collection because they are extracted from the real data. Multiple imputation methods and model-based procedures are used for producing partially synthetic data.

2.5.2 Data-generation steps

A basic data generation step may follow five major steps. The typical workflow of the data generation pipeline is shown in Fig. 2.5.

1. **Determine Objective:** The first step is to understand and establish the objectives for the planned synthetic-data generation and the techniques to be used later for data analysis. Further, one needs to understand organizational, legal, and other restrictions because data privacy is currently a major concern. For instance, if one needs clinical data for AIDS patients for analysis, confidentiality is a serious issue.
2. **Select Generator:** Next, one needs to select a data-generation model. For simulation, appropriate technical knowledge and adequate infrastructure are necessary. For example, generating artificial IoT traffic data using Monte Carlo simulation requires knowledge of various model hyperparameters to be tuned.

3. **Collect Sample Data:** Most synthetic data generators need real-data samples to learn the underlying probability distribution. The availability of carefully collected real-data samples determines the quality of the generated synthetic data.
4. **Train Model:** The selected data-generation model needs to be trained using collected real-data samples by tuning the hyperparameters. The target is to make the model understand well the sample data and create data close to real-data samples.
5. **Generate Data:** Once trained, synthetic datasets of any size can be generated using the trained model.
6. **Evaluate the Data:** Assessing the quality of generated synthetic data is important to ensure usability. The best way to evaluate this is to feed the data to data-analysis algorithms, learn a trained model, and test its quality with real-data samples. If the system performs according to expectation, the generated data can be used or archived. In the case of underperformance, the error may be fed back to the synthetic-data-generation model for further tuning.

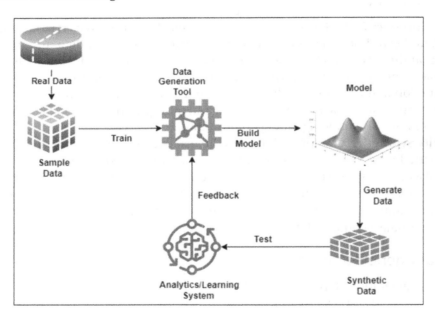

FIGURE 2.5 Synthetic-Data-Generation Workflow.

2.5.3 Generation methods

To generate synthetic data, the following techniques may be used.

- **Statistical Distribution:** The idea behind using a statistical distribution for synthetic-data generation [7] is first to learn the underlying probability distribution for real-data examples. The quality of synthetic data to be produced depends on sampling. If sampling is biased, naturally, the outcome may be ill-distributed and may not represent real data. There may be scenarios where no data samples are available. In such scenarios, expert judgment and experience matter greatly. A data scientist is able to generate random samples by following a probability distribution such as Gaussian, or Chi-square distribution.

- **Agent Modeling:** First, an agent data generator is modeled to behave as if in real scenarios [1]. Next, it uses a modeled agent for random fake-data generation. It may perform curve fitting to fit real data to a known distribution. Later, a suitable simulation method like Monte Carlo can be used for synthetic-data generation. A machine-learning model like a decision tree (discussed later in Chapter 6) can also be used as an alternative to fit the observed distribution. However, a proper selection of the machine-learning model is important as every model has its strengths and limitations for a good fit to the data distribution. A hybrid approach may also be adapted for the same. A part of the synthetic data may be produced using a statistical distribution approach, and the rest can be generated using agent modeling.

- **Deep Neural Networks:** In recent years, deep neural-network-based techniques (introduced in Chapter 7) have become popular and effective for generating bulk and close-to-real data. Advanced machine-learning approaches, including deep models, can learn relevant features for a variety of data. Neural networks are especially well suited for creating synthetic data through trial and error methods. They can learn to duplicate the data and generalize to create a representation that may have characteristics of real data. Two popular deep neural architectures include *Variational Autoencoders* (VAE) and *Generative Adversarial Networks* (GAN).

  VAEs [8] are members of the autoencoder family and represent an unsupervised technique. They are highly effective generative models that are able to learn the underlying distribution of the original data. The primary objective of a VAE is to guarantee that input and output data remain strikingly identical. GAN [3] is also an unsupervised technique that belongs to a generative family. It is an interesting model with two important components, a discriminator and a generator that work in an adversarial fashion. The generator is responsible for creating synthetic data, whereas the job of the discriminator is to detect the generated data as fake. By detecting synthetic data as fake, the generator gradually improves itself so that the synthetic-data examples become close to real-data examples, and undetectable by the discriminator as fake.

2.5.4 Tools for data generation

A plethora of synthetic data generators is available. They may be primarily categorized into two types: ready-to-use software and data-generator libraries. On the one hand, software tools are easy to use with limited or no knowledge of coding, but on the other hand, good, ready-to-use tools are usually not free. Moreover, other than parameter setting, no external control may be possible during data generation when using such a tool. In the case of software libraries, one needs the knowledge of coding so as to obtain better usability, customization, and control of synthetic-data generation. Discussed below are a few software tools or packages for generating data in different domains.

2.5.4.1 Software tools

A few examples of data-generation tools used in various domains where Data Science is commonly used are listed below.

- **Banking:** A good number of ready-to-use software tools are available to generate data for the banking, insurance, and fintech sectors. A few notable ones are *Hazy*,[3] *Datomize*,[4] *Mostly.AI*,[5] and *Facteus*.[6] The tools generate raw banking data for fintech industries. They provide alternative ways to generate banking customer data because real organizations in this sector have severe restrictions in terms of privacy and confidentiality in data storage and sharing. Synthetic data are generated by avoiding situations that pose the risk of fraud, as identified when gathering actual consumer data.
- **Healthcare:** Availability of patient data enables data scientists to design personalized care plans. Due to issues of privacy, security, and ownership, access to healthcare data is highly restricted. *MDClone*[7] is a software tool that provides a systematic approach to democratizing the availability of healthcare data for research, analytics, and synthesis while learning to code protecting sensitive data. It creates fake patient data based on real statistical characteristics of patients. Similarly, *SyntheticMass*[8] aims to statistically recreate data reflecting the actual population in terms of demographics, disease burdens, vaccinations, medical visits, and social determinants.
- **Computer Vision:** CVEDIA[9] offers a synthetic-data-generation tool for computer vision. CVEDIA's *SynCity*, allows users to create synthetic environments, generate synthetic data, and simulate scenarios for various use cases such as autonomous vehicles, robotics, and augmented and virtual reality. *Rendered.AI*[10] creates synthetic datasets for the satellite, autonomous vehicle, robotics, and healthcare industries. For example, using it one can recreate satellite-like images.

2.5.4.2 Python libraries

The Python language has become popular for Data Science research and development, and several Python-based data-generation libraries are freely available. It is important to choose the right Python library for the type of data that needs to be generated.

---

[3] https://hazy.com/.

[4] www.datomize.com.

[5] https://mostly.ai/.

[6] www.facteus.com.

[7] www.mdclone.com.

[8] https://synthea.mitre.org/.

[9] https://www.cvedia.com/.

[10] https://www.rendered.ai/.

- **Scikit-Learn:** *sklearn* is a popular library that can be used to generate fake data for various Data Science experiments. It can generate data for multiclass classification, clustering, regression analysis, and many other tasks. It can also create data examples similar to data samples drawn from real data.
- **Numpy:** This is another library that can be used to generate random data with various probabilistic distributions. *Numpy* and *sklearn* need to be used together for data-generation and -benchmarking activities.
- **Pydbgen:** This is a lightweight library to generate random data examples and export in popular data formats such as Pandas data frame, SQLite table, or MS Excel.
- **Sympy:** This library can be used to create datasets with complex symbolic expressions for regression and classification tasks.
- **Synthetic Data Vault (SDV):** This library can produce anonymous datasets based on the characteristics of real data. The key capabilities are modeling relational, time-series, single tables, and benchmarking data. The need for a sizable real dataset to train SDV models is a drawback that stands out.

Listed in Table 2.5 is a summary of various Python libraries for data generation. Like Python, R^{11} is also a commonly used programming language in Data Science. *bindata*, *charlatan*, *fabricatR*, *fakeR*, *GenOrd*, *MultiOrd*, and *SimMultiCorrData* are example of packages in R that can generate fake data.

Table 2.5   Python Synthetic-Data-Generation Libraries and capabilities.

Library	Activity
TimeseriesGenerator, SDV	Time-series data generator
Mesa	Framework for building, analyzing, and visualizing agent-based models
Zpy, Blender	Computer vision and 3D creation suites
Faker, Pydbgen, Mimesis	Fake database generation
DataSynthesizer	Simulates a given dataset

2.6 Summary

To start studying Data Science and learn about the use and practice of Data Science, it is of utmost importance to understand well the nature of data, data-storage formats, and sources of data availability. This chapter has presented a careful discussion of these topics. Real data are not always available for use due to various reasons. The solution is to artificially generate data that mimic real-life data. Various data-generation methods, basic steps of generation, and popular tools and libraries have also been discussed. The quality of data decides the outcome of data analytics. Quality data collection, preparation, and generation is an important, yet challenging task. The next chapter is dedicated to data-preparation processes.

---

[11] https://www.r-project.org/.

References

[1] Eric Bonabeau, Agent-based modeling: methods and techniques for simulating human systems, Proceedings of the National Academy of Sciences 99 (suppl_3) (2002) 7280–7287.

[2] Gregory Cohen, Saeed Afshar, Jonathan Tapson, Andre Van Schaik, EMNIST: extending MNIST to handwritten letters, in: 2017 International Joint Conference on Neural Networks (IJCNN), IEEE, 2017, pp. 2921–2926.

[3] Maayan Frid-Adar, Eyal Klang, Michal Amitai, Jacob Goldberger, Hayit Greenspan, Synthetic data augmentation using GAN for improved liver lesion classification, in: 2018 IEEE 15th International Symposium on Biomedical Imaging (ISBI 2018), IEEE, 2018, pp. 289–293.

[4] Václav Jirkovský, Marek Obitko, Semantic heterogeneity reduction for big data in industrial automation, in: ITAT, vol. 1214, 2014.

[5] Jure Leskovec, Rok Sosič, SNAP: a general-purpose network analysis and graph-mining library, ACM Transactions on Intelligent Systems and Technology (TIST) 8 (1) (2016) 1–20.

[6] Arsha Nagrani, Joon Son Chung, Weidi Xie, Andrew Zisserman, VoxCeleb: large-scale speaker verification in the wild, Computer Speech & Language 60 (2020) 101027.

[7] Trivellore E. Raghunathan, Synthetic data, Annual Review of Statistics and Its Application 8 (2021) 129–140.

[8] Zhiqiang Wan, Yazhou Zhang, Haibo He, Variational autoencoder based synthetic data generation for imbalanced learning, in: 2017 IEEE Symposium Series on Computational Intelligence (SSCI), IEEE, 2017, pp. 1–7.

3

Data preparation

3.1 Introduction

The quality of input data largely determines the effectiveness of Data Science tasks. If the quality of data is not adequate, even a brilliant exploratory or predictive Data Science technique might fail to produce reliable results. ***Data quality*** *refers to the state of the data in terms of completeness, correctness, and consistency.* Assessing and ensuring data quality by removing errors and data inconsistencies can improve the input-data quality and makes it fit for the intended purpose.

The availability of low-cost computing resources and high-speed data transmission has led to explosive data growth in many data-intensive computing domains. High-throughput experimental processes have resulted in the production of massive amounts of data, particularly in the biomedical field where experiments generate a wide variety of data such as mRNA, miRNA, gene expression, and protein–protein interaction data. Other sources of bulk data generation include the stock market, social networks, the Internet of Things (IoT), sensor networks, and business transactions. However, the effective storage and transmission of such data present a significant challenge. To address this challenge, efficient and scalable exploratory Data Science techniques are emerging as important tools for knowledge discovery from heterogeneous, multimodal data sources. Despite the plethora of data-exploration and machine-learning methods proposed in recent decades, the quality of input data remains a critical factor in producing reliable results. The heterogeneity of data due to different acquisition techniques and standards, as well as the presence of nonuniform devices in geographically distributed data repositories, make the task of producing high-quality data difficult in many cases.

In real-world scenarios, data are frequently of low quality and may not be suitable for use in advanced Data Science methods. Furthermore, these data may be incomplete or contain erroneous or misleading elements, known as outliers. Additionally, the same set of data can be represented in various formats, and values may have been normalized differently for different purposes. To use exploratory or predictive Data Science methods, it is necessary to adequately prepare the data beforehand through data preprocessing, which comprises several essential steps, including **Data Cleaning**, **Data Reduction**, **Data Transformation**, and **Data Integration**. Each of these steps is independent, equally crucial, and can be performed separately. The data and the tasks to be executed determine which steps need to be taken and when, and a domain expert may need to intervene to determine the appropriate steps for a particular dataset. It is also necessary to avoid using the steps as black boxes without understanding their underlying processes. A schematic overview of the overall workflow of preprocessing steps is illustrated in Fig. 3.1.

Fundamentals of Data Science. https://doi.org/10.1016/B978-0-32-391778-0.00010-7

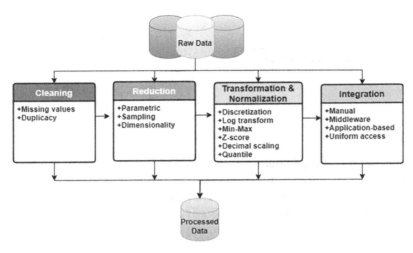

FIGURE 3.1 Data preprocessing steps and activities. The raw data may be single or multisourced, uniform or heterogeneous format.

Next, we discuss the major data-preprocessing steps in detail.

3.2 Data cleaning

The quality of data plays a significant role in determining the quality of the resulting output when using any algorithm. Creating high-quality data involves a step called data *cleaning* [16]. This step involves removing noise, estimating missing values, and correcting background information before normalizing the data. Data cleaning typically addresses two primary issues: missing values and data duplication. The latter issue can increase computation time without providing any additional value to the final result.

3.2.1 Handling missing values

Dealing with missing values is a crucial challenge in data preprocessing [20]. Missing values can have negative effects on the distribution of the dataset, introduce bias in results, or cause descriptive and predictive Data Science algorithms to fail. Thus they need to be handled with care to ensure optimal performance of the methods. Missing data may arise due to various technical reasons. Missing values in data can occur due to several reasons. For example, issues with the measuring equipment, such as malfunctioning or physical limitations, can result in certain values not being recorded. Similarly, missing values can occur during data transportation if there are interruptions in the communication media. In some cases, missing values may not have been addressed during the data-collection phase because they were considered insignificant. Additionally, some values may have been removed due to inconsistencies with other recorded data.

Different types of missing data can be classified into several categories [2]:

- **Missing Completely at Random (MCAR)**: This refers to the missing data, in which the absence of observation is unrelated to its actual value.
- **Missing at Random (MAR)**: When the probability of a missing observation is dependent on other aspects of the observed data, but not on the value of the missing observation itself.
- **Missing Not at Random (MNAR)**: This is the missing data that do not fit into either of the aforementioned two categories. In this type of missing value, the likelihood of a missing observation is correlated with its value.

To illustrate the different types of missing data values, let us consider the example of two variables, *a* and *b*, represented by vectors *A* and *B*, respectively. While all the values in vector *B* are recorded, there are some missing values in vector *A*. A missing value in *A* is considered MCAR if its probability of being missing is not dependent on the values in both *A* and *B*. Conversely, if the likelihood of a missing value in *A* depends on the values in *B* but not on the values in *A*, it is called MAR.

There are several ways to address missing data, and ignoring it altogether is one of them. However, this approach can have a detrimental impact on the results if a large proportion of the dataset is missing. In the case of MNAR missing values, parameter estimation is necessary to model the missing data. To handle missing data, four broad classes of approaches are available.

3.2.1.1 Ignoring and discarding data

One way to handle missing values is to ignore them, which is only appropriate for the MCAR situation. However, using this method for other types of missing data can introduce bias into the data and ultimately impact the results. There are two ways to discard missing data: complete-case analysis and attribute discarding. Complete-case analysis is the most common method and involves removing all instances with missing data and conducting further analysis only on the remaining data.

The second method involves determining the extent of missing data with respect to the number of instances and the relevance of the attributes in the dataset. Based on these considerations, instances that are not widespread or whose relevance to the dataset is not significant can be discarded. However, if an attribute with missing values is highly relevant, it must be retained in the dataset [5].

3.2.1.2 Parameter estimation

This method is appropriate for MNAR missing data and requires parameter estimation using statistical or supervised learning approaches such as probabilistic, KNN, or SVD models. To estimate the parameters of a model, the maximum-likelihood approach can be used, which involves a variant of the Expectation-Maximization algorithm [31]. The most common method for this approach is called BPCA [32], which involves three elementary steps: principal-components regression, Bayesian estimation, and an iterative Expectation-Maximization (EM) algorithm. This method is frequently used for biological data cleaning.

3.2.1.3 Imputation

Imputation refers to the actions taken to address missing data in a dataset. The objective of imputation is to replace missing data with estimated values by utilizing information from measured values to infer population parameters. Imputation may not provide the best prediction for the missing values, but it can enhance the completeness of the dataset [25]. Imputation approaches can be categorized into three broad classes: *Single-value imputation*, *Hot-deck and cold-deck methods*, and *Multiple imputations*.

1. **Single-value imputation**: This is a method of replacing a missing value with a single estimated value, which is assumed to be very similar to the value that would have been observed if the data example were complete. There are various ways to substitute a single value for the missing value in a data example. Some of these methods include:

    (i) Mean imputation: Replacing a missing attribute value in a data example with the mean of the existing values of the same attribute over all data examples is a common approach. However, this method can result in a poorer estimation of the standard deviation for the attribute in the resulting dataset because it reduces the diversity of the data [4].

    (ii) Regression imputation: Using a regression model to predict the missing value of an attribute in a data example based on the actual values of the attribute in other data examples is another method for imputing missing values. This approach preserves the variability in the dataset and avoids introducing bias in the results. Unlike mean imputation, which underestimates the correlation between attributes due to the loss of variability, this method tends to overestimate the correlation.

    (iii) Least-squares regression (LS impute): A special form of a regression model is used for imputing missing values, which involves developing an equation for a function that minimizes the sum of squared errors for the attribute's existing values across data examples from the model [6].

    (iv) Local least squares (LLS impute): This imputation method is similar to LS impute but involves a prior step that is known as KNN impute. KNN impute first identifies the K nearest neighbors that have high correlations, and then represents a target missing value in a data example using a linear combination of values in similar data examples [22].

    (v) K-Nearest-Neighbor imputation (KNN impute): The missing value for a particular attribute is estimated by minimizing the distance between the observed and predicted values using the known values of other examples. The K-nearest neighbors to the example with missing data are identified, and the final imputed value is estimated based on these neighbors.

2. **Hot-deck and cold-deck methods**: Hot-deck imputation is a method of filling in missing data on variables by using observed values from respondents within the same survey data set. It is also known as Last Observation Carried Forward (LOCF) and can be divided into random and *deterministic hot-deck prediction* [3], depending on the

donor with which the similarity is computed. This technique has the advantage of being free from parameter estimation and prediction ambiguities, and it uses only logical and credible values obtained from the donor pool set [3]. In contrast, cold-deck imputation selects donor examples from an external dataset to replace a missing attribute value in a data example.

3. **Multiple imputations**: To determine the missing attribute value, multiple imputation involves a sequence of computations. It replaces each missing value with a set of plausible values, which are then analyzed to determine the final predicted value. This technique assumes a monotonic missing pattern, where if one value for an attribute is missing in a sequence of values, all subsequent values are also missing. Two prevalent methods used to determine the set of plausible values are:

- Regression method: A regression model is constructed by utilizing the existing attribute values for attributes with missing values, and this process is carried out following the monotonic property. Subsequently, a new model is formed based on this regression model, and the process is repeated sequentially for each attribute that has missing values [34].
- Propensity-score method: In order to address missing values in a dataset, propensity-score imputation is used to calculate the conditional probability of assigning a certain value to a missing attribute, given a set of observed values. This method involves calculating a propensity score for each attribute with missing values, indicating the probability of the value being missing. Grouped data are then subjected to Bayesian bootstrap imputation based on the propensity score to obtain a set of values to replace the missing values [24].

3.2.2 Duplicate-data detection

The problem of storage shortage due to duplicate-data values in instances or attributes is known as data redundancy. The conventional method for addressing this issue involves identifying similar value chunks and removing duplicates from them. However, this approach is time intensive, and alternative techniques have been developed to handle redundant data.

3.2.2.1 Knowledge-based methods

One approach to address data duplication is to utilize domain-specific knowledge from knowledge bases in data-cleaning tasks. Intelliclean [28] is a tool that implements this approach by standardizing abbreviations to ensure consistency across records. For example, if one record abbreviates the word street as *St.*, and another abbreviates it as *Str*, while another record uses the full word, all three records are standardized to the same abbreviation. Then, domain-specific rules are applied to identify and merge duplicates, and to flag any other anomalies. This technique allows for more efficient and accurate duplication elimination, as compared to traditional methods that rely on identifying similar values and removing duplicates from them.

3.2.2.2 ETL method

The ETL method, which stands for extraction, transformation, and loading, has become the most widely used approach for eliminating duplicates [33]. This method involves three steps, namely extraction, transformation, and loading. It allows for two types of duplicate elimination: instance-level and schema-level processing. Instance-level processing aims to clean errors within the data itself, such as typos or misspellings. On the other hand, schema-level processing typically involves transforming the database into a new schema or data warehouse.

3.3 Data reduction

Large-scale data mining can be time consuming and resource intensive, making it impractical and infeasible. Using the entire dataset is not always necessary and may not contribute significantly to the quality of the results compared to using a smaller version of the data. Data-reduction techniques can reduce the size of the data by reducing its volume or the number of attributes without compromising the integrity of the data. There are several approaches available for data reduction, and an ideal method should be efficient and produce similar results to those obtained with the full dataset. Some of these methods are discussed briefly below.

3.3.1 Parametric data reduction

Parametric data-reduction techniques aim to reduce the volume of original data by using a more compact representation of the data. This is achieved by fitting a parametric model based on the distribution of the data and estimating the optimal values of the parameters required to represent that model. This approach involves storing only the model parameters and the outliers. Regression and nonlinear models are two popular examples of parametric data-reduction techniques.

On the other hand, nonparametric approaches do not rely on any specific model. Instead, they summarize the data using sample statistics, which will be discussed in more detail below.

3.3.2 Sampling

Dealing with a large dataset in its entirety can be a costly and time-consuming task. To avoid this, a smaller subset of representative data, consisting of k instances, is selected from a larger dataset of G instances, where k is less than or equal to G. This process is known as sampling. Sampling can be divided into the following categories:

- **Simple random sampling without replacement**: In this approach, a sample of size k is selected from a large dataset of size G based on a probability of selection (p), and the selected samples are not returned to the dataset. This is known as sampling without replacement, which can result in a nonzero covariance between the chosen samples, making the computations more complex. However, if the dataset contains a large number of instances, the covariance is likely to be negligible.

- **Simple random sampling with replacement**: In this type of sampling, the chosen sample is returned to the dataset, allowing for the possibility of duplicated data in the sample. This results in a zero covariance between two selected samples. When dealing with skewed data distributions, simple random sampling with replacement often yields poor results for any data-analysis technique. In such cases, sampling with or without replacement makes little difference. However, sampling without replacement typically produces more precise estimates than sampling with replacement. For unbalanced datasets with uneven sample distribution, adaptive sampling methods like stratified sampling and cluster sampling are more likely to improve performance.

- **Stratified sampling**: To perform stratified sampling, the original dataset is divided into subgroups or strata, from which a sample is generated using simple random sampling. On the other hand, in cluster sampling, the original dataset is divided into subgroups or clusters, and a random selection of n clusters is taken from the total N clusters. Then, the sample is generated by selecting elements from these selected clusters.

3.3.3 Dimensionality reduction

The process of dimensionality reduction involves eliminating attributes that are unnecessary or insignificant. Having irrelevant attributes in a dataset can make it difficult to extract valuable information and can also lead to erroneous analysis, often referred to as the *Curse of Dimensionality*. Data sets with a high number of dimensions are also likely to be sparse. This can cause issues with machine-learning techniques such as clustering, which rely on data density and distance, and can lead to inaccurate results in the case of sparse data. To enhance the performance of machine-learning algorithms with high-dimensional data sets, reducing dimensionality is crucial. The reduction process should preserve the information content of the original data to the greatest extent possible and ensure that the use of the reduced dataset does not affect the final result. One way to achieve dimensionality reduction is by selecting only the relevant attributes and ignoring the irrelevant ones.

One widely used technique for dimensionality reduction is Principal-Component Analysis (PCA), as proposed by Hotelling [19]. PCA is a statistical method that utilizes an orthogonal transformation to map a set of observations of possibly interrelated variables to a set of values of linearly independent variables known as principal components. It looks for k orthogonal vectors of n dimensions $(k < n)$ from the original data that best represent it. This process projects the original data onto linear subspaces, resulting in a low-dimensional approximation of the original data and, consequently, dimensionality reduction.

3.4 Data transformation

To analyze data that are collected from diverse sources, it is common to find them in various formats due to the differences in experimental setups, conditions, and variables of interest during data collection. Transformation of the data is often necessary and effective for analysis [11]. Transformation involves changing the format or structure of data from one to another to align better with the assumptions of statistical inference procedures or to enhance interpretability. The transformation function $y = f(x)$ is used to convert the data from one domain to another. Several techniques for data transformation are explained below.

3.4.1 Discretization

Discretization transforms the domain of values for each quantitative attribute into discrete intervals. Using discrete values during data processing offers a number of benefits, such as:

- Storing discrete data requires less memory compared to storing continuous data.
- Discrete data are likely to be simpler to visualize and understand.
- Discrete data are known to improve the accuracy of performance in some systems.

The following two concepts are commonly used when discussing discretization or interval generation:

- **Cutpoints**: The process of discretization involves identifying cutpoints, which are specific values within the range of a continuous attribute. These cutpoints divide the range of continuous values into successive intervals or bins. For instance, a continuous range such as $< a \cdots b >$ may be divided into intervals like $[a, c]$ and $(c, b]$, where c represents a cutpoint. This is described in [26].
- **Arity**: Arity [26] refers to the number of intervals generated for an attribute through discretization. It is determined by adding one to the number of cutpoints used in the process of discretization for that attribute. A high arity can lead to longer learning processes, while a very low arity may result in reduced predictive accuracy.

Any discretization method follows the enumerated steps below:

1. Arrange the continuous values in ascending or descending order for each attribute that needs to be discretized.
2. Determine the cutpoints for each of these attributes based on a certain criterion or method.
3. Generate intervals by dividing the range of values for each attribute using the computed cutpoints.
4. Map each value to its corresponding interval and represent it with a discrete value or category.

Below are a few measures that can be used by a discretizer to evaluate cutpoints and group different continuous values into separate discretized intervals:

- **Equal Width**: This method divides the range of values into equal intervals, each with the same width. The width of each interval is calculated as the difference between the maximum and minimum value of the attribute divided by the desired number of intervals.

- **Equal Frequency**: This method divides the range of values into intervals so that each interval contains the same number of observations. This method is based on the frequency of occurrence of values in the dataset.
- **Entropy-based**: This method is based on information theory and evaluates the quality of a cutpoint based on the reduction of entropy achieved by dividing the data into two intervals. The cutpoint that maximizes the reduction of entropy is selected.
- **Chi-square**: This method evaluates the quality of a cutpoint based on the difference in the distribution of values in the two intervals obtained by splitting the data. The cutpoint that maximizes the chi-square statistic is selected.

The choice of measure depends on the nature of the data and the objective of the analysis.

There are various techniques proposed in the literature for discretization, which are broadly classified into two categories: Supervised and Unsupervised techniques. Supervised techniques use class levels for discretization, whereas unsupervised techniques do not. Below are some of the techniques in each category.

3.4.1.1 Supervised discretization
This type of discretization technique uses the class labels of the original dataset to convert continuous data into discrete ranges [12]. There are two types of supervised discretization techniques.

1. **Entropy-Based Discretization Method**: The entropy measure is utilized in this approach to determine the boundaries for discretization. Chiu et al. [10] introduced a hierarchical technique that maximizes the Shannon entropy across the discretized space. The technique initiates with k partitions and applies hill climbing to refine the partitions, using the same entropy measure to obtain more precise intervals.
2. **Chi-Square-Based Discretization**: This technique of discretization employs the Chi-Square test to assess the probability of similarity between data in different intervals [21]. Initially, each unique value of the attribute is in its own interval, and the χ^2 value is calculated for each interval. The intervals with the smallest χ^2 values are merged, and the process is repeated until no further satisfactory merging is possible based on the χ^2 values.

3.4.1.2 Unsupervised discretization
The unsupervised discretization technique does not rely on any class information to determine the boundaries. Below are some examples of this type of discretization technique:

1. **Average and Midranged value discretization**: This is a binary discretization technique that uses the average value of the data to divide the continuous variable into two intervals. Values below the average are assigned to one interval, and values above the average are assigned to another interval. This technique does not use any class information to determine the boundaries. [12].
   To apply the average discretization method on a vector $A = [23.73, 5.45, 3.03, 10.17, 5.05]$, the first step is to calculate the average score, $\overline{A} == 9.88$. Based on this average, it dis-

cretizes the value using following equation.

$$D_i = \begin{cases} 1, & \text{if } A(i) >= \overline{A} \\ 0, & \text{otherwise.} \end{cases}$$

In our example, $\overline{A} = 9.486$, therefore, the discretized vector using average-value discretization is D=[1 0 0 1 0].

The midrange discretization method uses the middle value or midrange of an attribute in the dataset to determine the class boundary. To obtain the midrange value of a vector of values, the equation $M = (H + U)/2$ is used, where H is the maximum value and U is the minimum value in the vector. Discretization of values is carried out by assigning them to a class based on whether they fall below or above the midrange value.

$$D_i = \begin{cases} 1, & \text{if } A(i) >= M \\ 0, & \text{otherwise.} \end{cases}$$

For the vector A, we obtain M as 13.38. Therefore the corresponding discretized vector of values $D = [1 0 0 0 0]$. This approach is not of much use as it lacks robustness since outliers change the resulting vector significantly [12].

2. **Equal-Width Discretization**: This technique involves dividing the data into a specific number of intervals or bins, where the boundaries for each bin are determined by the range of the attribute being discretized [7]. The range is divided into equal parts, each corresponding to one of the intervals, and the boundaries for each interval are determined by the maximum (H) and minimum values (U) of the attribute.

   This approach divides the data into k equal intervals. The upper and lower bounds of the intervals are decided by the difference between the maximum and minimum values for the attribute of the dataset [7]. The number of intervals, k, is specified by the user, and the lower and upper boundaries, p_r and p_{r+1} of the class are obtained from the data (in particular, using the H and U values) in the vector according to the equation

$$p_{r+1} = p_r + (H - U)/k.$$

For our example, if $k = 3$, the group division occurs as given below

$$D_i = \begin{cases} 1, & \text{if } p_{r_0} <= A(i) < p_{r_1} \\ 2, & \text{if } p_{r_1} <= A(i) < p_{r_2} \\ 3, & \text{if } p_{r_2} <= A(i) <= p_{r_3}. \end{cases}$$

Therefore the corresponding discretized vector is $D = [3 1 1 2 1]$.

3. **K-means Discretization**: This technique divides the values of the variable into k intervals such that adjacent values are assigned to the same interval [29]. The number of intervals k is specified by the user, and the size of each interval is determined by the range of the data divided by k. Values are then assigned to the appropriate interval

based on their position within the range. For $k = 2$, discretization can be performed as follows.

$$D_i = \begin{cases} 1, & \text{if } A_{i+1} - A_i <= (H - U)/k \\ 0, & \text{otherwise.} \end{cases}$$

For our example, if $k = 2$, the discretized vector $D = [\, 3\ 1\ 1\ 2\ 1\,]$.

Some of the frequently used discretizers, out of the plethora of techniques available in the literature, are reported in Table 3.1.

Table 3.1   Some Discretizers and their Properties.

Discretizer	Measure Used	Procedure Adopted	Learning Model
Equal Width	Binning	Splitting	Unsupervised
Equal Frequency	Binning	Splitting	Unsupervised
ECD [1]	Equivalence class	Splitting	Unsupervised
Fixed-Frequency Discretization (FFD) [37]	Binning	Splitting	Unsupervised
ChiMerge [21]	Dependency	Merging	Supervised
Chi2 [27]	Dependency	Merging	Supervised
Ent-Minimum Description Length Principle (MDLP) [13]	Entropy	Splitting	Supervised
Zeta [18]	Dependency	Splitting	Supervised
Optimal Flexible Frequency Discretization (OFFD) [36]	Wrapping	Hybrid	Supervised
Class-Attribute Interdependence Maximization (CAIM) [23]	Dependency	Splitting	Supervised
Class-Attribute Dependent Discretization (CADD) [9]	Dependency	Hybrid	Supervised
Ameva [14]	Dependency	Splitting	Supervised

3.5 Data normalization

The term "normalization" is frequently used interchangeably with "transformation". Normalization refers to the process of transforming data so that the resulting distribution is approximately normal. The following paragraphs cover several techniques for normalization.

3.5.1 Min–max normalization

Given an attribute value, x, from the original dataset, it is transformed to a new value, x', through a mapping process.

$$x' = \frac{x - min}{max - min},$$

where min and max are the minimum and maximum values of the attribute in the dataset.

3.5.2 Z-score normalization

The transformation of variable values is determined by the mean and standard deviation of the variable values in the dataset. For instance, if X is a variable with values $x_1, x_2, ..., x_n$, the transformation can be achieved using the following formula:

$$x_i' = \frac{x_i - \bar{X}}{std_{dev}(\bar{X})},$$

where x_i' is the Z-score value of x_i, $\bar{X}$ is the mean of the values of X, and std_{dev} is the standard deviation given by

$$std_{dev}(\bar{X}) = \sqrt{\frac{1}{n-1} \sum_{i=1}^{n} (x_i - \bar{X})^2}.$$

3.5.3 Decimal-scaling normalization

Decimal scaling is a normalization technique that involves shifting the decimal point of an attribute's value. The number of decimal points shifted depends on the maximum absolute value of the attribute. To normalize a value v of an attribute to v', we divide v by 10^j, where j is the smallest integer such that $max(v') < 1$.

For example, if the values of an attribute range from -523 to 237, the maximum absolute value is 523. To normalize using decimal scaling, we would divide each value by 1000 (i.e., $j = 3$) so that -523 normalizes to -0.523 and 237 normalizes to 0.237.

3.5.4 Quantile normalization

Quantile normalization is a method used to make multiple distributions statistically identical. To normalize a column (which represents an attribute), each value in the column is ranked based on its position from lowest to highest. The column values are then rearranged in ascending order so that each column is in order from the lowest to the highest value. The average value for each row is then calculated using the reordered values. The average value calculated in the first row is the lowest value (rank 1) from every column, the second average value is the second-lowest (rank 2), and so on. Finally, the original values are replaced with the average values based on their assigned ranks. This results in each column having the same distribution of values.

Quantile normalization assumes that there are no global differences in the distribution of data in each column. However, if these assumptions are violated, it is not clear how to proceed with normalization. To address this issue, a recent technique called smooth quantile normalization (qsmooth) has been proposed. Qsmooth [17] is a generalization of quantile normalization that allows for differences in the distribution between groups by relaxing some of the assumptions made by quantile normalization.

3.5.5 Logarithmic normalization

In biomedical and psychosocial research, log-normalization is a popular technique used to address data skewness[1] and heteroskedasticity.[2] The purpose of log-normalization is to normalize the distribution of continuous data and make it more similar to a normal distribution, which reduces or eliminates input-data skewness [8]. This technique involves a simple scaling normalization called the log transformation, which replaces each value x with its logarithm, denoted by $x' = \log(x)$. The choice of the log base depends on the specific problem being addressed. For instance, the base 2 logarithm, denoted by $\log_2$, is commonly used for expression data normalization in microarray experiments in computational biology. Log-normalization is easy to apply and has proven to be beneficial for statistical modeling.

3.6 Data integration

If data are obtained from only a single source, it is often difficult to make a reliable decision. The use of multiple sources of data is preferred. Multisource data may be stored in various storage repositories such as databases, flat files, and multidimensional data cubes [15]. Often, the data structures and storage formats for the sources are different. To process, one needs to bring all data to a common format. Data integration is the process of analyzing data obtained from various sources, and obtaining a common format that can be used across all sources [35]. Such a format usually makes more accurate predictions.

Data integration can be achieved at three different levels: data level, processing level, and decision level [30]. Combining data from various sources and implementing a universal query system for all types of data involved is required to integrate data at the data level. This step is the most time consuming, as it requires considering various assumptions and experimental setup information before combining the data. The second level of data integration involves understanding and interpreting the datasets to identify associated correlations between the various datasets involved. The third level of data integration is at the decision level, where specific procedures are used to deal with different participating datasets and obtain individual results. These individual results are then mapped through a consensus process called "Ensembling".

Data integration involves processing steps like cleansing, sorting, enrichment, and other processes before porting or storage. The above preprocessing steps may be applied before or after porting the data to the destination storage, termed as *ETL* (Extract, Transform, Load) or *ELT* (Extract, Load, Transform). When data from different sources are ported to common destination entirely through the process of ETL/ELT, it is called *Tight Coupling*. Instead of porting, if the integration happens on-the-fly through query-based

---

[1] Skewness in a dataset refers to the degree of asymmetry in the distribution of its values, indicating whether the data is predominantly concentrated on one side of the mean.

[2] Heteroskedasticity in data occurs when the variability of errors or residuals in a statistical model is not constant across different levels of the independent variables.

data selection in the source storage for a consolidated view of the data, it is called *Loose Coupling*. Following are a couple of techniques for data integration.

3.6.1 Consolidation

Data consolidation is a tight coupling approach for data integration. It brings data from different sources to the centralized destination pool or data store. The centralized or consolidated storage is then used for downstream data-analytic tasks. An issue with the consolidation process is the *latency* in collecting updated data from multiple sources of storage to update the destination storage. Latency should be lower for more recent data in the centralized storage. Although the entire process of consolidation is expensive, with the advancement of integration technologies, it is now possible for near real-time upgradation.

3.6.2 Federation

Unlike consolidation, federation does not involve any physical movement of data. Rather, a virtual data storage is used to unify multiple data sources using a variety of data models. It is a type of loose coupling integration. It is an on demand process of integration based on user query. When, a user issues a query, data are fetched from heterogeneous data sources by splitting the query into versions understandable by the candidate source storage. Consolidation of data sources in advance is avoided in a federation data-integration model.

3.7 Summary

To deal with the incomplete and imperfect nature of real-life data, data-preparation steps are necessary. These steps are crucial to handle different forms of incompleteness and imperfection, including missing values, noise, inconsistencies, and the curse of dimensionality. However, the nature of real data is not static and data distributions vary over time, making it challenging to apply state-of-the-art preprocessing techniques. This challenge is particularly pronounced when data are rapidly produced. Therefore the Big Data community is interested in conducting an empirical study to assess the suitability and efficiency of preprocessing techniques for handling dynamic data.

References

[1] Dhrubajit Adhikary, Swarup Roy, A new equivalence class based approach for discretizing quantitative data using point shift mechanism, in: 2015 International Symposium on Advanced Computing and Communication (ISACC), IEEE, 2015, pp. 174–180.
[2] Paul D. Allison, Handling missing data by maximum likelihood, in: SAS Global Forum, vol. 23, 2012.
[3] Rebecca R. Andridge, Roderick J.A. Little, A review of hot deck imputation for survey non-response, International Statistical Review 78 (1) (2010) 40–64.
[4] Amanda N. Baraldi, Craig K. Enders, An introduction to modern missing data analyses, Journal of School Psychology 48 (1) (2010) 5–37.

[5] Gustavo E.A.P.A. Batista, Maria Carolina Monard, A study of k-nearest neighbour as an imputation method, in: International Conference on Health Information Science, 2002, https://api.semanticscholar.org/CorpusID:37493644.

[6] Trond Hellem Bø, Bjarte Dysvik, Inge Jonassen, LSimpute: accurate estimation of missing values in microarray data with least squares methods, Nucleic Acids Research 32 (3) (2004) e34–e34.

[7] Jason Catlett, On changing continuous attributes into ordered discrete attributes, in: European Working Session on Learning, Springer, 1991, pp. 164–178.

[8] Feng Changyong, Wang Hongyue, Lu Naiji, Chen Tian, He Hua, Lu Ying, et al., Log-transformation and its implications for data analysis, Shanghai Archives of Psychiatry 26 (2) (2014) 105.

[9] John Y. Ching, Andrew K.C. Wong, Keith C.C. Chan, Class-dependent discretization for inductive learning from continuous and mixed-mode data, IEEE Transactions on Pattern Analysis and Machine Intelligence 17 (7) (1995) 641–651.

[10] David K.Y. Chiu, Benny Cheung, Andrew K.C. Wong, Information synthesis based on hierarchical maximum entropy discretization, Journal of Experimental and Theoretical Artificial Intelligence 2 (2) (1990) 117–129.

[11] Rajashree Dash, Rajib Lochan Paramguru, Rasmita Dash, Comparative analysis of supervised and unsupervised discretization techniques, International Journal of Advances in Science and Technology 2 (3) (2011) 29–37.

[12] James Dougherty, Ron Kohavi, Mehran Sahami, et al., Supervised and unsupervised discretization of continuous features, in: Machine Learning: Proceedings of the Twelfth International Conference, vol. 12, 1995, pp. 194–202.

[13] Usama M. Fayyad, Keki B. Irani, Multi-interval discretization of continuous-valued attributes for classification learning, in: International Joint Conference on Artificial Intelligence, 1993, https://api.semanticscholar.org/CorpusID:18718011.

[14] L. Gonzalez-Abril, Francisco Javier Cuberos, Francisco Velasco, Juan Antonio Ortega, Ameva: an autonomous discretization algorithm, Expert Systems with Applications 36 (3) (2009) 5327–5332.

[15] Jim Gray, Surajit Chaudhuri, Adam Bosworth, Andrew Layman, Don Reichart, Murali Venkatrao, Frank Pellow, Hamid Pirahesh, Data cube: a relational aggregation operator generalizing group-by, cross-tab, and sub-totals, Data Mining and Knowledge Discovery 1 (1) (1997) 29–53.

[16] Katherine G. Herbert, Jason T.L. Wang, Biological data cleaning: a case study, International Journal of Information Quality 1 (1) (2007) 60–82.

[17] Stephanie C. Hicks, Kwame Okrah, Joseph N. Paulson, John Quackenbush, Rafael A. Irizarry, Héctor Corrada Bravo, Smooth quantile normalization, Biostatistics 19 (2) (2018) 185–198.

[18] K.M. Ho, P.D. Scott, Zeta: a global method for discretization of continuous variables, in: Proceedings of the 3rd International Conference on Knowledge Discovery and Data Mining, 1997, pp. 191–194.

[19] Harold Hotelling, Analysis of a complex of statistical variables into principal components, Journal of Educational Psychology 24 (6) (1933) 417.

[20] Hyun Kang, The prevention and handling of the missing data, Korean Journal of Anesthesiology 64 (5) (2013) 402–406.

[21] Randy Kerber, ChiMerge: discretization of numeric attributes, in: Proceedings of the Tenth National Conference on Artificial Intelligence, 1992, pp. 123–128.

[22] Hyunsoo Kim, Gene H. Golub, Haesun Park, Missing value estimation for DNA microarray gene expression data: local least squares imputation, Bioinformatics 21 (2) (2005) 187–198.

[23] Lukasz A. Kurgan, Krzysztof J. Cios, CAIM discretization algorithm, IEEE Transactions on Knowledge and Data Engineering 16 (2) (2004) 145–153.

[24] Philip W. Lavori, Ree Dawson, David Shera, A multiple imputation strategy for clinical trials with truncation of patient data, Statistics in Medicine 14 (17) (1995) 1913–1925.

[25] Roderick J.A. Little, Donald B. Rubin, Statistical Analysis with Missing Data, John Wiley & Sons, 2014.

[26] Huan Liu, Farhad Hussain, Chew Lim Tan, Manoranjan Dash, Discretization: an enabling technique, Data Mining and Knowledge Discovery 6 (2002) 393–423.

[27] Huan Liu, Rudy Setiono, Chi2: feature selection and discretization of numeric attributes, in: Proceedings of 7th IEEE International Conference on Tools with Artificial Intelligence, IEEE, 1995, pp. 388–391.

[28] Wai Lup Low, Mong Li Lee, Tok Wang Ling, A knowledge-based approach for duplicate elimination in data cleaning, Information Systems 26 (8) (2001) 585–606.

[29] James MacQueen, et al., Some methods for classification and analysis of multivariate observations, in: Proceedings of the Fifth Berkeley Symposium on Mathematical Statistics and Probability, Oakland, CA, USA, vol. 1, 1967, pp. 281–297.

[30] Luciano Milanesi, Roberta Alfieri, Ettore Mosca, Federica Viti, Pasqualina D'Ursi, Ivan Merelli, Sys-bio gateway: a framework of bioinformatics database resources oriented to systems biology, in: Sandra Gesing, Jano I. van Hemert (Eds.), Proceedings of the International Workshop on Portals for Life Sciences, IWPLS 2009, Edinburgh, UK, September 14-15, 2009, in: CEUR Workshop Proceedings, vol. 513, CEUR-WS.org, 2009.

[31] Todd K. Moon, The expectation-maximization algorithm, IEEE Signal Processing Magazine 13 (6) (1996) 47–60.

[32] Shigeyuki Oba, Masa-aki Sato, Ichiro Takemasa, Morito Monden, Ken-ichi Matsubara, Shin Ishii, A Bayesian missing value estimation method for gene expression profile data, Bioinformatics 19 (16) (2003) 2088–2096.

[33] Erhard Rahm, Hong Hai Do, Data cleaning: problems and current approaches, IEEE Data Engineering Bulletin 23 (4) (2000) 3–13.

[34] Donald B. Rubin, Multiple Imputation for Nonresponse in Surveys, vol. 81, John Wiley & Sons, 2004.

[35] Maria Victoria Schneider, Rafael C. Jimenez, Teaching the fundamentals of biological data integration using classroom games, PLoS Computational Biology 8 (12) (2012) e1002789.

[36] Song Wang, Fan Min, Zhihai Wang, Tianyu Cao, OFFD: optimal flexible frequency discretization for naive Bayes classification, in: International Conference on Advanced Data Mining and Applications, Springer, 2009, pp. 704–712.

[37] Ying Yang, Geoffrey I. Webb, Discretization for naive-Bayes learning: managing discretization bias and variance, Machine Learning 74 (1) (2009) 39–74.

4

Machine learning

4.1 Introduction

Data Science aims to identify novel patterns in large collections of data. The discovered patterns often help in quick and smart decision making. For instance, being a frequent user of any email service such as Gmail, one receives numerous important and unimportant emails daily. To deal with a large volume of incoming emails, one may opt to use a facility that filters unwanted emails automatically as Spam without explicit human intervention. Various marketing and social-media-related intended and unintended communications that are pushed can be screened automatically to keep the inbox clean. Automated email recognition and categorization guarantee efficient time utilization for users who need to apportion time to other activities. Likewise, in healthcare, smart MRI analysis can detect brain tumors automatically, or a smartwatch can analyze health status in real time and raise an alarm in advance to warn about possible health issues.

In conventional programming, programs make decisions based on algorithms that programmers create a priori. However, it has been found impossible to write a general program to recognize credit-card fraud based on rules obtained by analyzing how someone uses the card. Similarly, just by analyzing network traffic based on rules or algorithms written a priori, it is impossible to recognize whether it is normal or abnormal. This is because the number and/or complexity of rules or algorithms are context dependent and become unwieldy. We need a program that automatically learns contextual rules or parameters (say, individualized thresholds) for recognizing fraud, network attacks, or spam emails. The program itself attempts to identify patterns from past instances or experiences for the environment. In other words, the data are used to identify the relevant patterns in what is often known as data-driven decision making. Machine Learning is a subdomain of Artificial Intelligence (AI) that deals with developing algorithms that let a program automatically learn inherent patterns in the given data. Mathematically, it produces an implicit or explicit function that is used for recognition or prediction in the future. During the learning phase, the algorithm enhances its own performance on a particular activity as it is exposed to more data or experiences without being explicitly programmed.

Let us try to understand the difference between explicit (conventional) programming and machine learning with the help of Fig. 4.1. Consider writing a program whose algorithm uses rules to produce output given an input. In such a case, in explicit programming, the coder writes the program to map input data to the expected output. The coder must decide the program structure and the rules or algorithms. She corrects the code if her code does not give the expected outcome. In contrast, in the case of machine learning, a prebuilt program accepts several inputs and corresponding expected outcomes. It generates rules

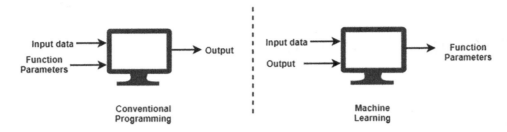

FIGURE 4.1 Illustration of explicit programming vs. Learning Programs.

that constitute the decision-making algorithm by trying to faithfully map the input to the output. Through an iterative refinement (learning) process, parameters (which describe components of the rules or algorithms) are updated automatically when the program fails to give the correct output for an input–output pair. Instead of depending exclusively on predetermined rules and instructions, the main objective of machine learning is to enable computers to recognize patterns, make judgments, and solve problems based on the data they have been exposed to.

The first formal definition of machine learning was given by the renowned computer scientist Tom M. Mitchell [12].

Definition 4.1.1. A computer program is said to learn from experience E concerning some class of tasks T and performance measure P if its performance at task T, as measured by P, improves with experience E.

Let us take the example of the Spam-Filtering task to understand the above definition. The task T is to recognize and classify spam emails when experience (E), i.e., known examples of spam emails and nonspam (legitimate) emails, are given for learning. Finally, how many previously unseen spam and legitimate emails it can successfully detect is assessed to evaluate its performance (P). Learning proceeds iteratively until performance is satisfactory.

This first phase of learning from the examples is known as *Training*. The output prediction for future input data based on learned parameters is called *Testing*. Such a guided learning paradigm based on a known set of example input and output pairs is called *Supervised Learning*. When learning needs to be done in the absence of known input–output pairs, it is called *Unsupervised Learning*. The names, supervised and unsupervised, relate to the presence or absence of outputs associated with inputs as a guide for learning suitable values of the parameters.

4.2 Machine Learning paradigms

Depending on the type of experience data the underlying learning model uses, machine-learning approaches can be categorized into three types: supervised, unsupervised, and semisupervised. Let us briefly introduce each of these three learning approaches.

4.2.1 Supervised learning

Consider the illustration in Fig. 4.2 (the rightmost part of the figure). The task is to correctly recognize the class of an uncategorized (unlabeled) object instance, given some instances, each with an assigned category or class. In this case, the class of an object corresponds to its geometric shape. The example is a 4-class problem since the shape may be any of four possibilities: circle, triangle, rectangle, or pentagon. A set of instances, along with their labels called *class labels*, are provided to train a learning model. Note that the instances may or may not be given as drawings but in terms of some attributes or features whose values may or may not be numeric. Examples of such features may include the number of sides, lengths of sides, internal angles, and the color of the shape. The features may or may not be relevant to the task at hand. Supervised machine learning (more precisely *classification* task) can be defined as follows.

Definition 4.2.1 (Supervised Learning). The learning algorithm learns a function, $\mathcal{F}$, from a set of (labeled) training instances, I_{train}, to map each test instance $X_i \in I_{test}$ to an output or target variable y. This mapping is determined based on the n features of $X_i = \{x_1, x_2, \cdots, x_n\}$ and the best probable corresponding labels based on the learning from I_{train} during training:

$$\mathcal{F} : I_{test} \implies y.$$

The algorithm uses the labeled training data to learn patterns and relationships between input features and target outputs, enabling it to predict the labels of new, unseen test instances with the knowledge gained from the labeled training data. During testing, labels of the test instances are kept hidden from the learning algorithm to match how accurately the algorithm can predict the labels based on its prior learning.

To understand better, let us consider a simpler two-class or binary class (yes/no) problem, where $y \in \{0, 1\}$. We need to learn (i.e., find patterns in the training instances) to predict the label of an unlabeled instance as either 'circle' ($y = 1$) or 'not circle' ($y = 0$). A two-class classification is called binary classification and is often easier to learn than multiclass classification [3]. That is why, even though we started with a multiclass classification problem, we reduced it to a binary classification problem for discussion. Assume that the class of objects with the class label 'circle' includes all circular objects. On the other hand, the objects included in the class 'not circle' are instances of other shapes, namely, triangle, rectangle, and pentagon. We train the model with labeled instances. In particular, during training, we train the model with salient features of the classes, such as the number of straight-line segments or the number of corners in exemplar objects. There may be many features, but we may be better off considering only the features that help distinguish a class of objects from others, i.e., classify the objects with high accuracy. Once the prediction model has been trained, the next step is to test its effectiveness using a set of unlabeled test instances. The task is to predict the class labels of unlabeled test instances

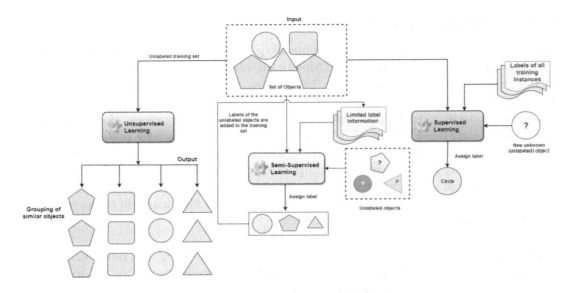

FIGURE 4.2 The three learning approaches: Unsupervised, Supervised, and Semisupervised.

using the model learned from the labeled examples, which summarizes prior knowledge. Since we consider the task as a 2-class problem for any given test object, if the learned model assigns the class label 'circle', the object should ideally have a circular shape. If the model does not assign the label 'circle', it should ideally belong to one of the three other shapes. A supervised-learning method attempts to map the test instances in this way to their respective true classes. The effectiveness of such a method depends on (i) the quality and quantity of training instances, (ii) the quality and quantity of the attributes or features used when learning to classify, and (iii) the type of function the learning algorithm is restricted to learn to discriminate among the classes.

To start the learning process, the raw rows or instances gathered or captured are preprocessed and usually split into three subsets, training, validation, and testing, as shown in Fig. 4.3. The training set is used for training the model, and the validation set is used for evaluating the performance of the trained model to improve the trained model. The ultimate purpose of a supervised-learning method is to predict a class instance with the best possible accuracy for an unlabeled test instance. The validation set is used throughout the model's training phase to evaluate the model's performance and fine tune the model's parameters. On the other hand, the test set is utilized once the model has been fully trained to evaluate the model's performance on data that has never been seen before. Once it achieves satisfactory performance, the learned model prepares to predict the category of unseen objects. There is always scope for fine tuning the model's training by tuning the parameters of the machine-learning algorithm or by restarting the whole process with different combinations of steps. Fig. 4.3 presents the overall process of a

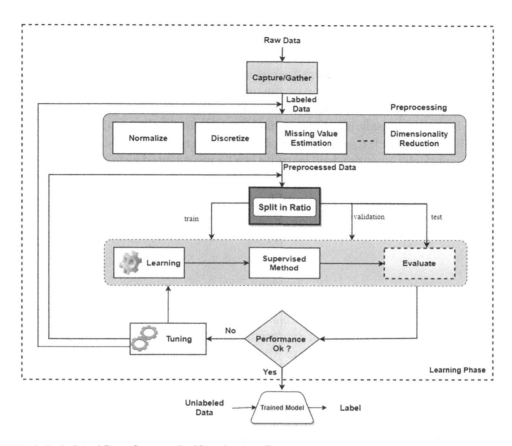

FIGURE 4.3 Typical workflow of a supervised-learning paradigm.

supervised-learning method. There are five major steps: (i) Capturing or gathering raw data, (ii) Preprocessing to make the data ready for machine-learning algorithms, (iii) Splitting the preprocessed data into three parts: training, validation, and testing, (iv) Learning the classification model from the training data and fine tuning on the validation data, and finally (v) Evaluating the trained model on the testing set, to check the performance of the method. If the performance is not satisfactory in terms of accuracy, further parameter tuning may be performed. In some situations, reperforming all or some of the prior steps with different combinations of alternatives may improve the prediction performance. Once the model performs satisfactorily, it may be deployed for real-world application. However, a good prediction system should be able to adapt to dynamic real-world scenarios by updating itself. For example, spam-email characteristics may change with time or a change of context to make the detectors malfunction. Hence, a regular updating of a detection system is essential to adjust to the changing scenarios. We present detailed discussions on supervised-learning techniques in subsequent chapters (Chapters 5 and 6).

4.2.2 Unsupervised learning

Unlike supervised learning, no prior knowledge (in terms of labeled data) is available [2,13] during unsupervised learning. An unsupervised method attempts to learn patterns from the features that can be used to group or cluster unlabeled data instances in some natural or useful way without the intervention of human expert(s). It is also known as *exploratory data analysis*. More precisely, it involves a machine-learning algorithm that learns patterns from an unlabeled dataset, $I_{unlabeled}$, in the absence of corresponding output or target variables. Mathematically, unsupervised learning can be represented as follows.

Definition 4.2.2 (Unsupervised Learning). Let $\mathcal{I}_{unlabeled}$ be the input space, which contains all input instances X, where $X = \{x_1, x_2, \cdots, x_n\}$, without any output label. The algorithm seeks to find a function $\mathcal{F}$ that maps the input space $\mathcal{I}_{unlabeled}$ to a set of representations or clusters $\mathcal{Z}$:

$$\mathcal{F} : \mathcal{I}_{unlabeled} \implies \mathcal{Z}.$$

The algorithm explores the inherent relationships and similarities within the data, aiming to discover clusters, patterns, or representations that can provide valuable insights into the underlying structure of the data. Unlike supervised learning, where labeled data guides the learning process, unsupervised learning relies solely on the input data to autonomously identify meaningful patterns and extract valuable information.

Applications of unsupervised-learning methods, such as social-media community finding [13] and biological complex finding [9,11,15,16], are suitable for exploratory data analysis. As shown in Fig. 4.2, the input to an unsupervised method is a set of objects of various shapes, usually given in terms of some descriptive features—without any labels (the pictures may be given in some cases). The task is to group similar objects so that the intragroup similarity among the members of a group is high, whereas, for a pair of instances from two different groups, the similarity is significantly less. It may lead to grouping such as (i) shapes with three corners, (ii) shapes with four corners, (iii) shapes with five corners, and (iv) shapes with no corners, or the groups may be unexpected. For example, if similarly shaped objects had different colors, the objects may get grouped by color and not by shape. In designing an unsupervised method, proximity measures play a crucial role. The precision of cluster analysis given by an unsupervised method depends on the effectiveness (or expressiveness) of the proximity measure used [4]. The success of such measures is also highly influenced by the data types, dimensionality, number of instances, and purpose of use (e.g., whether to capture trend or proximity). Input-order independence, effective noise handling, border object handling, the ability to recognize clusters of any shape and low dependence on algorithmic parameters are desired qualities of such methods [14]. To learn more about the unsupervised-learning technique, clustering, please refer to Chapter 9.

4.2.3 Semisupervised learning

Obtaining many labeled instances for all the classes is difficult in real life. When only a limited number of labeled training instances are available, how to develop a prediction model that ensures the best possible accuracy remains a challenging issue. Another issue with supervised learning is that it requires hand labeling a large number of training instances by domain experts, incurring large costs [5]. Unsupervised learning groups instances, but the obtained groups may not be consistent with actual classes present in the data, and hence may not be very useful [8]. To address these issues, semisupervised learning has been found to be useful. Semisupervised learning holds an intermediate position between supervised and unsupervised learning. Such approaches attempt to learn how to predict the class labels for unlabeled instances using labeled and unlabeled training instances [19,21]. Usually, the semisupervised approach is useful when unlabeled instances are more numerous than labeled instances. A small set of labeled instances are used to train any supervised model and, later, the same trained model is used to label a large pool of unlabeled instances.

Definition 4.2.3 (Semisupervised Learning). Given a set of (limited) labeled training instances, I_{train}, and a larger set of unlabeled instances, $I_{unlabeled}$, the algorithm learns a function, $\mathcal{F}$, from I_{train} that maps each instance $X_i \in I_{unlabeled}$ to an output or target variable y.

In Fig. 4.2, a trained semisupervised model is used to label unlabeled circles, triangles, or any other shapes using semisupervised learning. In one iteration, labeled objects with high confidence scores are added to the training input for continued learning and model improvement. The process is iterated until as many unlabeled instances as possible can be utilized for the overall performance improvement of the model. Although the idea of semisupervised learning is effective, when the amount of labeled data is extremely limited, there is a high probability that the model will overfit the training data and generate inaccurate labels. This can result in the entire model being highly erroneous. Therefore, it becomes crucial to establish a confidence threshold to determine which labeled instances should be included in the retraining process.

To distinguish among the three learning approaches, we can use the following metaphor. Supervised learning takes place when a learner is given ample learning supervision by a tutor, although the learner also uses her own analysis during learning to generalize from presented labeled examples. In unsupervised learning, the learner learns by self-study, performing her own analysis without the support of a tutor. Finally, in semisupervised learning, the learner is expected to have the ability to revise or generalize the learned content by self-study while receiving limited guidance from the tutor.

4.3 Inductive bias

A supervised machine-learning algorithm of the classifier type learns a model of the data from the training instances and uses this model to classify previously unseen examples.

This model may be an explicit mathematical function of the attributes or features or can be implicit. Since the number of possible functions to learn is infinite, depending on the algorithm, it may limit itself to a particular class of functions that it seeks to learn from the training examples. This is called the *Inductive Bias* of an algorithm. Thus, a classification algorithm may have some bias to start with. For example, it is possible that a certain classifier is biased from the beginning to draw a straight line or a plane or a hyperplane in a high dimension to separate examples of one class from another. We see such a scenario in Fig. 4.4(a). Here, the small squares represent examples from one class, and the small circles represent examples of a second class. In mathematical terms, the function such a classifier can learn is simply a straight line or a plane or a hyperplane in higher dimensions. Each training example has say two features, x and y. x can be income in dollars, and y can be savings in dollars, and the two classes are *loan-approved* (circles) and *loan-not-approved* (squares). The function learns to approve or not approve loans based on prior approval data, where each prior applicant is described in terms of income and savings.

A second example of inductive bias in a classifier may be that the learner can learn to draw two separating lines that are parallel to the axes. A pair of such lines are used to separate a class's examples from those of another class. Fig. 4.4(b) depicts such a situation. Such a classifier has to learn the coordinates of the point where the two lines meet and the direction of the two axes. A third type of bias associated with a classifier may be that it learns the description of a rectangular area that separates the two classes such that the border lines are parallel to the axes. Such a classifier has to learn four values, the x and y values for the lower corner of the rectangle, and the height and width of the rectangle (or, equivalently, the x and y values for the diagonally opposite corner). Fig. 4.4(c) shows such a rectangle that includes the square classes. A final example of inductive bias is to find a separator region bounded by margins on two sides, with the actual separator line or hyperplane exactly in the middle of the region. It also seeks to find the widest margin such that the classes are maximally separated with a "*no-man's land*" region between them. Fig. 4.4(d) shows such a maximum-margin classifier. Depending on the classifier, the inductive bias may be complex and difficult to explain. A more complex inductive bias may be that the classifier is quadratic (circular or ellipsoidal), or that it is translation invariant, meaning that the presence of a certain feature (e.g., a line or a circle) in an image matters, but not its position (in image classification). Complex machine-learning approaches like artificial neural networks have no inductive bias in the sense discussed here; they are known to be universal function approximators. Unsupervised algorithms such as clustering algorithms may have learning biases also. For example, some clustering algorithms are designed in such a way that they can discover clusters that are circular (spherical or hyperspherical if the data are high dimensional) or ellipsoidal (or, hyperellipsoidal) in nature. A very popular clustering algorithm called K-means has such a bias. Certain other clustering algorithms are grid based. Such algorithms work in spaces where the clustering algorithm needs the data to be divisible in grid configurations, and discovered clusters are grid aligned. There are other clustering algorithms that can find clusters of any random but "natural" shape and even clusters of shapes that are enclosed within other shapes (refer to Fig. 9.1 in Chapter 9).

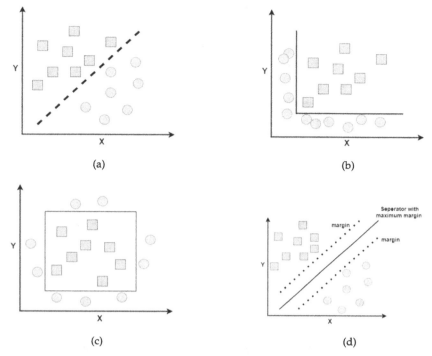

FIGURE 4.4 Different types of inductive bias. (a) A classifier may be biased to draw a straight line separating examples of the two classes. (b) The inductive bias may draw two lines that start at a point and are parallel to the two axes (c) The inductive bias may be that the classifier can draw rectangles around the space required by examples of a class. (d) The inductive bias is to draw a separator line or hyperplane like in (a), but with maximum separation of the two classes.

4.4 Evaluating a classifier

A crucial part of machine learning is evaluating the performance of the learning algorithm. We discuss here how supervised machine-learning algorithms, particularly classifiers, are evaluated. Evaluation of another popular supervised machine-learning approach called regression will be discussed in the next chapter. The evaluation process for clustering algorithms is discussed in Chapter 9.

Assessing the performance of classifiers is a fundamental aspect of machine learning. Once we have trained a classification algorithm and performed classification tasks on examples the classifier has not been trained, we need to know how good our classifier is. Evaluation of classification is also important to compare various classifiers we may have as candidates for a task at hand. It involves quantifying how well a classifier distinguishes between different classes in a dataset. Researchers and practitioners can gain insights into a classifier's performance by employing a range of evaluation metrics. This evaluation process aids in making informed decisions about model selection, parameter tuning, and understanding the classifier's behavior under different scenarios.

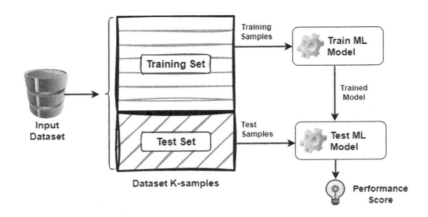

FIGURE 4.5 A typical Training and Testing workflow of a classifier on the split dataset.

4.4.1 Evaluation steps

First, we need to train our classifier. To train a classifier, we need labeled data or a dataset. For many problems we want to solve, datasets are available and can be found on the web for download. For other tasks, we may need to collect data ourselves. Quite frequently, we may be able to test our classifier on several datasets that are available for download from a website like the UCI Machine Learning Repository[1] or Kaggle.[2] Both sites contain hundreds of datasets that researchers from around the world have collected, performed experiments with, and then submitted so that others can perform experiments and compare results in a consistent manner. Fig. 4.5 illustrates the situation. In many datasets, the training and test sets or splits are clearly identified. In such cases, one trains the model on the training set and tests the trained model on the test set, and reports the results. In many machine-learning competitions, only the training set is given. The competition organizers keep the testing set hidden and run the (trained) model uploaded by the participants to obtain results on the testing set and rank participants. Since everyone tests the same way, machine-learning algorithms developed by different individuals or groups can be easily compared.

Performance evaluation should be carried out during the validation and testing phases of a machine-learning model development. These two phases serve distinct purposes in the model-development process.

4.4.1.1 Validation

After the model has been trained using the training dataset, the validation step begins. A separate validation dataset (also known as a development or holdout dataset) is used during this phase to fine tune hyperparameters, select the optimum model architecture,

---

[1] https://archive.ics.uci.edu/.

[2] https://www.kaggle.com/datasets.

and prevent overfitting (discussed below). *Hyperparameters* are the external parameters of the machine-learning model that control how a learning model learns from data. Hyperparameter tuning refers to choosing the ideal settings for hyperparameters to improve a model's performance and capacity to work accurately for unseen data samples. Typically, tuning is carried out by evaluating different combinations on a validation dataset. During the validation phase, performance evaluation assists in making decisions that improve the model's capability to handle new, previously unknown data. Effective performance evaluation ensures that machine-learning models are dependable, resilient, and suitable for real-world scenarios.

4.4.1.2 Testing
The final assessment occurs during the testing phase when the model has been tuned through validation. The test set should be different from both the training and validation sets. The test set acts as an unbiased measure of the model's performance in the real world. The goal of the trained model testing is to determine how well the model will function on entirely fresh, untested data. Measurements made using the testing dataset clearly show the model's actual performance and capacity for new instances. Note that once a model has been trained on the training data and improved iteratively on the validation data, the final trained model is obtained. The final trained model is then tested once on the testing data.

4.4.1.3 K-fold crossvalidation
Often, a dataset is available without any identified training and testing subsets. In such cases, the usual approach is to randomly divide the dataset into K equal groups or parts or folds. In crossvalidation, training and testing sets are not prespecified. The entire dataset is divided randomly into K parts. One part is set aside for testing, and the other $K - 1$ parts are used for training. This is one execution. For this execution, results are recorded. Experiments are executed considering each of the K parts or folds as a testing set and the rest as a training set. Thus, K experiments are executed. The results from all the executions are averaged and reported. How experiments are performed in crossvalidation is shown in Fig. 4.6. The purpose of performing crossvalidation is to remove accidental regularities that may occur in a dataset by coincidence and gain some feel for how a machine-learning algorithm is likely to behave with unseen examples, or in other words, gauge its predictive capability. Since we randomly pick the examples in the folds, the assumption is that any random or accidental associations or correlations among data items are at least partially removed from consideration.

4.4.2 Handling unbalanced classes

Frequently in a dataset, the examples from the various classes may be unbalanced. For example, suppose that a dataset has examples from two classes c_1 and c_2, and if 95% of the examples are from one class, say c_1, and the classifier does not learn anything and simply says that every example is from class c_1, we obtain 95% accuracy. When we work with such

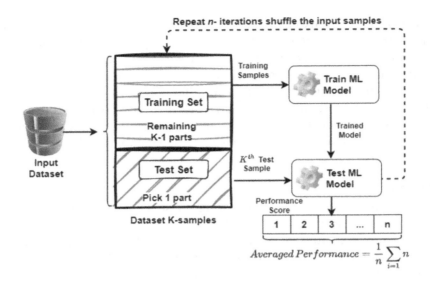

FIGURE 4.6 K-crossvalidation framework for training and testing.

unbalanced datasets, we need to take extra care to make sure the results are valid. There are several ways to handle such a situation. One approach is to oversample the smaller class. In other words, when we randomly pick examples to go into our folds, we pick examples from the smaller class with replacement, i.e., we pick the same example several times. Another approach is to create synthetic examples. In this case, for the smaller class, we generate new examples by making small random changes to some of the attribute values in the real examples. When we do so, we must make sure that the new examples created after perturbation are valid examples. For example, if the dataset is about elementary-school students, maybe the age of the students should never be over 12. We also should know that certain algorithms perform reasonably well on unbalanced datasets compared to others. For example, decision-tree approaches may work well directly on an unbalanced dataset without data augmentation. There are advanced artificial neural techniques such as Variational Autoencoders (VAE) [10] and Generative Adversarial Networks (GAN) [7] that can be used to generate realistic artificial examples for data augmentation, especially when dealing with images.

4.4.3 Model generalization

Building models in machine learning is not just about achieving impressive results on training data. It is about creating models that work well on never encountered, real-world data. The ability of a trained model to perform accurately on new, unseen data that it has not encountered during the training phase is referred to as *generalization*. When a model generalizes successfully, it means that it has discovered in the training data significant patterns and correlations that it can use to interpret new, previously unexplored

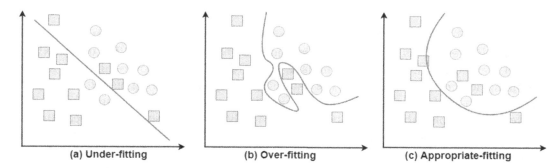

FIGURE 4.7 For a binary class problem, three trained classifiers create different separation boundaries during training based on training examples. The separation plane in (a) shows underfitting, (b) overfitting when the fitted function is very complex, and (c) appropriate fitting when the fitted function is somewhere in between, leading to better generalization.

data. A model will not perform well on new data if it merely memorizes the training data without grasping the underlying patterns in the general population of data examples, not just the examples that have been sampled for training. Therefore, generalization is crucial. A *memorizing model* merely recalls the training instances it has seen without truly understanding the underlying principles. Instead of generalizing from the data to understand the underlying relationships and trends, a memorizing model essentially memorizes the individual instances in the training data. It should be clear now why we need different training and test sets. If we train on test data, we have no idea whether the learning model is correctly generalizing or simply memorizing the training examples. However, achieving generalization for a learning algorithm is hindered by two major pitfalls, Underfitting and Overfitting.

4.4.3.1 Underfitting

A machine-learning algorithm is considered to be *underfitting* when it fails to understand the fundamental patterns or relationships within the data. It performs poorly on both the training data and fresh, previously unseen data because it struggles to grasp the underlying patterns and intricacies present in the data. This typically occurs when there are insufficient training data. Even when we attempt to develop a linear model with insufficient nonlinear data, underfitting results. To understand this better, let us consider a two-class classification problem to classify circles and squares, as shown in Fig. 4.7(a). The model is so simple that it tries to create a linear boundary or straight line to separate samples of the two classes. This leads to poor performance during both training and testing. In these situations, the machine-learning model will likely produce a number of incorrect predictions since its learned rules are too simple and naive to be applied to such sparse data. More data can be used to prevent underfitting, while feature selection can be used to reduce the number of features. As is clear, the simplicity of the fitted model may also be a significant reason for underfitting.

4.4.3.2 Overfitting

When a model achieves excellent prediction performance during training but fails to produce reliable predictions on test data, it is said to be *Overfitted*. Overfitting occurs when a model, trained with an abundance of data, learns not only the meaningful patterns but also the noise and errors present in the dataset. Consequently, during testing, the model displays a high level of variability. Since the model captured patterns corresponding to excess details and noise, the model struggles to properly classify unseen data. Overfitting is particularly noticeable in nonparametric and nonlinear machine-learning algorithms, as they possess greater flexibility to construct models from the dataset. This can inadvertently result in models that diverge from reality. For example, during training for our two-class classification problem, the model creates a separation curve or boundary that perfectly separates all the training instances (Fig. 4.7(b)) into true classes. This means the model achieves zero error during training. Such a complex boundary reduces the model's generalization capability, as the separation boundary is created with the involvement of noisy samples or random fluctuations in the training examples. As a result, when this overfitted model encounters new or unseen data, it might struggle to classify accurately because it has incorporated too many characteristic details of a flawed training set. Thus, if the learned model has a very high capacity, it may also overfit. An example of a high-capacity model may be an elaborate artificial neural network with a very large number of weights or parameters.

When hyperparameters are tuned using training data, this might result in overfitting, reducing the generalization capability of the model by making it too specifically customized to the training data's nuances. To put it another way, optimizing hyperparameters purely based on how much they enhance performance on the training set may result in a model that underperforms on fresh data. To overcome this issue, it is common practice to split the available data into three sets: training, validation, and testing. The training set is used to learn new parameters, the validation set aids in hyperparameter tuning, and the testing set evaluates the generalization capability of the model. Data splitting is one of the measures to control overfitting. Using K-fold crossvalidation creates K distinct training and testing sets, making generalization a little more achievable.

In addition to data splitting, early stopping is another measure to counter overfitting. The training is stopped early if the model's performance degrades during validation. This prevents the model from continuing to learn intricate details of the training data that may not generalize well to new data. Regularization is another popular method to control overfitting. These techniques introduce penalties to discourage the model from adopting extreme parameter values.

4.4.3.3 Accurate fittings

The ideal scenario is to have the model with appropriate fitting (Fig. 4.7(c)), such that the model performs relatively well during both training and testing. Achieving accurate fitting involves creating a model that can effectively capture the underlying patterns and relationships within the data, both in the training phase and when applied to new, unseen data. To achieve accurate fittings, we must choose a model that is neither highly complex (overfitting) nor too simplistic (underfitting). It is critical to strike the correct balance between these two extremes. Complex models can capture intricate patterns in the training data, but they may have difficulty generalizing, while overly simple models may overlook significant trends. Intuitively, when we try to deal with a nonlinearly separable classification problem, a relatively simple nonlinear boundary (Fig. 4.7(c)) may achieve better generalization. A simple nonlinear boundary focuses on capturing the major trends and general shapes of the data distribution without getting caught up in every tiny fluctuation. This prevents the model from becoming too specialized to the training data and helps it generalize better to unseen instances. Hence, allowing a certain degree of misclassification during training is often beneficial for achieving balanced and accurate fitting. While it might seem counterintuitive, striving for zero misclassification on the training data can lead to overfitting.

4.4.4 Evaluation metrics

We can use several metrics to quantify the performance of a classifier. The commonly used metrics are Accuracy, Precision, Recall, and F-measure. We provide the definitions below. Assume that there are only two classes, i.e., it is a binary classifier. To understand the terminology below, consider the first class as the *Positive* class (e.g., a class called *DOG*) and the second class as the *Negative* class (e.g., a class called *NOT-DOG*).

- *True Positives (TP)*: The set of positive examples in the test set that are classified as positive by the trained program. This is the set of true positives, i.e., positives classified as positive. Let TP be the size of the set of true positives.
- *False Negatives (FN)*: The set of positive examples in the test set that are classified as negative by the trained program. This is the set of false negatives, i.e., positives classified as negative. FN is the number of elements in the set of false negatives.
- *True Negatives (TN)*: The set of negative examples in the test set that are classified as negative by the trained program. This is the set of true negatives, i.e., negatives classified as negative. TN is the size of the set of true negatives.
- *False Positives (FP)*: The set of negative examples in the test set that are classified as positive by the trained program. This is the set of false positives, i.e., negatives classified as positive. FP is the size of the set of false positives.

As an example, if the Positive class corresponds to *DOG* and the Negative class corresponds to *NOT-DOG*, TP is the number of actual examples of dogs in the test set that are classified (or predicted) as dogs by a trained classifier. TN is the number of actual test examples of non-dogs that are classified as non-dogs by the trained classifier. FP is the number of tested non-dogs that are classified as dogs, and FN is the number of tested dogs that are classified as non-dogs. The trained classifier performs correctly in the case of true positives and true negatives, and it performs incorrectly in the case of false positives and false negatives.

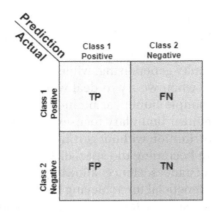

FIGURE 4.8 A confusion matrix displays numeric results of a classification experiment in one place.

4.4.4.1 Confusion matrix

If the classifier is binary, i.e., there are only two classes, the four values—TP, TN, FP, and FN—can be represented as a 2×2 matrix. Such a matrix is called a Confusion Matrix (see Fig. 4.8).

In the binary confusion matrix, the first row contains numbers corresponding to the actual Positive class, and the second row contains numbers corresponding to the Negative class. The first column contains numbers corresponding to the predicted (or, classified by the trained classifier) Positive class, and the second column contains numbers corresponding to the predicted Negative class. Combining the row and column designations, the $< 1, 1 >$ cell contains the number of actual positive test examples predicted as positive (TP). The $< 1, 2 >$ cell contains the number of actual positive test examples classified as negative (FN)—these are error cases. Similarly, the $< 2, 1 >$ cell contains the number of actual negative test examples classified as positive (FP)—these are also error cases. The $< 2, 2 >$ cell contains actual negative test examples classified as negative (TN).

If there are more than two classes, a confusion matrix can also be populated in such a case. The values in row i provide the number of actual examples of class i that have been classified by the trained model as examples of classes $1, 2, \cdots, n$, respectively. The values in column j provide the number of predicted examples of class j that actually belong to classes $1, 2, \cdots n$, respectively. Thus, the value is cell $< i, j >$ is the number of test examples of actual class i that are classified by the trained classifier as belonging to class j. The value in cell $< i, i >$ is a case of true positive classification, but the entries in all other cells in row i are cases of error. Thus, the number in cell $< i, i >$ is TP_i for class i. It is possible to obtain FN_i and FP_i values for class i ($i = 1, \cdots, n$) from the values in the table. The metrics can be computed for each class i separately in this case.

Chapter 4 • Machine learning   63

4.4.4.2 Accuracy

Accuracy is defined as the percentage of testing examples classified correctly by a classifier. In other words, for binary classification:

$$Accuracy = \frac{TP + TN}{TP + TN + FN + FP}. \qquad (4.1)$$

The numerator is the total number of examples classified correctly, whether positive examples or negative. The denominator is the total number of examples in the testing dataset. As alluded to earlier, it may not always be a good metric, especially if the classes are unbalanced. If accuracy is 100%, all the examples have been classified correctly. However, as noted earlier, if the classes are very unbalanced, a high accuracy may not indicate that the results are actually good.

Thus, in addition to or in lieu of accuracy, it is common to calculate other metrics for the evaluation of classifiers. Precision, Recall, and F-measure are a few of the many popular metrics. We discuss them below.

4.4.4.3 Precision and recall

Kent et al. [1] were the first to use the concepts of Recall and Precision, albeit the term precision did not come into usage until later in the context of information retrieval. We can compute precision and recall for each of the two classes, *Positive* and *Negative* separately, assuming it is a case of binary classification. The definitions are given below for the *Positive* (+) class:

$$Precision_+ = \frac{TP}{TP + FP}, \qquad (4.2)$$

$$Recall_+ = \frac{TP}{TP + FN}. \qquad (4.3)$$

$Precision_+$ gives a measure of precision for the *Positive* class. Precision is measured by dividing the number of test examples that are correctly classified as positive by the classifier by the number of test examples that are classified as positive (either correctly or wrongly) by the classifier. Precision for a class c is the proportion of all test examples that have been classified as belonging to class c that are true members of class c. For example, suppose our trained classifier classifies 4 testing examples as belonging to the *Positive* class. Suppose 3 of these are correctly classified and 1 is wrongly classified. That is, 3 *Positive* examples are classified as *Positive* by the classifier; in addition, 1 *Negative* test example is classified as *Positive* by the classifier. In this situation, the precision for the *Positive* class is 3/4 or 75%.

Recall measures the proportion of test examples belonging to a certain class that are classified correctly as belonging to the class. The formula given above computes recall for the positive class. Assume we have a total of 10 examples of the *Positive* class in all of the test set. Of these, our trained classifier classifies 8 as belonging to the *Positive* class. Then, our recall for the *Positive* class is 8/10=0.8 or 80%.

For a binary classifier, it is normally sufficient to provide the values of $Precision_+$ and $Recall_+$, and simply call them *Precision* and *Recall*. However, it is possible to compute precision and recall for the *Negative* class also:

$$Precision_- = \frac{TN}{TN + FN},$$ (4.4)

$$Recall_- = \frac{TN}{TN + FP}.$$ (4.5)

These definitions also assume we have a binary classifier. We show how precision and recall are computed for the *Negative* class because the idea can be generalized if we have multiclass classification, say with m classes. If we have m number of classes in total, we compute these two metrics, precision and recall for each class. In other words, we compute $2m$ values.

4.4.4.4 F-measure
The F-measure (also known as the F1-score) is a popular performance metric that was introduced by Van Rijsbergen [20]. It combines both precision and recall into a single score, providing a balanced evaluation of the model's performance. The F-measure for a class is the harmonic mean of the recall and precision for the class. The harmonic mean is used since the two metrics being combined are in opposition to each other—when one goes up, the other goes down. The F-measure is defined as follows.

$$\frac{1}{F} = \frac{1}{2}\left(\frac{1}{Precision} + \frac{1}{Recall}\right), \text{ or}$$ (4.6)

$$F = 2\frac{Precision \times Recall}{Precision + Recall}.$$

The F1-score is a number between 0 and 1, with 1 representing perfect precision and recall and 0 indicating poor performance. A higher F1-score reflects a better balance of precision and recall, as well as a good trade-off between false positives and false negatives. The F1-score is very helpful when working with datasets that are unbalanced and have one class outnumbering the other in terms of instance count. It is a useful indicator for gauging the general effectiveness of a binary classification model since it considers both false positives and false negatives, allowing for a more thorough evaluation.

4.4.4.5 Macroaveraging
If we have several classes, we can compute the three metrics: precision, recall, and F-measure for each class separately. However, how do we report the overall results considering how the trained classifier performs over all the classes? In macroaveraging, the values of precision, recall, and F-measure are calculated separately for each class. Macroaveraged precision is simply the average of the precision for all classes [17]. If we have m classes, we compute the sum of the precision values for all classes and divide the sum by m to obtain

the macroaveraged precision:

$$\text{Precision}_{i_{\text{macro}}} = \frac{1}{m} \sum_{i=1}^{m} \frac{TP_i}{TP_i + FP_i}. \qquad (4.7)$$

Similarly, macroaveraged recall and macroaveraged F-measure can be calculated as follows:

$$\text{Recall}_{i_{\text{macro}}} = \frac{1}{m} \sum_{i=1}^{m} \frac{TP_i}{TP_i + FN_i}, \qquad (4.8)$$

$$\text{F1}_{i_{\text{macro}}} = \frac{1}{m} \sum_{i=1}^{m} \frac{2 \times \text{Precision}_{i_{\text{macro}}} \times \text{Recall}_{i_{\text{macro}}}}{\text{Precision}_{i_{\text{macro}}} + \text{Recall}_{i_{\text{macro}}}}. \qquad (4.9)$$

4.4.4.6 Microaveraging

A microaveraged performance metric can also be used in multiclass classification tasks. It gives equal importance to each instance in the dataset and is suitable when there are class imbalances, as it treats all instances equally regardless of the class distribution. The microaverage is computed by considering the overall count of true positives (TP), false positives (FP), true negatives (TN), and false negatives (FN) for each class. These counts are used to calculate the precision, recall, and F1-score. The microaverage places greater emphasis on the majority class and proves beneficial when there is an imbalance among the classes. For each class i (where $i = 1 \cdots m$), the microaveraged precision is calculated as the ratio of the sum of true positives across all classes to the sum of true positives and false positives across all classes:

$$\text{Precision}_{i_{\text{micro}}} = \frac{\sum_{i=1}^{m} TP_i}{\sum_{i=1}^{m} (TP_i + FP_i)}. \qquad (4.10)$$

Similarly, the microaveraged recall and F-score can be calculated as follows:

$$\text{Recall}_{i_{\text{micro}}} = \frac{\sum_{i=1}^{m} TP_i}{\sum_{i=1}^{m} (TP_i + FN_i)}, \qquad (4.11)$$

$$\text{F1}_{i_{\text{micro}}} = \frac{2 \times \text{Precision}_{i_{\text{micro}}} \times \text{Recall}_{i_{\text{micro}}}}{\text{Precision}_{i_{\text{micro}}} + \text{Recall}_{i_{\text{micro}}}}. \qquad (4.12)$$

For crossvalidation, classification experiments are repeated several times, as discussed earlier. If we repeat the experiment K times, we obtain the average precision, recall, and F1 values over all the repeated runs. These are the values that are used to evaluate a classification algorithm. Often just a single metric, the average F1-score over all runs is used to evaluate a classification algorithm. The higher the average F1-score over all runs of an experiment, the better the trained classifier.

4.4.5 Hypothesis testing

Imagine having a dataset and conducting two machine-learning experiments using either two different algorithms or the same algorithm with two distinct configurations. From these experiments, we acquire two result sets, one for various classes and another for the overall dataset. The question arises: are these two result sets significantly different from a statistical standpoint? Also, we often perform experiments with the same learning algorithm but with different datasets or different-sized samples from the same dataset and want to know if the results are different from one another. Hypothesis testing provides a rigorous statistical framework to address this question.

Consider that we have two sets of data, such as accuracy or F1-scores, from two machine-learning experiments. Algorithm A with Configuration 1 might be used in the first experiment, while Algorithm A with Configuration 2 or Algorithm B might be used in the second experiment. Additionally, we have individual performance measures for each class in the classification task as well as an aggregate statistic that compiles data from all classes.

One can use statistical tests like the *t-test* [18] and *ANOVA* [6] to determine whether the two sets of results are statistically different from one another. These tests assist in determining whether any observed variations in the performance indicators are the result of random chance or statistically significant.

4.4.5.1 t-test

The t-test is used to compare the means of two groups. In the context of machine-learning experiments, it can be employed to compare the means of performance metrics (e.g., accuracy and F1-score over several runs) between the two sets of results. The t-test is applicable when the data follow a normal distribution and the variances of the two groups are assumed to be equal.

Suppose we have two sets of experimental results, denoted as Group 1 and Group 2, with their respective performance metrics (e.g., accuracy and F1-score). Let $\bar{x}_1$ and $\bar{x}_2$ be the sample means of the performance metrics, considering examples in the dataset, for Group 1 and Group 2, respectively, and let s_1 and s_2 be the corresponding sample standard deviations considering examples in the dataset. The t-statistic can be computed as follows:

$$t = \frac{(\bar{x}_1 - \bar{x}_2)}{\sqrt{\frac{s_1^2}{n_1} + \frac{s_2^2}{n_2}}}, \tag{4.13}$$

where n_1 and n_2 are the sample sizes of Group 1 and Group 2, respectively. The t-test allows us to determine if there is a statistically significant difference between the means of the two groups. If the t-statistic falls within the critical region, the null hypothesis is rejected, indicating that there is a significant difference between the two experimental results. Otherwise, if the t-statistic falls outside the critical region, there is no significant evidence to reject the null hypothesis, suggesting that the two sets of results are not significantly different.

4.4.5.2 ANOVA (analysis of variance)

ANOVA is used to compare means among three or more groups. In the context of machine learning, ANOVA can be employed when there are more than two sets of results to compare, such as when multiple algorithms or configurations are being evaluated on different datasets or sample sizes. ANOVA helps determine if there are significant differences among the means of the different groups.

Suppose we have k groups of experimental results, denoted as Group 1, Group 2, ..., Group k, with their respective performance metrics (e.g., accuracy and F1-score). Let $\bar{x}_1, \bar{x}_2, \cdots, \bar{x}_k$ be the sample means of the performance metrics for each group, and let $n_1, n_2, \cdots, n_k$ be the corresponding sample sizes. Additionally, let $\bar{x}$ be the overall sample mean obtained by pooling all data points from all groups. The significance (also known as F-statistics) can be calculated as follows:

$$F\text{-}statistics = \frac{\text{Between-Group Variability}}{\text{Within-Group Variability}}. \tag{4.14}$$

The between- and within-group variabilities can be calculated as follows:

$$\text{Between-Group Variability} = \sum_{i=1}^{k} n_i \cdot (\bar{x}_i - \bar{x})^2, \tag{4.15}$$

$$\text{Within-Group Variability} = \sum_{i=1}^{k} \sum_{j=1}^{n_i} (x_{ij} - \bar{x}_i)^2. \tag{4.16}$$

If the F-statistic is greater than the critical value for a chosen significance level (e.g., 0.05) the null hypothesis is rejected, indicating that there are statistically significant differences among the experimental results. Otherwise, it suggests that the means of the groups are not significantly different.

4.5 Summary

This chapter has introduced the field of machine learning and discussed three different categories of learning paradigms. In supervised learning, it is necessary to have access to labeled data examples. Labeling is usually performed by humans and is expensive in terms of time and money. In unsupervised machine learning, data examples are not labeled. Unlabeled examples are usually more abundant, and easier and cheaper to collect. However, results from unsupervised learning are often exploratory in the sense that unsupervised approaches provide initial insights into a dataset that has been collected or is available, and further experiments and exploration are necessary to gather more insights and useful information. With plentiful unlabeled data and a limited amount of labeled data, it is also possible to combine the two approaches using what is called semisupervised learning to obtain the best of both worlds.

This chapter also discussed how machine-learning experiments should be run and the results should be reported. The major concerns related to model generalization have also been discussed thoroughly. In addition, the chapter presented a detailed discussion of evaluation metrics or measurements that can be computed to assess the performance of supervised-learning models. This chapter sets the stage for the next several chapters where a number of machine-learning algorithms of various types are discussed in detail.

References

[1] Kent Allen, Madeline M. Berry, Fred U. Luehrs Jr., James W. Perry, Machine literature searching VIII. Operational criteria for designing information retrieval systems, American Documentation (pre-1986) 6 (2) (1955) 93.

[2] Horace B. Barlow, Unsupervised learning, Neural Computation 1 (3) (1989) 295–311.

[3] Dhruba Kumar Bhattacharyya, Jugal Kumar Kalita, Network Anomaly Detection: A Machine Learning Perspective, CRC Press, 2013.

[4] M. Emre Celebi, Kemal Aydin, Unsupervised Learning Algorithms, vol. 9, Springer, 2016.

[5] Pádraig Cunningham, Matthieu Cord, Sarah Jane Delany, Supervised learning, in: Machine Learning Techniques for Multimedia: Case Studies on Organization and Retrieval, Springer, 2008, pp. 21–49.

[6] R.A. Fisher, Statistical methods for research workers, Proceedings of the Royal Society of London. Series A, Containing Papers of a Mathematical and Physical Character 106 (738) (1925) 150–174.

[7] Ian Goodfellow, Jean Pouget-Abadie, Mehdi Mirza, Bing Xu, David Warde-Farley, Sherjil Ozair, Aaron Courville, Yoshua Bengio, Generative adversarial networks, Communications of the ACM 63 (11) (2020) 139–144.

[8] Gareth James, Daniela Witten, Trevor Hastie, Robert Tibshirani, Jonathan Taylor, Unsupervised learning, in: An Introduction to Statistical Learning: with Applications in Python, Springer, 2023, pp. 503–556.

[9] Monica Jha, Swarup Roy, Jugal K. Kalita, Prioritizing disease biomarkers using functional module based network analysis: a multilayer consensus driven scheme, Computers in Biology and Medicine 126 (2020) 104023.

[10] Diederik P. Kingma, Max Welling, Auto-encoding variational Bayes, arXiv preprint, arXiv:1312.6114, 2013.

[11] Hazel Nicolette Manners, Swarup Roy, Jugal K. Kalita, Intrinsic-overlapping co-expression module detection with application to Alzheimer's disease, Computational Biology and Chemistry 77 (2018) 373–389.

[12] Tom M. Mitchell, Machine Learning, McGraw-Hill Series in Computer Science, McGraw-Hill, 1997.

[13] Keshab Nath, Swarup Roy, Sukumar Nandi, InOvIn: a fuzzy-rough approach for detecting overlapping communities with intrinsic structures in evolving networks, Applied Soft Computing 89 (2020) 106096.

[14] Swarup Roy, Dhruba K. Bhattacharyya, An approach to find embedded clusters using density based techniques, in: International Conference on Distributed Computing and Internet Technology, Springer, 2005, pp. 523–535.

[15] Swarup Roy, Dhruba K. Bhattacharyya, Jugal K. Kalita, CoBi: pattern based co-regulated biclustering of gene expression data, Pattern Recognition Letters 34 (14) (2013) 1669–1678.

[16] Pooja Sharma, Hasin A. Ahmed, Swarup Roy, Dhruba K. Bhattacharyya, Unsupervised methods for finding protein complexes from PPI networks, Network Modeling Analysis in Health Informatics and Bioinformatics 4 (2015) 1–15.

[17] Marina Sokolova, Guy Lapalme, A systematic analysis of performance measures for classification tasks, Information Processing & Management 45 (4) (2009) 427–437.

[18] Student, The probable error of a mean, Biometrika 6 (1) (1908) 1–25.

[19] Jesper E. Van Engelen, Holger H. Hoos, A survey on semi-supervised learning, Machine Learning 109 (2) (2020) 373–440.

[20] C. Van Rijsbergen, Information retrieval: theory and practice, in: Proceedings of the Joint IBM/University of Newcastle upon Tyne Seminar on Data Base Systems, vol. 79, 1979.

[21] Xiaojin Jerry Zhu, Semi-Supervised Learning Literature Survey, Technical Report, Computer Sciences, University of Wisconsin-Madison, No. 1530, 2005.

Regression

5.1 Introduction

Often, we are given a dataset for supervised learning, where each example is described in terms of a number of features, and the label associated with an example is numeric. We can think of the features as independent variables and the label as a dependent variable. For example, a single independent variable can be a person's monthly income, and the dependent variable can be the amount of money the person spends on entertainment per month. The person is described in terms of one feature, income; the person's label is the amount of money spent on monthly entertainment. In this case, the training dataset consists of a number of examples where each person is described only in terms of monthly income. Corresponding to each person there is a label, which is the person's entertainment expense per month. Usually, the example is written as x and the label as y. If there are m examples in the training set, we refer to the ith example as $x^{(i)}$ and its corresponding label as $y^{(i)}$.

The training examples can be more complex if each person is described in terms of a number of features such as the person's age, income per month, gender, number of years of education, marital status, and the amount of money in the bank. In this case, the example is a vector x (also denoted as $\vec{x}$), and the label is still y. In a training set, the ith example can be referred to as $x^{(i)}$, and the corresponding label is $y^{(i)}$. The example $x^{(i)}$ is described in terms of its features. Assume there are k features, for example. The ith example is written as

$$x^{(i)} = [x_1^{(i)}, \cdots x_j^i \cdots x_k^{(i)}],$$

where $x_j^{(i)}$ is the jth feature's value for the ith example.

Table 5.1 shows a dataset from 1968, obtained from a University of Florida website[1] that presents the concentration of LSD in a student's bodily tissue and the score on a math exam. The data are based on [15]. The independent variable is LSD concentration and the dependent variable is the score. Table 5.2 provides another dataset, obtained from the same University of Florida website, based on [8]. Each row corresponds to an immigrant nationality group such as Irish, Turkish, Dutch, or Slovak. The independent variables are the percentage of an immigrant group that speaks English, the percentage that is literate, and the percentage that has been living in the US for more than 5 years. The data are from 1909. The dependent variable is the average weekly wage in dollars. For discussion purposes, ignore the first column that spells out the nationality of an immigrant group. In

---

[1] http://www.stat.ufl.edu/~winner/datasets.html.

Fundamentals of Data Science. https://doi.org/10.1016/B978-0-32-391778-0.00012-0

Table 5.1 LSD level in tissue and performance on a math exam.

LSD	Math
78.93	1.17
58.20	2.97
67.47	3.26
37.47	4.69
45.65	5.83
32.92	6.00
29.97	6.41

other words, assume only each nationality group is described by three features, and the label or dependent variable is Income. This chapter's examples deal with these and other similar datasets.

5.2 Regression

In regression, given a training dataset like the ones discussed above, a machine-learning program learns a function or a model that describes the trend in the values of the dependent variable (i.e., the label) in terms of the values of the independent variables, i.e., the feature values. The goal is to be able to predict the value of the dependent variable when a new unseen example is presented to the learned model. The regression function or model learned should be such that it generalizes well from the data, which is likely to have errors. In other words, the model learned should not overfit the training data, i.e., it should not just remember the training data or learn incidental patterns in the training data, but should predict values for unseen examples, and do so well without making too many errors.

5.2.1 Linear least-squares regression

Consider the first dataset given above. Suppose the goal is to fit a straight line that describes the relationship between the amount of drug in the tissue of an exam taker and the score received on a math test. A scatter plot for the dataset is shown in Fig. 5.1(a). Viewing the scatter plot shows a potentially straight-line relationship going from the top left corner to the bottom right corner. A regression line may look like the one shown in Fig. 5.1(b). However, there are many possible lines we can draw as the linear fit to the data. Which line should we draw?

Visually, a generic training dataset can be thought of like the matrix or table as given in Table 5.3. Note that the first column is for illustration only; it should not be considered a part of the data table for regression.

In the Drugs and Grades dataset, the goal is to fit a line defined by the equation $y = mx + c$ to the data points given, since each data example is specified as a single scalar x. The assumption in regression or machine learning, in general, is that the data are not per-

Table 5.2 Shows the percentage speaking English, the percentage literate, the percentage in the US for 5 or more years, and the income for immigrant groups in the US 1909.

Nationality	English	Literacy	Residency	Income
Armenian	54.9	92.1	54.6	9.73
Bohemian/Moravian	66.0	96.8	71.2	13.07
Bulgarian	20.3	78.2	8.5	10.31
Canadian (French)	79.4	84.1	86.7	10.62
Canadian (Other)	100.0	99.0	90.8	14.15
Croatian	50.9	70.7	38.9	11.37
Danish	96.5	99.2	85.4	14.32
Dutch	86.1	97.9	81.9	12.04
English	100.0	98.9	80.6	14.13
Finnish	50.3	99.1	53.6	13.27
Flemish	45.6	92.1	32.9	11.07
French	68.6	94.3	70.1	12.92
German	87.5	98.0	86.4	13.63
Greek	33.5	84.2	18.0	8.41
Hebrew (Russian)	74.7	93.3	57.1	12.71
Hebrew (Other)	79.5	92.8	73.8	14.37
Irish	100.0	96.0	90.6	13.01
Italian (Northern)	58.8	85.0	55.2	11.28
Italian (Southern)	48.7	69.3	47.8	9.61
Lithuanian	51.3	78.5	53.8	11.03
Macedonian	21.1	69.4	2.0	8.95
Hungarian	46.4	90.9	44.1	11.65
Norwegian	96.9	99.7	79.3	15.28
Polish	43.5	80.1	54.1	11.06
Portuguese	45.2	47.8	57.5	8.10
Roumanian	33.3	83.3	12.0	10.90
Russian	43.6	74.6	38.0	11.01
Ruthenian	36.8	65.9	39.6	9.92
Scottish	100.0	99.6	83.6	15.24
Servian	41.2	71.5	31.4	10.75
Slovak	55.6	84.5	60.0	11.95
Slovenian	51.7	87.3	49.9	12.15
Swedish	94.7	99.8	87.4	15.36
Syrian	54.6	75.1	45.3	8.12
Turkish	22.5	56.5	10.0	7.65

fect, because the data have flaws due to reasons such as observation errors, record-keeping errors, and transcription errors. Thus, when we fit a regression line, each point may actually be not sitting exactly on the line. Thus, the fit of each point to the regression line may have an error in it. Let the error associated with the fit of the ith point be $\epsilon^{(i)}$. Since the

Table 5.3 A generic dataset for regression.

No.	$x_1 \cdots x_n$	y
1	$\cdots$	.
2	$\cdots$	.
$\vdots$	$\vdots$	$\vdots$
i	$x_1^{(i)} \cdots x_n^{(i)}$	$y^{(i)}$
$\vdots$	$\vdots$	$\vdots$
n	$\cdots$	.

dataset has n points in it, the total cumulative error of fit of a regression line to a training dataset can be written as

$$E = \sum_{i=1}^{n} \epsilon^{(i)} \tag{5.1}$$

if all the errors across the dataset are simply added. Thus, we can write an objective function to obtain the regression line as

Find line $y = mx + c$ such that it minimizes the cumulative error $E = \sum_{i=1}^{n} \epsilon^{(i)}$.

However, the problem with this approach is that some of the errors are positive and some of the errors are negative, and since negative errors cancel positive errors, the regression line that minimizes the direct sum of errors may actually turn out to be a bad fit.

An alternative may be to compute the cumulative error as

$$E = \sum_{i=1}^{n} |\epsilon^{(i)}|, \tag{5.2}$$

where $|.|$ is the absolute value. This representation of cumulative error is good, but absolute values are usually difficult to deal with in mathematics. As a result, an alternative approach that is commonly used is that the cumulative error is the sum of the squares of individual errors, given by

$$E = \sum_{i=1}^{n} \left\{ \epsilon^{(i)} \right\}^2. \tag{5.3}$$

Since squares of both positive and negative errors are positive, the total error does not vanish unless there is no cumulative error at all. If there is error, depending on the magnitude of the individual errors (less than 1 or more than 1), the error can be squashed or magnified. The modified objective function to obtain the regression line becomes

Find line $y = mx + c$ such that it minimizes the cumulative error $E = \sum_{i=1}^{n} \left\{ \epsilon^{(i)} \right\}^2$.

Since the error $\epsilon^{(i)} = y^{(i)} - mx^{(i)} - c$, we can write the cumulative error expression as

$$E = \sum_{i=1}^{n} \left\{ y^{(i)} - mx^{(i)} - c \right\}^2.$$

In this expression, the $x^{(i)}$ and $y^{(i)}$ values are known from the training dataset. The values for m and c that minimize E need to be obtained. To find the equation of the line that minimizes this cumulative error, we obtain its partial derivatives with respect to the two "variables" m and c, set them to 0 and solve for the values of m and c:

$$0 = \frac{\partial E}{\partial m}$$

$$= \frac{\partial}{\partial m} \sum_{i=1}^{n} \left\{ y^{(i)} - mx^{(i)} - c \right\}^2$$

$$= 2 \sum_{i=1}^{n} \left\{ y^{(i)} - mx^{(i)} - c \right\} \frac{\partial}{\partial m} \sum_{i=1}^{n} \left\{ y^{(i)} - mx^{(i)} - c \right\}$$

$$= 2 \sum_{i=1}^{n} \left\{ y^{(i)} - mx^{(i)} - c \right\} x^{(i)}$$

$$= \sum_{i=1}^{n} \left\{ y^{(i)} - mx^{(i)} - c \right\} x^{(i)}.$$

This gives us a linear equation in m and c as shown below:

$$\sum_{i=1}^{n} y^{(i)} - \left(\sum_{i=1}^{n} \left\{ x^{(i)} \right\}^2 \right) m - \left(\sum_{i=1}^{n} \left\{ x^{(i)} \right\} \right) c = 0. \qquad (5.4)$$

Similarly, we set the partial derivative of E w.r.t. c to 0 as well:

$$0 = \frac{\partial E}{\partial c}$$

$$= \frac{\partial}{\partial c} \sum_{i=1}^{n} \left\{ y^{(i)} - mx^{(i)} - c \right\}^2$$

$$= 2 \sum_{i=1}^{n} \left\{ y^{(i)} - mx^{(i)} - c \right\} \frac{\partial}{\partial c} \sum_{i=1}^{n} \left\{ y^{(i)} - mx^{(i)} - c \right\}$$

$$= -2 \sum_{i=1}^{n} \left\{ y^{(i)} - mx^{(i)} - c \right\}$$

$$= \sum_{i=1}^{n} \left\{ y^{(i)} - mx^{(i)} - c \right\}.$$

This can be rewritten as another linear equation in m and c:

$$\sum_{i=1}^{n} y^{(i)} - \left(\sum_{i=1}^{n} \left\{x^{(i)}\right\}\right) m - \sum_{i=1}^{n} c = 0$$

$$\sum_{i=1}^{n} y^{(i)} - \left(\sum_{i=1}^{n} \left\{x^{(i)}\right\}\right) m - nc = 0. \tag{5.5}$$

Given two linear equations (Eqs. (5.4) and (5.5)) in m and c, we can solve for them and obtain the equation of the line.

The solutions are

$$m = \frac{n\left(\sum_{i=1}^{n}\left\{x^{(i)}y^{(i)}\right\}\right) - \left(\sum_{i=1}^{n}\left\{x^{(i)}\right\}\right)\left(\sum_{i=1}^{n} y^{(i)}\right)}{n\left(\sum_{i=1}^{n}\left\{x^{(i)}\right\}^2\right) - \left(\sum_{i=1}^{n}\left\{x^{(i)}\right\}\right)^2} \tag{5.6}$$

$$c = \frac{1}{n}\sum_{i=1}^{n} y^{(i)} - m\frac{1}{n}\sum_{i=1}^{n} x^{(i)}$$

$$= \bar{y} - m\bar{x}. \tag{5.7}$$

In Eq. (5.7), $\bar{y}$ and $\bar{x}$ are means.

The equations can be programmed directly using a programming language of choice. We can also perform linear least-squares regression using a wide variety of tools. If the dataset is two-dimensional, it is a good idea to perform a scatter plot of the data first. A scatter plot for the Drugs and Grades dataset is seen in Fig. 5.1. A linear regression fit obtained using the programming language R can be seen in Fig. 5.1. The equation of the line given by R is

$$y = -0.09743473x + 9.21308462. \tag{5.8}$$

5.3 Evaluating linear regression

There are various metrics that are used to evaluate how well a trained model fits the data. A few of them are discussed below.

5.3.1 Coefficient of determination R^2

A common way to determine how well a linear regression model fits the data is by computing the coefficient of determination, R^2. If $\hat{y}^{(i)}$ is the predicted value for $x^{(i)}$ using the regression line, whereas $y^{(i)}$ is the actual value of the dependent variable, the coefficient of determination is given as follows:

$$R^2 = 1 - \frac{\sum_{i=1}^{n}\left(y^{(i)} - \hat{y}^{(i)}\right)^2}{\sum_{i=1}^{n}\left(y^{(i)} - \bar{y}\right)^2}. \tag{5.9}$$

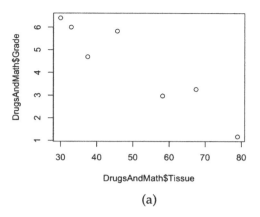
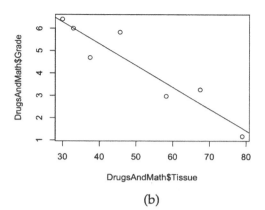

(a)   (b)

FIGURE 5.1 Scatter Plot and Linear-Regression Line for the Drugs and Grades dataset.

If the predicted values are close to the actual values, R^2 should be close to 1. If the predictions are not close to the actual values, $R^2 \approx 0$. The value of R^2 is always between 0 and 1, and a value close to 1 is better, although if we are comparing two fitted models, one that has a higher R^2 value is not necessarily better.

The R^2 metric describes the proportion of variation in the predicted variable explained by the regression model. The R^2 value is 0.8778 for the Drugs and Grades dataset, signifying that 87.78% of the variation in the math grade is captured by the model.

5.3.2 Standard error of regression and F-statistic

The Sum of Squared Errors, SSE, is defined as

$$SSE = \sum_{i=1}^{n} \left(\hat{y}^{(i)} - \overline{y} \right)^2. \tag{5.10}$$

The Mean-Squared Error, MSR, is defined as

$$MSR = \frac{SSE}{n-q}, \tag{5.11}$$

where q is the number of coefficients in the model. $q = 2$ for linear regression in two dimensions. Thus, for linear regression

$$MSR = \frac{SSE}{n-2}. \tag{5.12}$$

This definition reflects the fact that to begin with, we have n variables or degrees of freedom in the description of the data. Since there are 2 coefficients, two degrees of freedom are covered by the linear regression line.

The Standard Error for linear regression is defined as

$$StdError = \sqrt{MSE},$$ (5.13)

where MSE is the mean of the squared errors. Smaller values of $StdError$ are better. A value closer to 0 is good for $StdError$.

F-statistic is defined as

$$F\text{-}statistic = \frac{MSR}{MSE}.$$ (5.14)

A higher value of F-statistic is better than a lower value.

5.3.3 Is the model statistically significant?

It is customary to compute what is called a p-value when performing many statistical computations to indicate if the results obtained are statistically significant, i.e., they have any (statistical) merit. That is, whether the results can be trusted and are meaningful, and can really be used. P-values are computed in different ways for different situations.

When we compute a p-value, there is always an associated Null Hypothesis and an Alternate Hypothesis. For linear regression, the Null Hypothesis is that the coefficients associated with the independent variables (i.e., features) are all 0. In other words, the Null Hypothesis says there is no relationship of any significance between the dependent and independent variables. The Alternate Hypothesis is that the coefficients are not equal to 0, i.e., there is actually a linear relationship between the dependent variable and the independent variable(s).

To compute the p-value, we have to assume a value for what is called a t-value in turn. A larger t-value, written simply as t, indicates that it is less likely that the coefficient is not equal to 0 purely by chance. Hence, a higher t-value is better.

The p-value is defined as $probability(> |t|)$, i.e., the probability that the t-value computed is high or higher than the observed value when the Null Hypothesis is true. Hence, if $probability(> |t|)$ is low, the coefficients are significant (significantly different from 0). If $probability(> |t|)$ is high, the coefficients are not significant.

In practice, we can assume a significance level for p-value as something like 0.05. If the computed p-value is less than the significance level (here, 0.05), we can reject the Null Hypothesis that the coefficients are 0. In other words, the linear regression model we computed is meaningful and can be used for predictive purposes.

5.4 Multidimensional linear regression

So far, we have discussed regression where we have a single independent variable and a single dependent variable. However, it is quite likely that we have several independent variables or features that are needed to describe each example in a training dataset. Thus, the ith example is given as

$$\boldsymbol{x}^{(i)} = [x_1^{(i)}, \cdots x_j^{(i)} \cdots x_k^{(i)}],$$

where $x_j^{(i)}$ is the jth feature's value for the ith example.

In such a case, when we fit a regression model, the equation of the model is

$$y = \theta_0 + \theta_1 x_1 + \cdots + \theta_j x_j + \cdots + \theta_n x_n, \tag{5.15}$$

where

$$\boldsymbol{\theta} = [\theta_0, \theta_1, \cdots, \theta_j, \cdots, \theta_n]$$

is the vector of coefficients in the fitted line. In other words, they are the parameters that need to be found to do the fitting.

For the ith point $[x_1^{(i)}, \cdots, x_n^{(i)}]$, its y value $y^{(i)}$ is given as follows:

$$y^{(i)} = \theta_0 + \theta_1 x_1(i) + \cdots + \theta_j x_j(i) + \cdots + \theta_n x_n(i) + \epsilon^{(i)},$$

where $\epsilon^{(i)}$ is the error in prediction by the fitted hyperplane. In other words, the error in fit of the ith training data point is given as

$$\epsilon^{(i)} = y^{(i)} - \theta_0 - \theta_1 x_1^{(i)} - \cdots - \theta_j x_j^{(i)} - \cdots - \theta_n x_n^{(i)}. \tag{5.16}$$

Just like fitting a regression line when we had scalar x, we find the equation of the hyperplane in n dimensions, for which the sum of the squares of the individual errors for each training data point is minimized.

Thus the objective function to obtain the regression hyperplane becomes:

Find hyperplane $y = \theta_0 + \theta_1 x_1 + \cdots + \theta_j x_j + \cdots + \theta_n x_n$ such that it minimizes the cumulative error $E = \sum_{i=1}^{n} \left\{ \epsilon^{(i)} \right\}^2$.

Like before, we have to compute $\frac{\partial E}{\partial \theta_0} \cdots \frac{\partial E}{\partial \theta_j} \cdots \frac{\partial E}{\partial \theta_n}$ and set each one to 0.

Let us just compute $\frac{\partial E}{\partial \theta_0}$ here and leave the others as an exercise for the reader:

$$
\begin{aligned}
0 &= \frac{\partial E}{\partial \theta_0} \\
&= \frac{\partial}{\partial \theta_0} \left\{ \epsilon^{(i)} \right\}^2 \\
&= \frac{\partial}{\partial \theta_0} \left\{ y^{(i)} - \theta_0 - \theta_1 x_1^{(i)} - \cdots - \theta_j x_j^{(i)} - \cdots - \theta_n x_n^{(i)} \right\}^2 \\
&= 2 \left\{ y^{(i)} - \theta_0 - \theta_1 x_1^{(i)} - \cdots - \theta_j x_j^{(i)} - \cdots - \theta_n x_n^{(i)} \right\} \\
&\qquad \frac{\partial}{\partial \theta_0} \left\{ y^{(i)} - \theta_0 - \theta_1 x_1^{(i)} - \cdots - \theta_j x_j^{(i)} - \cdots - \theta_n x_n^{(i)} \right\} \\
&= -2 \left\{ y^{(i)} - \theta_0 - \theta_1 x_1^{(i)} - \cdots - \theta_j x_j^{(i)} - \cdots - \theta_n x_n^{(i)} \right\} \\
&= \left\{ y^{(i)} - \theta_0 - \theta_1 x_1^{(i)} - \cdots - \theta_j x_j^{(i)} - \cdots - \theta_n x_n^{(i)} \right\}. \tag{5.17}
\end{aligned}
$$

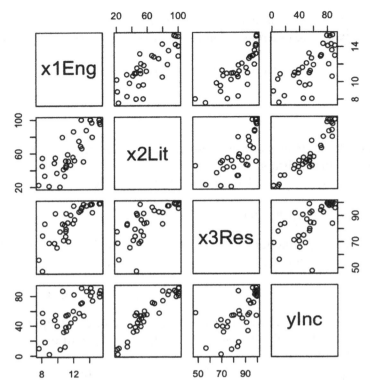

FIGURE 5.2 A plot of the Immigrants dataset. For each plot in a row, the indicated variable in the row is on the X-axis, with the indicated variable from another row on the Y-axis.

By partially differentiating E with respect to $n + 1$ variables, we obtain $n + 1$ equations in $n + 1$ unknowns, $\theta_0 \cdots \theta_n$. We can solve them using any method for solving a system of linear equations and obtain the equation of the hyperplane that is the regression model that fits the training dataset the best.

A plot of the multidimensional Immigrants and Incomes dataset, discussed earlier, is shown in Fig. 5.2. A linear regression fit using R is given as

$$y = -0.2318x_1 + 1.0436x_2 - 0.1568x_3 + 7.4864, \tag{5.18}$$

with a p-value of $1.307\ e^{-14}$, a very small value, making the fit line statistically acceptable. x, y, and z are the three independent variables.

5.5 Polynomial regression

Instead of fitting a linear function, we can also fit a polynomial of a higher degree to our data. We assume that we have only one independent variable. The independent variable is x, and the dependent variable is y. In such a case, just by looking at the data, we may be able to surmise that a linear relationship does not exist between the dependent and independent variables. In such a situation, we can attempt to fit polynomials of several degrees to see what fits the data the best.

In general, we can fit an nth-degree polynomial:

$$y = \theta_n x^n + \cdots + \theta_1 x + \theta_0 \tag{5.19}$$

to a dataset. As an illustrative example, let us try to fit a second-degree polynomial

$$y = \theta_2 x^2 + \theta_1 x + \theta_0. \tag{5.20}$$

Since there is an error of fit at every point, we can write for the ith point,

$$y^{(i)} = \theta_2 \left(x^{(i)} \right)^2 + \theta_1 x^{(i)} + \theta_0 + \epsilon^{(i)}.$$

Then, the error of fit at the ith training point is

$$\epsilon^{(i)} = y^{(i)} - \theta_2 \left(x^{(i)} \right)^2 - \theta_1 x^{(i)} - \theta_0.$$

As usual, the problem to solve becomes:

Find $y = \theta_n x^n + \cdots + \theta_1 x + \theta_0$ *such that it minimizes the cumulative error* $E = \sum_{i=1}^{n} \left\{ \epsilon^{(i)} \right\}^2.$

Like before, we have to compute $\frac{\partial E}{\partial \theta_0}$, $\frac{\partial E}{\partial \theta_1}$, and $\frac{\partial E}{\partial \theta_2}$ and set each one to 0. Let us just compute one of these partial derivatives here: $\frac{\partial E}{\partial \theta_2}$.

$$
\begin{aligned}
0 &= \frac{\partial E}{\partial \theta_2} \\
&= \frac{\partial}{\partial \theta_2} \left\{ \epsilon^{(i)} \right\}^2 \\
&= \frac{\partial}{\partial \theta_2} \left\{ y^{(i)} - \theta_2 \left(x^{(i)} \right)^2 - \theta_1 x^{(i)} - \theta_0 \right\}^2 \\
&= 2 \left\{ y^{(i)} - \theta_2 \left(x^{(i)} \right)^2 - \theta_1 x^{(i)} - \theta_0 \right\} \\
&\qquad \frac{\partial}{\partial \theta_2} \left\{ y^{(i)} - \theta_2 \left(x^{(i)} \right)^2 - \theta_1 x^{(i)} - \theta_0 \right\} \\
&= -2 \left\{ y^{(i)} - \theta_2 \left(x^{(i)} \right)^2 - \theta_1 x^{(i)} - \theta_0 \right\} \left(x^{(i)} \right)^2 \\
&= \left\{ y^{(i)} - \theta_2 \left(x^{(i)} \right)^2 - \theta_1 x^{(i)} - \theta_0 \right\} \left(x^{(i)} \right)^2. \tag{5.21}
\end{aligned}
$$

Similarly, we can obtain two other equations by setting the other two partial derivatives to 0. Thus, we have three linear equations in three unknowns, θ_2, θ_1, and θ_0. This gives us the second-degree function we want to fit into the training dataset.

Table 5.4 Heat Capacity of Hydrogen Bromide and Temperature in Kelvin.

Capacity	Temp
10.79	118.99
10.80	120.76
10.86	122.71
10.93	125.48
10.99	127.31
10.96	130.06
10.98	132.41
11.03	135.89
11.08	139.02
11.10	140.25
11.19	145.61
11.25	153.45
11.40	158.03
11.61	162.72
11.69	167.67
11.91	172.86
12.07	177.52
12.32	182.09

In a similar way, we can find the polynomial of any degree that we want to fit to a dataset. Consider a dataset (see Table 5.4) called the Heat Capacity dataset obtained from the University of Florida website on regression data. Three regression models, linear, quadratic, and cubic were fit to this dataset, considering the independent variable, capacity. The three regression models that were obtained are given below:

$$t = -348.277 + 43.761c \tag{5.22}$$

$$t = -3403.875 + 576.460c - 23.174c^2 \tag{5.23}$$

$$t = -12993.483 + 3078.252c - 240.513c^2 + 6.287c^3. \tag{5.24}$$

For the linear fit, an R implementation using a library finds an R^2 value of 0.9457, and F-statistic value of 278.8 with a p-value of $1.516e^{-11}$. For the quadratic fit, a similar R implementation finds an R^2 value of 0.9886, and F-statistic value of 648 with a p-value of $2.747e^{-15}$. For the cubic fit, R implementation finds an R^2 value of 0.9892, and F-statistic value of 426.5 with a p-value of $5.437e^{-14}$. Based on the R^2 values, the linear fit explains 94.57% of the variance in the data, the quadratic fit explains 98.86% of the variance, and the cubic fit explains 98.92% of the variance. Thus, the quadratic and cubic fits are almost the same in quality. Based on F-statistic values, the quadratic fit is the best since the value is the highest. The p-values for the F-statistics are low, and as a result, all three are statistically acceptable. A comparison of the three models over a scatter plot of the data is given in

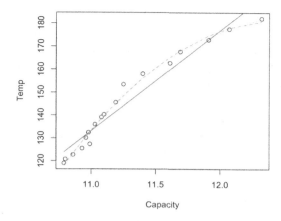

FIGURE 5.3 Scatter Plot, and Plot of three Polynomial Fits on the Heat-Capacity dataset.

Fig. 5.3. In this figure, the quadratic fit is dashed black, and the cubic fit is red dashed (mid gray in print version). Based on the F-statistic values, in this case, the quadratic fit should arguably be considered the best fit among the three. Moreover, a simpler fit (explanation) is usually better than a more complex fit.

5.6 Overfitting in regression

Overfitting happens when there are coincidental or accidental patterns in the training data. These accidental patterns could arise due to the presence of noise in the data, or because the size of the training dataset is small, or can happen just randomly.

One of the main purposes of a machine-learning algorithm is to be able to generalize, i.e., look past such accidental or random patterns and be able to perform well when dealing with previously unseen data. In the case of regression, it would mean that we would like to be able to predict well the value of the dependent variable given the values of the independent variables. A regressor that fits very well to the data, i.e., has a very low sum of squared errors, may in fact, be a bad generalizer; we say that such a regressor has *overfit* the data. A regressor that does not fit at all well to the data, i.e., has a very high sum of squared errors, is also bad; we say that such a regressor *underfits* the data. The goal in machine learning, and regression, in particular, is to neither underfit nor overfit.

5.7 Reducing overfitting in regression: regularization

There are several approaches to reducing overfitting. A common method is called regularization.

When a function is fitted using ordinary least-squares linear regression, the coefficients are easily understood and as a result, the fit is easy to interpret. However, it is likely that some predictor variables not related to the dependent variable occur in the fitted equation, making it unnecessarily complex. Even if a predictor is unrelated to the dependent variable, the corresponding coefficients are unlikely to be zero in an ordinary least-squares fit. There may also be wild swings in the fitted curve if the degree of the polynomial is high, i.e., we fit a polynomial instead of a linear function.

Thus, a preferred solution to the regression problem may fit a linear (or maybe even higher order) function, but constrain the sizes of the coefficients in some ways. Such methods are called *shrinkage* methods. Shrinkage methods suppress the values of the coefficients, preventing the fitted equation from fitting too closely to the dataset (also, from having wild swings if it is a polynomial fit), i.e., providing stability to the predictions and a measure of generalization. Some shrinkage methods can also reduce certain coefficients to zero, performing a type of feature reduction or selection.

We discuss three methods of shrinkage regression here: Ridge regression, Lasso regression, and Elastic-Net regression.

5.7.1 Ridge regression

Ridge Regression [7,12,13] attempts to dampen the coefficients of a least-squares linear regression fit. Ridge regression adds an extra component to the error function, which is also called a loss function. It is denoted by L here. The objective of optimization (minimization) becomes

$$L = \sum_{i}^{N} \left(y^{(i)} - \theta_0 \sum_{j=1}^{n} \theta_i x_j^{(i)} \right)^2 + \lambda \sum_{j=1}^{n} \theta_j^2 \qquad (5.25)$$

$$= LSE + \lambda \sum_{j=1}^{n} \theta_j^2, \qquad (5.26)$$

where LSE is the Least-Squared Error we have discussed before. $\lambda \sum_{j=1}^{n} \theta_j^2$ is called the Shrinkage Penalty. When the value of shrinkage-penalty coefficient λ is set arbitrarily high, the coefficients $\theta_0, \theta_1, \cdot, \theta_n$ can be forced to be arbitrarily low. Therefore such an optimization objective has the effect of shrinking the coefficients toward 0. The values of the θ_is can be positive or negative.

λ is the tuning parameter that controls or regulates the amount of shrinkage of the regression coefficients. For example, if $\lambda = 0$, there is no shrinkage; it is simply the linear least-squares regression. If $\lambda \to \infty$, the values of the θ_js can be made arbitrarily small. The fitted equations that come out of Ridge Regression are not unique and depend on the value of λ. Thus, it is important to obtain good values of λ. This is usually done using the approach called crossvalidation.

θ_0 is the intercept on the Y-axis of the model. If θ_0 is shrunk or penalized, we may force the model to always have a low intercept. Thus, it is recommended that θ_0 be estimated separately. The developers of a library called *glmnet* [5,7] that performs Ridge Regression

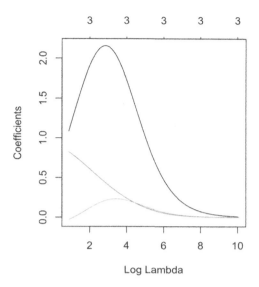

FIGURE 5.4 Values of the coefficients as λ changes in Ridge Regression.

in R recommend that

$$\theta_0 = \bar{y} = \frac{1}{N} \sum_{i=1}^{N} y^{(i)}.$$

The remaining parameters are obtained using an optimization technique.

It is to be noted that λ is a parameter in Ridge Regression and potentially can take any value. When using a programming library such as *glmnet* to perform Ridge Regression in a language like R, it is possible to change the value of λ and plot the coefficients of the fitted polynomial as well as the mean-squared error. Based on these plots, the user can decide upon the value of λ to use.

Fig. 5.4 shows a graph with coefficients on the Y-axis and log of the λ values on the X-axis, obtained by running the implementation of Ridge Regression in the *glmnet* library in R on the Immigrants and Incomes dataset discussed earlier. The log plot allows for a large range of λ values to be considered. It shows the coefficients as y values obtained for least-squares linear regression as λ is varied. It also shows that values of two of the coefficients first grow and then fall toward 0 as λ increases. For high values of λ, all the coefficients become very low, close to 0. The value of the third coefficient is the highest for the smallest value of λ, and then falls all the way as the values of λ become larger.

The plot of crossvalidation, which shows how the mean-squared error changes with respect to changes in the log of λ is given in Fig. 5.5. The actual values of λ fitted vary between $log(0)$, which is 1 and $log(2^{10})$, which is 1,204, in this case. The plot also shows the value of λ (λ_{min}) where the best value of squared error occurs. It occurs on the left vertical line. The other vertical line is the λ (λ_{1se}) value at one standard deviation away from where

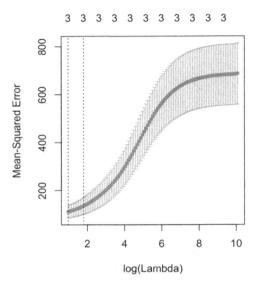

FIGURE 5.5 Ridge-Regression crossvalidation: Mean-squared error as λ changes.

the smallest sum of least-squared errors occurs. The authors of *glmnet* [5] recommend the λ_{1se} for better results. The plot has the number 3 written on top several times, indicating that there are 3 nonzero parameters or coefficients in the model that result as λ changes. That is, the number of nonzero parameters is always 3, no matter what value of λ is picked, in this case.

5.7.2 Lasso regression

In Ridge Regression, as the value of the hyperparameter λ becomes larger, the parameters $\theta_0 \cdots \theta_n$ become smaller and smaller, but the way it has been designed, the parameters never become 0 although they could become quite small. Thus, all the features are likely to matter in the final regression equation no matter what value of λ we choose, although some may become marginal if the corresponding coefficients become really small in absolute value. The fact that all independent variables remain at the end, may not matter much in this example with three independent variables or features, but if we have a lot of features, to begin with, say tens or hundreds or even more, Ridge Regression will keep them all even though the coefficients may become tiny. Making some of the coefficients exactly 0 so that they do not matter at all in the final equation may improve the interpretability of the regression equation since the number of dependent variables will become smaller. For example, if there are 50 independent variables to begin with and we are left with only 10 at the end, it becomes much easier to visualize or understand the relationships between the ten independent variables and the dependent variable. Lasso Regression [9,11] has been designed to achieve this type of reduction in the number of independent variables that

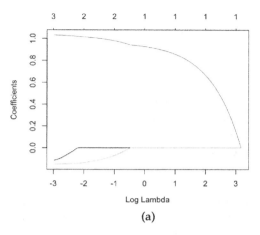
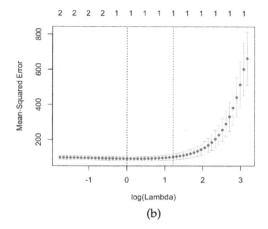

(a)   (b)

FIGURE 5.6 (a) Values of the coefficients as λ changes in Lasso Regression, (b) Mean-Squared Error as λ changes in Lasso Regression.

matter as λ becomes larger. The effect of removing independent variables from consideration is called feature selection.

Lasso Regression is quite similar to Ridge Regression in formulation, but instead of an L_2-loss in Ridge Regression, it uses absolute values of the parameters in the shrinkage component or squared loss:

$$L = \sum_{i}^{N} \left(y^{(i)} - \theta_0 - \sum_{j=1}^{n} \theta_i x_j^{(i)} \right)^2 + \lambda \sum_{j=1}^{n} \left| \theta_j \right| \tag{5.27}$$

$$= LSE + \lambda \sum_{j=1}^{n} \left| \theta_j \right|. \tag{5.28}$$

Here, λ is a hyperparameter, and like in Ridge Regression, it is used to increase the generalization and prediction accuracy of regression in addition to providing better interpretability. For our example with three dependent variables, Fig. 5.6 shows how the coefficients (θ_0, θ_1, and θ_2) vary as $log(\lambda)$ changes from -3 to slightly over 3 when running *glmnet*'s implementation of Lasso Regression. The integers on the top of the two plots show the number of nonzero coefficients in the resulting regression equation as the value of λ changes. Fig. 5.6(a) shows that there are 3 nonzero coefficients when $log(\lambda) = -3$, but the number goes down to 1 when $log(\lambda) \geq 0$. Fig. 5.6(b) shows that the mean-squared error is minimum at $log(\lambda) = 0$ or $\lambda = 1$.

5.7.3 Elastic-net regression

Ridge Regression pushes the values of the coefficients toward 0 but does not make them exactly 0, as the hyperparameter λ rises in value. Lasso Regression pushes the values of coefficients to 0 as the value of λ becomes large, partially eliminating features as λ rises. Thus, both regression techniques produce a family of learned models, not just one. The proper choice of λ that leads to the regression equation we finally use, needs a search through crossvalidation. The authors of Ridge and Lasso regression software in the library called glmnet [5] recommend using a value where the mean-squared error is the lowest, and also alternatively, a value of λ one standard deviation away in the positive direction of λ. The ultimate choice is the user's based on real-world considerations outside of mathematics.

A problem both Ridge and Lasso regressions try to solve is how to handle features that are related to each other linearly. It has been observed that if there are two linearly related features, the coefficient of one may become zero much before the other does in Lasso Regression, and thus, some similar features are left in whereas others are eliminated. Thus, related features are not treated similarly, and why this happens to which feature is not systematic or easily explainable. Ridge regression pushes coefficients of all of the related features down in a more predictable manner, although not down to 0.

Elastic-Net Regression [16] attempts to combine the effects of both Lasso and Ridge regressions. In other words, Elastic-Net Regression treats related features similarly as coefficients are shrunk, and also helps in feature selection. The objective function for Elastic-Net Regression minimizes the loss function given below:

$$L = \sum_{i}^{N} \left(y^{(i)} - \theta_0 - \sum_{j=1}^{n} \theta_i x_j^{(i)} \right)^2 + \lambda \left((1-\alpha) \sum_{j=1}^{n} \theta_j^2 + \alpha \sum_{j=1}^{n} \left| \theta_j \right| \right) \tag{5.29}$$

$$= LSE + \lambda \left((1-\alpha) \sum_{j=1}^{n} \theta_j^2 + \alpha \sum_{j=1}^{n} \left| \theta_j \right| \right). \tag{5.30}$$

Thus, the two losses, one used in Ridge and the other used in Lasso, are both used, and which one is predominant depends on the value of α. The λ hyperparameter exerts the chosen amount of shrinkage pressure on the coefficients used in the regression. Ideally, one should search over all possible values of both λ and α to find the combination that works best. Therefore a grid search or some intelligent search in both parameters may be helpful. Note that in glmnet, setting alpha to 0 performs Ridge Regression, and setting alpha to 1 performs Lasso Regression. Any other value of alpha makes it Elastic-Net Regression.

In the run whose results are reported in Fig. 5.7, $\alpha = 0.5$ so that both constraints, Ridge and Lasso, are equally weighted. glmnet performs crossvalidation to find the right value of λ that minimizes the sum of squared errors for a given value of α. Fig. 5.7(a) shows the search over values of $log(\lambda)$, and Fig. 5.7(b) shows the change in mean-squared error as $log(\lambda)$ changes.

5.8 Other approaches to regression

Regression is concerned with predicting a dependent numeric value, given the values of a number of associated independent variables. So far, this chapter has assumed that all

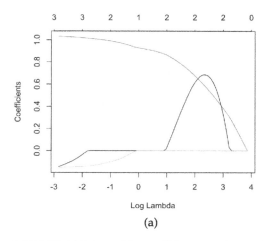 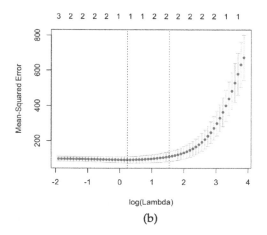

FIGURE 5.7 (a) Values of the coefficients as λ changes in Elastic-Net Regression, (b) Mean-Squared Error as λ changes in Elastic-Net Regression.

the independent variables are numeric; this does not have to be the case always. The approaches to regression presented in the previous sections generate explicit regression functions, which are then used to obtain regressed or predicted values for an unseen example with a new set of independent variables.

Other approaches to regression are also commonly used. A simple but often effective idea is to predict the value of a dependent variable for an unseen example by finding k previously seen examples from the training dataset that are neighbors of the unseen example, and predicting that the value associated with the unseen example is an average of the values for the neighbors. For the approach to be efficient, given a dataset with potentially a large number of examples, the algorithm should be able to find k nearest neighbors quickly. This approach is called k-Nearest-Neighbor (kNN) Regression [1,4].

Approaches to regression also include ones that perform regression using decision trees or ensembles of decision trees, such as Random Forests and boosted trees. These are discussed in Chapter 6 of this book. Such regressors are able to predict the value of the dependent variable in a piecewise manner, instead of the methods seen so far, which produce a continuous mathematical function for prediction.

There is a very successful approach to traditional machine learning called Support-Vector Machines [3,14] that, at its simplest, produces separating lines or (high-dimensional) planes that separate examples of various classes. This approach produces separators among classes that are maximally separated. In other words, the separators are not simple lines or hyperplanes, but have widths to them, and are optimized to be as wide as possible. There is an approach to producing regressors based on this method, called Support-Vector Regression [2,10]. Support-Vector Machines will be discussed in Chapter 6.

Among the many and varied methods for performing regression, one that is commonly used these days uses Artificial Neural Networks, or Deep Learning [6]. For Deep Learning to work well for regression, the amount of training data needed is usually quite high, if the neural network is trained from scratch. However, if a large amount of related data is available, the neural network can first be pretrained on the available data. Many such pretrained neural networks are available for free download. It can then be trained on a smaller amount of data (but still often substantial in the order of thousands or more examples) on the actual problem, and it is quite likely that it will still obtain very good regression results. If the amount of pretraining data is extremely large, pretrained networks may be able to perform regression for new problems with only a very small amount of task-specific data, usually just a few and in the tens; this is called Few-Shot Learning. Usually, few-shot learning results are not great, but a pretrained neural network being able to solve a variety of regression problems using few-shot learning is quite valuable. Going further, zero-shot learning attempts to perform regression or other tasks with no new examples of the new and specialized task at all. Artificial Neural Networks are discussed in Chapter 7.

5.9 Summary

This chapter has provided a comprehensive overview of regression analysis and its applications, including techniques for evaluating and improving linear-regression models, and approaches to addressing overfitting in regression models. We begin by introducing regression analysis and explaining the basic concepts of linear regression. We then move on to discuss techniques for evaluating linear regression models, including the coefficient of determination R^2, standard error of regression, and F-statistic. We also cover the importance of assessing the statistical significance of a model.

Next, we explore multidimensional linear regression and polynomial regression, which involve modeling relationships between multiple independent variables and a dependent variable. We also discuss the challenges of overfitting in regression models and techniques for reducing overfitting, including regularization methods such as ridge regression, lasso regression, and elastic-net regression. The techniques discussed in this chapter are fundamental to the field of statistical modeling and are widely applicable in a range of industries and disciplines.

References

[1] Naomi S. Altman, An introduction to kernel and nearest-neighbor nonparametric regression, American Statistician 46 (3) (1992) 175–185.

[2] Mariette Awad, Rahul Khanna, Support vector regression, in: Efficient Learning Machines, Springer, 2015, pp. 67–80.

[3] Corinna Cortes, Vladimir Vapnik, Support-vector networks, Machine Learning 20 (3) (1995) 273–297.

[4] Evelyn Fix, Joseph Lawson Hodges, Discriminatory analysis. Nonparametric discrimination: consistency properties, International Statistical Review/Revue Internationale de Statistique 57 (3) (1989) 238–247.

[5] Jerome Friedman, Trevor Hastie, Rob Tibshirani, Regularization paths for generalized linear models via coordinate descent, Journal of Statistical Software 33 (1) (2010) 1.

[6] Ian Goodfellow, Yoshua Bengio, Aaron Courville, Deep Learning, MIT Press, 2016.

[7] Trevor Hastie, Robert Tibshirani, Jerome H. Friedman, Jerome H. Friedman, The Elements of Statistical Learning: Data Mining, Inference, and Prediction, vol. 2, Springer, 2009.

[8] Robert Higgs, Race, skills, and earnings: American immigrants in 1909, The Journal of Economic History 31 (2) (1971) 420–428.

[9] Fadil Santosa, William W. Symes, Linear inversion of band-limited reflection seismograms, SIAM Journal on Scientific and Statistical Computing 7 (4) (1986) 1307–1330.

[10] Alex J. Smola, Bernhard Schölkopf, A tutorial on support vector regression, Statistics and Computing 14 (3) (2004) 199–222.

[11] Robert Tibshirani, Regression shrinkage and selection via the lasso: a retrospective, Journal of the Royal Statistical Society, Series B, Statistical Methodology 73 (3) (2011) 273–282.

[12] Andreï Tikhonov, Solutions of Ill-Posed Problems, revised edition, Winston & Sons, Washington, D.C., 1977.

[13] Andrei Nikolaevich Tikhonov, A.V. Goncharsky, V.V. Stepanov, Anatoly G. Yagola, Numerical Methods for the Solution of Ill-Posed Problems, vol. 328, Springer Science & Business Media, 1995.

[14] Vladimir N. Vapnik, An overview of statistical learning theory, IEEE Transactions on Neural Networks 10 (5) (1999) 988–999.

[15] John G. Wagner, George K. Aghajanian, Oscar H.L. Bing, Correlation of performance test scores with "tissue concentration" of lysergic acid diethylamide in human subjects, Clinical Pharmacology & Therapeutics 9 (5) (1968) 635–638.

[16] Hui Zou, Trevor Hastie, Regularization and variable selection via the elastic net, Journal of the Royal Statistical Society, Series B, Statistical Methodology 67 (2) (2005) 301–320.

6

Classification

6.1 Introduction

Machine learning begins with a collection of data examples of interest to explore. It applies a variety of algorithms to discover hidden regularities or patterns in the collection in the form of a model, and uses the learned model for the purposes at hand.

There are primarily two types of machine-learning algorithms: *supervised* and *unsupervised* (see Chapter 4). Supervised algorithms use a collection of data where each example has been examined and labeled by an "expert" in some way. For example, many supervised algorithms perform what is called *classification*. The idea of *class* is fundamental to classification. We think of a class as a group of objects that share some properties or *attributes* or *features*, and can be thought of as a "natural" category, relevant to a problem at hand. Each example is described in terms of a number of features. For example, if our examples are from the domain of vehicles, each example may have features such as height, length, width, weight, color, runs on gas or diesel or electricity, and the number of seats. The associated classes may be the names of the manufacturers of the vehicles, such as Ford, General Motors, and BMW. The purpose of supervised learning in this case is to learn the patterns from a dataset of examples where the manufacturers are known, so that the learned patterns can be used to "predict" or identify the manufacturers of a vehicle for which the manufacturer is unknown, based on the feature values.

Thus, for supervised learning, we need a dataset of examples, where each example is described in terms of its features and also, its class. Table 6.1 presents a few simple but real datasets that have been frequently used in the traditional machine-learning literature. We provide detailed information about the third dataset in Table 6.2. The details have been culled from the UCI Machine Repository site.

Table 6.1 A few datasets from the UCI Machine Learning Repository.

Dataset	Description	#Classes	#Features	#Examples
Covertype	Predict forest cover type, e.g., pine,	7	54	581 012
Glass Identification	Predict the type of glass such as window, tableware, headlamps	7	10	214
Iris	Predict the type of iris as one of setosa, versicolor, and virginica	3	4	150
Lenses	Whether a patient should be fitted with contact lenses (soft or hard) or not	3	4	24
Seeds	Whether a kernel is one of three types of wheat: Kama, Rosa, and Canadian	3	7	210

Fundamentals of Data Science. https://doi.org/10.1016/B978-0-32-391778-0.00013-2

Table 6.2   Attributes in the Iris dataset from the UCI Machine Learning Repository.

Description: Predicting the type of Iris flower based on four attribute values. The dataset is derived from a paper a British biologist and statistician published in 1936 [8].[a] The classes are Iris Setosa, Iris Versicolor, and Iris Virginica.

Attribute Information:

Name	Data Type	Measurement
Sepal length	quantitative	cm
Sepal width	quantitative	cm
Petal length	quantitative	cm
Petal width	quantitative	cm

[a] https://archive.ics.uci.edu/ml/datasets/iris.

Table 6.3   Attributes in the Covertype dataset from UCI Machine Learning Repository.

Description: Predicting forest cover type from cartographic variables. The actual forest cover type for a given observation (30 x 30 meter cell) was determined from US Forest Service data. Independent variables were derived from US Geological Survey and USFS data. The classes are spruce or fir (type 1), lodgepole pine (type 2), Ponderosa pine (type 3), cottonwood or willow (type 4), aspen (type 5), Douglas-fir (type 6), and Krummholz (type 7).

Attribute Information:

Name	Data Type	Measurement
Elevation	quantitative	meters
Aspect	quantitative	azimuth
Slope	quantitative	degrees
Horizontal_Distance_To_Hydrology	quantitative	meters
Vertical_Distance_To_Hydrology	quantitative	meters
Horizontal_Distance_To_Roadways	quantitative	meters
Hillshade_9am	quantitative	0 to 255
Hillshade_Noon	quantitative	0 to 255
Hillshade_3pm	quantitative	0 to 255
Horizontal_Distance_To_Fire_Points	quantitative	meters
Wilderness_Area (4 binary columns)	qualitative	0 or 1
Soil_Type (40 binary columns)	qualitative	0 or 1
Cover_Type (7 types)	integer	1 to 7

Given a dataset where each example is described in terms of features and a class label, the task of supervised learning is to discover patterns to build a "model" for distinguishing among the classes involved. A model, in this case, is a way to put together a description of the found patterns in such a way that it can be used to tell apart unseen examples of the classes. The model may not be explicit, but stored implicitly. Each machine-learning algorithm has a learning bias, and the model formed depends on the learning bias.

Unsupervised learning also attempts to discover patterns in a dataset. Datasets used in unsupervised learning have examples like datasets used in supervised learning, where examples are described in terms of features. However, examples used in unsupervised learning lack the class label. A common form of unsupervised learning, called *clustering*, attempts to put examples in various groups, where examples inside a group are similar to each other as much as possible, and dissimilar to examples in other groups as much as possible. To be able to perform such grouping, we need to use what is called a *similarity measure* or *similarity metric* to compute how similar or dissimilar are two examples. On the flip side, we can use a *distance measure* or *distance metric* also. It may also be useful to measure the similarity or dissimilarity between two groups of examples.

Researchers and practitioners have developed a large number of algorithms for supervised and unsupervised learning. In this chapter, we start the discussion on supervised learning. Among the most commonly used supervised learning or classification algorithms are nearest-neighbor algorithms, tree-based algorithms, support-vector machines, and neural networks. We discuss some of these algorithms in this chapter. We present artificial neural networks in Chapter 7. It is also commonplace to use several classifiers that complement each other to perform a number of separate classifications, and then use a group decision in terms of voting or consensus, to make the final classification. Such classifiers are called *ensemble* classifiers. Ensembles are discussed in Chapter 10.

Developing a classifier algorithm, training it on examples, and using it on unseen examples (also, called *testing*) are the main things we do in supervised learning. The purpose of using a supervised learning algorithm is to be able to train it on a dataset and use the trained model to classify unseen examples in a test setting or in the real world. To decide which algorithm is best suited for a certain supervised machine-learning task, we need to evaluate the performance of the algorithm in terms of metrics. We discuss the most common evaluation metrics for classifiers in this chapter.

6.2 Nearest-neighbor classifiers

These are among the simplest of classifiers, but also at the same time quite effective. There is an adage in English "A man is known by the company he keeps." Imagine that we have a number of data examples for which we know the classes they belong to. That is, each example is described in terms of its features and also has a class label assigned to it. Now, suppose a new example comes along and we have to decide which class it belongs to. First, we need to situate the new unclassified example e among the already labeled examples considering the features we use to describe examples. For example, if each example is described in terms of just two numeric features, say x and y, we can place the new example e among the already classified examples in a 2D space described by two dimensions x and y, assuming the two features are orthogonal to each other. To assign a class to the new example, we can look at its neighbors. In general, we can look at k neighbors. If k is equal to one, we look at just one neighbor, the nearest neighbor n of e, and assign e the class that this neighbor n belongs to. Fig. 6.1(a) shows a situation where the unknown example e is

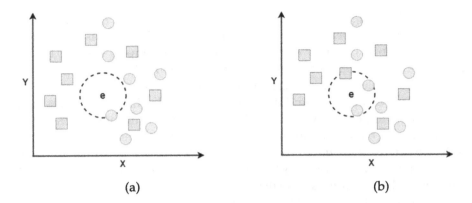

FIGURE 6.1 k-NN classifier, with $k = 1$ and $k = 3$.

closest to a known example of the "circle" class. Thus, the unknown example is assigned to be in the circle class as well. If k is equal to 3, we look at 3 nearest neighbors, n_1, n_2, and n_3, and assign e the class that a majority of these three belong to. Fig. 6.1(b) shows an example where of the three nearest neighbors of the unknown example e, two belong to the circle class and one belongs to the "square" class. Thus, if we go by majority vote, the unknown example is assigned the circle class by a two-to-one vote of its three nearest neighbors. Similarly, for a more general value of k, the k nearest neighbors can, in some sense vote and assign a class to the new example e. This is the essence of how nearest-neighbor classifiers work.

Fig. 6.2 shows a graph of the results of experiments with the Iris dataset. The dataset was randomly divided into two parts: one with 60% of the 150 examples for training, and the rest of the examples for testing. Values of $k = 1$ through $k = 15$ were tried to determine what value works the best for this dataset. For each value of k, the figure shows the error in training as well as testing. The training-error values are shown in the dashed graph in blue. The test results are in the graph with straight lines between two neighboring points. The figure shows that $k = 6$ or $k = 7$ produces the least test error and hence, any of these two values can be used.

6.2.1 Storing classified examples

In the case of nearest-neighbor classifiers, we do not process the classified examples or the input training set at all. We simply record them in a manner that makes it efficient to look up neighbors of a new unknown example. This recording may involve indexing the examples by their feature values in terms of binary trees or more advanced data structures like K-D trees where there may be a large number of branches at each node. When a new example comes along, by traversing the index structure, we are able to find the appropriate number of neighbors quickly. The neighbors' classes are retrieved, and frequently, a ma-

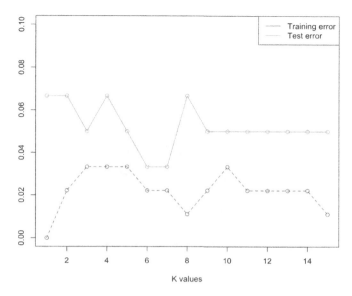

FIGURE 6.2 Running k-NN classifier with several values of k. The figure shows misclassification errors for training and testing. The top plot shows test error, and the lower one training error.

jority vote is taken to produce a class assignment for the new example e. Using an efficient data structure is important if the number of data examples is large.

A K-D tree is a binary tree in which every node is a k-dimensional point, in our case a k-dimensional training example. An interior node can be thought of as implicitly generating a splitting hyperplane that divides the space into two parts, known as half-spaces. Points to the left of this hyperplane are represented by the left subtree of that node and points right of the hyperplane are represented by the right subtree. Because each training example has k features or dimensions, any of the dimensions may be chosen at an internal node for splitting. The dimension to split and the values to split at for the dimension can be chosen in many different ways depending on the algorithm to build and traverse the tree. We do not discuss the construction, updating, and traversal of the index-data structure in this book.

6.2.2 Distance measure

Thus, to find neighbors of an example, we have to traverse the index structure to obtain potential neighbors of example e. Next, we have to compute the pairwise distance or similarity between e and the potential neighbors to find the k nearest neighbors. We must use a similarity or distance metric to implement a nearest-neighbor algorithm.

To measure the distance between two training examples, the most common measure used is Euclidean distance. Assume that a data example or point has m dimensions. If a and b are two data examples, we can write $a = \langle a_1 \cdots a_m \rangle$ and $b = \langle b_1 \cdots b_m \rangle$, where a_i and

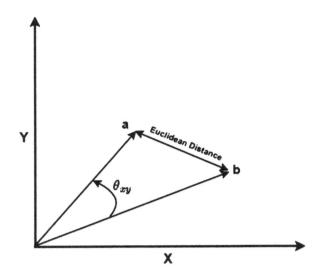

FIGURE 6.3 Euclidean and Cosine distances between two data points.

b_i are attribute values of the data points. The Euclidean distance between two examples is given as

$$d_{Euclidean}(a, b) = \sqrt{(a_1 - b_1)^2 + \cdots + (a_m - b_m)^2} = \sqrt{\sum_{i=1}^{m}(a_i - b_i)^2}. \qquad (6.1)$$

The Euclidean distance between two vectors is nonnegative, but potentially unbounded. Another common way to compute the distance between two data examples is to compute an angular distance between the two examples or points since each example is a vector. The traditional way is to compute the cosine of the angle between the two vectors representing the two points:

$$d_{Cosine}(a, b) = cos(\theta_{ab}) = \frac{\sum_{i=1}^{m} a_i b_i}{\sqrt{\sum_{i=1}^{m} a_i^2 \sum_{i=1}^{m} b_i^2}}, \qquad (6.2)$$

where θ_{ab} is the angle between the two vectors a and b. The cosine distance is always between -1 and 1, with -1 meaning the two vectors are exactly opposite and 1 meaning the two vectors point in the same direction. Two data examples are considered closest when the cosine distance is 1 and farthest when the cosine distance is -1. Fig. 6.3 shows two examples and the Euclidean and cosine distances between them.

6.2.3 Voting to classify

Once k nearest neighbors have been identified, these neighbors vote to determine the class of the new and unclassified example e. Often, majority voting is used, but in some cases consensus may be necessary. In majority voting, the class that occurs most frequently becomes the class of the new example. The class that a neighboring example belongs to is considered its vote. If two classes receive the same number of votes, it is usually broken randomly. In consensus voting, all of the k neighbors have to belong to the same class to assign a class to the new example. If this is not the case, the new example remains unclassified.

In an advanced version of the algorithm, it is possible to weigh the votes by distance from e, with nearer neighbors getting a greater amount of say compared to neighbors that are farther away. If there are k nearest neighbors, and d is the distance metric used, the amount of vote given to one of these neighbors x_j can be computed using the following formula:

$$v_j = \frac{\frac{1}{d(x_j,e)}}{\sum_{i=1}^{k} \frac{1}{d(x_i,e)}}. \tag{6.3}$$

For each of the k nearest neighbors, the inverse of the distance from the unknown example e is computed and these inverse values are added together to form the denominator of the formula. The amount of vote example x_j receives is obtained by dividing the inverse of its distance from the unknown example by the denominator. As a result, a neighbor close to the unknown example receives a high vote compared to an example that is farther away. If the inverse of the distance were not computed and the distance was used as is, a nearer neighbor would have been given smaller weight compared to a neighbor that is farther away. The value v_j is normalized to be between 0 and 1 and all the values add up to 1, and as a result, v_j can be thought of as a probability as well. Fig. 6.4 shows how the misclassification rates change during training and testing when Euclidean distance is used with voting as discussed. In this example, the use of voting makes the performance of the classifier during testing more stable as the value of k increases, but the performance actually degrades for higher values.

The inductive bias of a nearest-neighbor classifier is a little difficult to explain. Each classified example can be thought of as having a "halo" around it in the form of what is called a Voronoi diagram. In 2D, it is the area that describes its neighborhood. This neighborhood is never built explicitly by a nearest-neighbor algorithm, but used only for understanding how it works. Different neighbors of a new example e have their own neighborhoods, and these neighborhoods together decide which class a new example is assigned.

6.3 Decision trees

Often, as human, when we make decisions regarding classification, we use a tree-like structure to do so. The nodes of such a tree have queries, and the branches lead to different paths based on answers to the queries. We start from the top node, which poses a query,

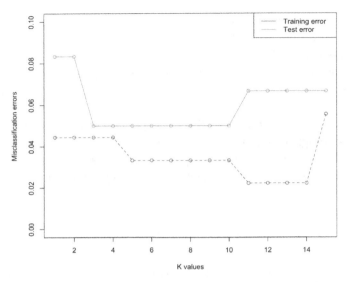

FIGURE 6.4 Running a k-NN classifier with several values of k. The figure shows misclassification errors for training and testing when the Euclidean distance is used as discussed above for voting. The top plot shows test error, and the lower one training error.

and based on the answer, we take a path down the tree, and we answer another query at the second node, and take another path down the tree based on the answer. We traverse the tree in such a manner to a leaf node where we know to which class the example belongs (see a sample tree in Fig. 6.5).

Decision trees are built using supervised learning. In other words, we build a decision tree from a number of labeled examples given to us. Table 6.4 (attribute description) and Table 6.5 show one of the popularly used labeled dataset, *Lenses*,[1] from UCI Repository. Here, we describe how a decision tree can be built from such a dataset and used after it has been constructed. Many different decision trees can be constructed from this dataset. Fig. 6.5 shows a decision tree that has been built from this dataset.

The Lenses dataset, like any other dataset, is given in terms of features or attributes for the examples. Each example in the table has four features. All four are descriptive, giving us details of an example. The *class distribution* in Table 6.4 gives the idea that the number of examples belong to a particular class. This is also called the *label* and is assigned to an example by an "expert" or "supervisor". In Table 6.5 the class or label column is separated from the descriptive features by a vertical line in the table. The descriptive features may take only categorical values, where the number of distinct values is small, or depending on the dataset also numeric values, where the number of possible values is potentially infinite. The class label usually takes a small number of values.

---

[1] https://archive.ics.uci.edu/dataset/58/lenses.

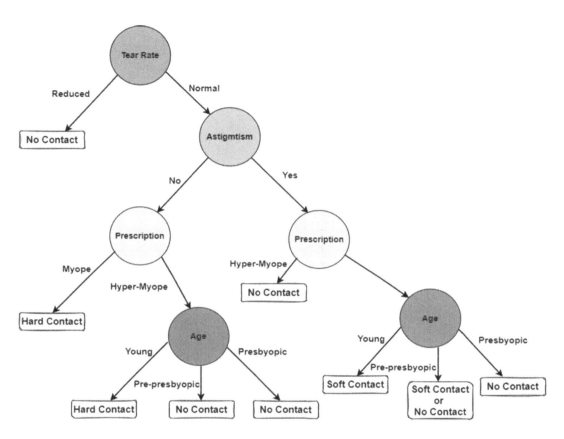

FIGURE 6.5 An example of a decision tree built from the Lenses dataset.

Table 6.4 Description of the Lenses Dataset from the UCI Machine-Learning Repository.

Number of Instances:	24
Number of Attributes:	4 (all nominal)
Number of classes:	3
	1 : the patient should be fitted with hard contact lenses,
	2 : the patient should be fitted with soft contact lenses,
	3 : the patient should not be fitted with contact lenses.
Attribute 1	age of the patient: (1) young, (2) prepresbyopic, (3) presbyopic
Attribute 2	spectacle prescription: (1) myope, (2) hypermetrope
Attribute 3	astigmatic: (1) no, (2) yes
Attribute 4	tear-production rate: (1) reduced, (2) normal
Class Distribution:	
	1. hard contact lenses: 4
	2. soft contact lenses: 5
	3. no contact lenses: 15

Table 6.5 Lenses Dataset from the UCI Machine-Learning Repository.

No.	Att 1	Att 2	Att 3	Att 4	Class
1	1	1	1	1	3
2	1	1	1	2	2
3	1	1	2	1	3
4	1	1	2	2	1
5	1	2	1	1	3
6	1	2	1	2	2
7	1	2	2	1	3
8	1	2	2	2	1
9	2	1	1	1	3
10	2	1	1	2	2
11	2	1	2	1	3
12	2	1	2	2	1
13	2	2	1	1	3
14	2	2	1	2	2
15	2	2	2	1	3
16	2	2	2	2	3
17	3	1	1	1	3
18	3	1	1	2	3
19	3	1	2	1	3
20	3	1	2	2	1
21	3	2	1	1	3
22	3	2	1	2	2
23	3	2	2	1	3
24	3	2	2	2	3

6.3.1 Building a decision tree

The task of building a decision tree involves deciding which one among the descriptive features should be used to construct the first query. A query involves a feature or attribute and a value for the attribute, along with a comparison operator, as seen in Fig. 6.5. The first question to ask is which feature among the ones available should be the feature for the question, i.e., on what basis should it be selected? The next question is what value of the feature is relevant to ask the question and what the comparison operator should be. The comparison operator is usually $=$ for text or a categorical attribute, and $>$, $<$ or $\geq$, or $\leq$ for a numerical attribute.

The objective is to build a compact decision tree that is consistent with all the examples from which it is built. The process is called training the decision tree. If the goal is to build the most compact tree, the process becomes exponential in time since the number of possible trees that can be built from n examples having k features with each feature taking a number of values exponential in number. Thus, in practice, there is no time to create all possible trees and choose the most compact or optimal tree. Therefore, the approach to build a decision tree is necessarily heuristic and expedited, deciding on a feature to use in a query in a greedy fashion, a decision that cannot be undone once made. Thus, it is quite likely that a decision tree built by one of the commonly used methods is only locally optimal and is not the globally optimal one. However, even such locally optimal trees often happen to be quite good classifiers.

6.3.2 Entropy for query construction

The choice of a feature or attribute to construct a query at a certain level of the tree is usually made using the concept of entropy (and, sometimes another similar concept called the Gini Index). Entropy measures the amount of chaos or dissimilarity in a dataset, or how mixed up or inhomogeneous a dataset is. Homogeneity is measured with respect to the class(es) the examples in the dataset belong to. For example, if a dataset has n examples, and all examples belong to the same class c, the entropy of the dataset is 0 (see Fig. 6.6(a)). In this case, there are 10 examples in the dataset included inside the big dashed circle, and each data example is a small rectangle, and thus the dataset is homogeneous. If a dataset has n examples, and $\frac{n}{2}$ examples belong to class c_1 and $\frac{n}{2}$ examples belong to another class c_2, it can be said that the dataset is completely mixed up and its entropy is 1. See Fig. 6.6(b), where the dataset of 10 contains 5 examples of the circle class and 5 from the rectangle class.

The entropy of a binary dataset D containing n examples, with n_1 examples from class c_1 and n_2 examples from class c_2 is

$$Entropy(D) = -\frac{n_1}{n} \, log\frac{n_1}{n} - \frac{n_2}{n} \, log\frac{n_2}{n} \tag{6.4}$$

$$= -p_1 \, log \, p_1 - p_2 \, log \, p_2 \tag{6.5}$$

$$= \sum_{i=1}^{2} -p_i \, log \, p_i, \tag{6.6}$$

where $n_1 + n_2 = n$, $p_1 = \frac{n_1}{n}$, and $p_2 = \frac{n_2}{n}$. p_1 is the probability of a data example being in class c_1, and p_2 is the probability of a data example being in class c_2. If the dataset contains examples from k classes $C = \{c_1 \cdots c_k\}$, the entropy of the dataset is

$$Entropy(D) = \sum_{i=1}^{k} -p_i \, log \, p_i, \tag{6.7}$$

where $p_i = \frac{n_i}{n}$ with n_i being the number of examples in class c_i in the dataset, with $\sum_{i=1}^{k} p_i = 1$ and $\sum_{i=1}^{k} n_i = n$. Thus, the entropy of a dataset considering the classes of examples in it lies between 0 and 1. If a dataset is not as mixed as being equally divided between two (or more) classes, the entropy value is less than one. Therefore the entropy curve looks like the one given in Fig. 6.7, assuming we have two classes to choose from. The X-axis shows the proportion of examples from one of the two classes in the dataset

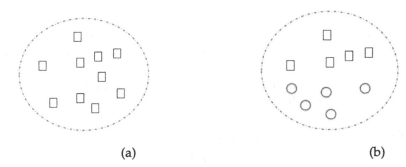

FIGURE 6.6 (a) A dataset with all examples from the same class, with a dataset entropy of 0. (b) A dataset with half the examples from one class and the other half from another, with an entropy of 1.

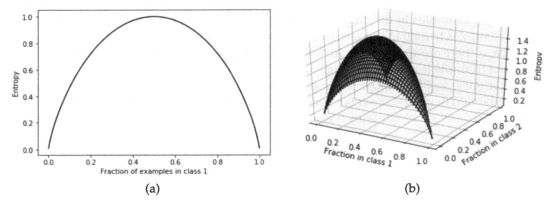

FIGURE 6.7 Entropy Graph. (a) The X-axis shows the fraction of the dataset in a class where there are two classes. The Y-axis shows the entropy of the dataset. (b) The case with three classes.

and the Y-axis shows the entropy of the dataset. If there are two classes, the maximum value of entropy is 1 and the minimum value is 0. If there are k classes in the dataset, the maximum value of entropy is $log_2 n$ and the minimum value remains 0.

To select a feature to create a query, it is not necessary to look at the entropy of a dataset directly, but compute how the entropy will change when the different features are used to separate out the dataset into subsets. If a feature f is used to separate the dataset into disjoint subsets, one can compute the before-separation entropy of the entire set, and the after-separation weighted sum of entropies of the subsets, and then the change in entropy when the feature f is used for the process:

$$\Delta Entropy(D, f) = Entropy(D) - \sum_{i=1}^{nv(f)} \frac{|D_{fi}|}{|D|} Entropy(D_{fi}), \qquad (6.8)$$

where D is the original dataset, $nv(f)$ is the number of distinct values feature f can take, $|D|$ is the number of examples in dataset D, and $|D_{fi}|$ is the number of total examples in the

ith branch of the tree, i.e., the number of examples in D that have the ith unique value of f. If f can take k possible discrete values, there will be k branches, as shown in Fig. 6.8(a). As a specific example, if Att_1 is the attribute chosen to use as a query, there will be three branches, as shown in Fig. 6.8(b). If one choose Att_2 as the query feature, there will be two branches, as shown in Fig. 6.8(c).

If Att_1 is the attribute at the top node, there will be three branches, one for each value of Att_1, namely, 1, 2, and 3. At the top node, the entire *Lenses* dataset is associated with the node. There will be a subset associated with each of the branches, the specific subset depending on the value of Att_1. Let the subset associated with the value 1 of attribute 1 be called $Lenses_{11}$, the subset associated with value 2 of attribute be called $Lenses_{12}$, and the subset associated with value 3 of attribute be called $Lenses_{13}$. It is possible to compute the entropy of each of these subsets, and then the change in entropy when the *Lenses* dataset is divided with $Attr_1$ as follows:

$$Entropy(Lenses) = -\frac{4}{24} log_2 \frac{4}{24} - \frac{5}{24} log_2 \frac{5}{24} - \frac{15}{24} log_2 \frac{15}{24} = 1.3261$$

$$Entropy(Lenses_{11}) = -\frac{2}{8} log_2 \frac{2}{8} - \frac{2}{8} log_2 \frac{2}{8} - \frac{4}{8} log_2 \frac{4}{8} = 1.5000$$

$$Entropy(Lenses_{12}) = -\frac{1}{8} log_2 \frac{1}{8} - \frac{2}{8} log_2 \frac{2}{8} - \frac{5}{8} log_2 \frac{5}{8} = 1.3844$$

$$Entropy(Lenses_{13}) = -\frac{1}{8} log_2 \frac{1}{8} - \frac{1}{8} log_2 \frac{1}{8} - \frac{6}{8} log_2 \frac{6}{8} = 1.0613.$$

Now, it is possible to compute the change in entropy when the dataset is divided into three subsets using $Attr_1$ as follows:

$$\Delta Entropy = Entropy(Lenses) - \frac{1}{3} \left(\sum_{i=1}^{3} Entropy(Lenses_{1i}) \right)$$

$$= 1.3261 - \frac{1}{3}(1.5000 + 1.3844 + 1.0613)$$

$$= 0.01086.$$

The change in entropy can be computed for all the features of an example, and choose the one that produces the highest amount of entropy change. In other words, one choose the feature that separates the mixed dataset into less mixed subsets of data. This is because the ultimate objective in building a decision tree is to obtain a tree where all the leaves have homogeneous subsets of data, i.e., each leaf node corresponds to a small number of examples corresponding to only one of the possible classes.

After selecting the first attribute to ask a question at the top node of the decision tree, one builds edges out of this node taking into account the distinct values the relevant categorical attribute takes. Thus, if an attribute takes 3 values (say, *yes*, *no*, and *maybe*), there will be three edges out of this top node, corresponding to each of the values. All the data

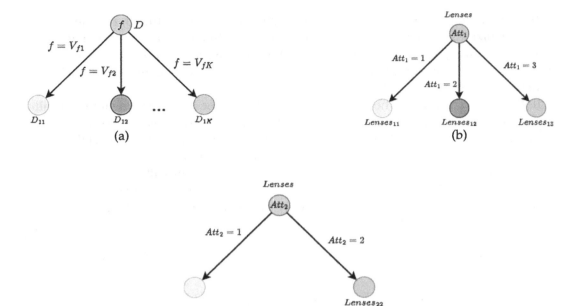

FIGURE 6.8 (a) Branching dataset D on feature f that has k possible values. (b) Branching dataset *Lenses* on feature Att_1 that has 3 possible values. (b) Branching dataset *Lenses* on feature Att_2 that has 2 possible values.

examples are associated with the top node, and examples associated with the appropriate feature values are passed along down the edges. Thus, there are three children nodes of the top node, each associated with a smaller number of examples. One repeats the process of building a subtree by selecting a head node for the subtree using entropy-change computation as described above. Building a subtree stops below a node if all the data examples associated with it become homogeneous, i.e., their class becomes the same.

There are many issues that arise when building a useful decision tree that can be used in practice. The basic algorithm can be modified to address these issues. However, to keep our discussions brief, we present only a few of them, and these all too briefly.

6.3.3 Reducing overfitting in decision trees

One problem that is important to discuss is how to handle *overfitting*. The purpose of a machine-learning algorithm is to *generalize* from the limited number of examples by eliminating any accidental similarities that may exist so that the trained algorithm can be used on unseen examples or released to the public to use on whatever dataset they may have. Thus, a machine-learning algorithm must generalize to the right amount and try to avoid overfitting, which is difficult to do. In the context of decision trees, it is possible to build a large decision tree with many levels and many nodes to fit the training data exactly. Assume that a decision tree with 100 nodes has been built to fit the dataset exactly. However, it is likely that such a tree is too detailed and in addition to generalizing the information contained in the dataset, it also has built nodes to cater to accidental regularities that are present in this particular dataset only and may not be present in another similar dataset that may be collected for the same phenomena. Thus, it is possible that when put into action after training, a decision tree with 20 nodes will perform as well as or even better than a decision tree with 100 nodes.

Fig. 6.5 shows a fairly detailed tree for the Lenses dataset with only 24 training examples. Perhaps a smaller tree may work as well, or even if it performs a little worse on the training data, it is likely to work better on unseen data that it needs to classify. In this specific tree, not only do the data fit the tree exactly, there is also a problem where one cannot decide what class it should be in one of the leaf nodes. How to build a decision tree with the right number of nodes so as not to overfit, but learn the requisite generalization involves pruning the decision tree. There are several ways to prune a decision tree, and the preferred way is to build a complete decision tree with as many nodes as necessary to produce homogeneous leaf nodes, but then remove as many nodes as possible so as to remove nodes created to cater to accidental regularities. One may simply pick and delete a random node and all nodes below it, and test the new reduced tree using the same dataset that was used to test the entire tree, and if the new reduced tree's performance is not worse than the original tree, keep the reduced tree. The process of pruning is repeated a number of times, and a node is removed only if performance does not degrade or degrades just a little. It has been seen that reduction of 50–70% or even more nodes may keep the tree's performance almost as good as it was for the original tree. Therefore, it may be hypothesized that the nodes eliminated from the original tree covered only nonessential (maybe duplicated) and possibly accidental regularities and their removal actually not only does not degrade performance, but makes the performance more efficient (since fewer nodes need to be tested for an unseen example) and at the same time possibly a better performer on unseen examples. Thus, a pruned tree like the one shown in Fig. 6.9 may work better than the tree in Fig. 6.5 although the more complex tree has 6 nodes compared to just 3 nodes in the pruned tree.

Another way to prune converts the decision tree to a set of if-then rules. Each path from the root to a leaf node of a decision tree can be described by writing a corresponding if-then rule. All the rules corresponding to all the paths can be written as a set of rules or a *ruleset*. Fig. 6.10 shows three of the rules that can be written to describe the decision tree in Fig. 6.5. Converting to a set of rules removes the distinction between nodes that occur at upper levels of the tree and those that occur below because all of a rule's conditions are considered to be at the same level. In addition, since a node, especially at an upper level occurs several times in the rules, the node's influence becomes distributed and as a result, the effect of pruning a node is distributed, since a node now occurs in several rules. Once there is a set of rules, pruning simply picks a random rule and picks a random antecedent and removes the antecedent. If after removing the antecedent, the ruleset's performance does not decrease, the reduced rule is kept in the ruleset instead of the original rule. If the

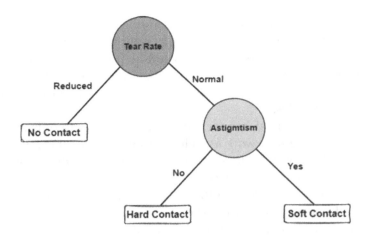

FIGURE 6.9 An example of a pruned decision tree built from the Lenses dataset.

```
If (Tear Rate = reduced)
                  then No contact
If (Tear Rate = normal)  and (Astigmatism = no) and
      (Prescription = myope)
                  then Hard contact
If (Tear Rate = normal)  and (Astigmatism = no) and
      (Prescription = hypermyope) and (age=young)
                  then Hard contact
```

FIGURE 6.10 Three if-then rules obtained from the decision tree shown in Fig. 6.5 for the Lenses dataset.

ruleset's performance decreases that specific antecedent is not removed, but another one is tried for removal and the process is repeated. In general, it is possible to remove many antecedents and even complete rules and obtain a simplified ruleset.

6.3.4 Handling variety in attribute types

Another issue of importance in building decision trees and also in many other machine-learning algorithms is that the values of the features may be categorical, or mixed between numeric and categorical, or that all values of all features are numeric. A numerical attribute simply may take values that are continuous numbers, e.g., age and weight are such attributes. However, there are other attributes whose values are from a finite set, even though they are numeric integer values. For example, the value of a feature *year* in a college setting may take only 4 values: 1, 2, 3, and 4, or *freshman, sophomore, junior,* and *senior,* alternatively. All the features in the Lenses dataset take categorical values, which can be denoted using a limited number of integers or words. The decision-tree algorithm discussed so far works well with discrete categorical values, and cannot handle continuous numeric values directly. The approach to handle such cases is to automatically (or by inspection) create intervals that span the range of values for the attribute. For example, a feature called *age*, whose value may lie in the continuous range 0–100, may be divided into a few intervals such as *infant* (0–4), *child* (5–12), *teenager* (13–19), *young adult* (20–29), *middle aged* (30–55), *senior* (56–70), *older adult* (71–100), or something similar. This obtains 7 possible discrete values instead of an infinite number of numbers. All features in the Covertype dataset, described in Table 6.3 are numeric, and all but the last three can take possibly an infinite number of values. For example, the three Hillshade attributes take values in the range 0–225. We may divide the values into three ranges with a name assigned to each range: *low* (0–63), *medium* (64–127), and *high* (128–255). We may divide the values into five ranges instead of three if it suits our purpose. There are also published algorithms that analyze the values that actually occur and decide how many ranges may be most appropriate. Software-generated names can then be assigned to these ranges.

6.3.5 Missing values in decision tree construction

A final issue to discuss in the context of decision trees is the presence of examples where some feature values are missing. Since feature values are needed to decide what feature would go on a node and what question to ask, the absence of a value makes it difficult to place the example on a tree-node's children. As a result, if an attribute is suitable to become a node in the tree, one must make up values for missing attributes unless such examples are completely thrown out from consideration. Making up a missing value is called *imputation*. A simple way to impute a missing value is to find the most common value for the attribute associated with examples at the node, and replace the missing value with that common value. Another way to impute may be to look at the values of other features and find a value that goes best with the values that are there in the other fields considering the cooccurrence or correlations among values. There are other sophisticated ways to impute as well.

6.3.6 Inductive bias of decision trees

The inductive bias of Decision-Tree classifiers can be thought of as giving preference to building trees that are short and compact, among all the trees that can be built consistent with the data. This is the so-called Occam's Razor or Occam's Principle. The Decision-Tree building algorithm can be thought of as searching greedily through the space of such trees to hone in on the one that is optimal, but as mentioned earlier only locally optimal. At another level, the inductive bias of Decision-Tree classifiers can be thought of as dividing the search space into subspaces corresponding to the classes by drawing lines parallel to the axes. In a 2D space, the subspaces may look like what we see in Fig. 6.11. The subspaces corresponding to a single class need not be disjoint; in other words, Decision Trees are able to find classes that occupy noncontiguous spaces. In higher-dimensional spaces, the subspaces are higher-dimensional rectangles or hyperrectangles.

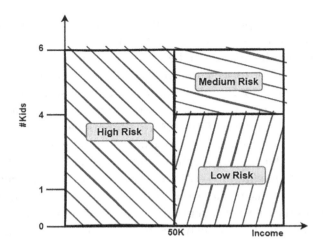

FIGURE 6.11 A decision tree produces rectangular disjoint regions for a class.

6.4 Support-Vector Machines (SVM)

Assume that there is a set of examples that belong to two classes. As usual, each example is described in terms of a number of features. Consider the number of features to be two for now. Each example also has a label indicating which class the example belongs to. If it is possible to find a linear function of the features, i.e., a line or a plane that can separate examples of one class from those of the other class completely, there exists a linear classifier. As a simplistic example, consider a set of training examples each of which is described in terms of two features x_1 and x_2. Let the examples be denoted in 2D space as seen in Fig. 6.12(a). One class is depicted as containing a number of rectangles, and the other class contains a number of circles. In the following discussions, we sometimes refer to the rectangle class as the $+$ class and the circle class as the $-$ class. A line that separates one class from another is shown in the diagram. This line is a linear separator between the two classes. In fact, it is possible to obtain infinitely many linear separators between the two classes. The goal in binary linear classification is to learn the best linear separator or discriminator given a number of labeled points belonging to the two classes in a dataset.

6.4.1 Characterizing a linear separator

Let a linear separator function be called $f(\vec{x})$, where $\vec{x}$ is a vector representing the features of a data example. If the two classes are linearly separable (i.e., the data examples in the two classes can be separated with a linear function without any errors), the function to learn is a straight line like that shown in Fig. 6.12(a). In 3D, the function is a plane, and

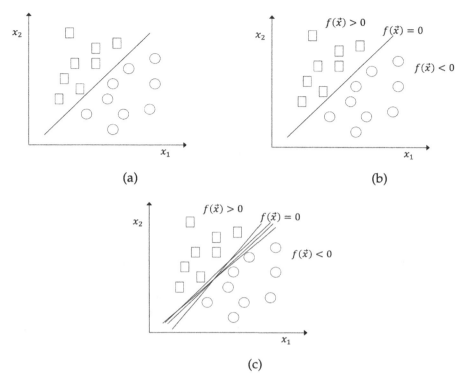

FIGURE 6.12 A linear discriminator or separator between two classes. (a) Shows one discriminator, (b) Shows how the discriminator function divides the space of data examples, (c) Shows that an infinite number of discriminator functions are possible consistent with two linearly separable classes.

in higher dimensions, it is a separating hyperplane. Since it is a linear separator, $f(\vec{\mathbf{x}})$ is a linear function of the features of the example $\vec{\mathbf{x}}$:

$$
\begin{aligned}
f(\vec{\mathbf{x}}) &= \theta_0 + \theta_1 x_1 + \theta_2 x_2 + \cdots + \theta_n x_n \\
&= \vec{\theta} \cdot \vec{x} + \theta_0 \\
&= \vec{\theta}^T \vec{x} + \theta_0,
\end{aligned} \tag{6.9}
$$

where n is the number of features or dimensions in the example $\vec{\mathbf{x}}$, and $\cdot$ represents the dot product (or component-wise product) of the two vectors. The superscript T indicates the transpose of the column vector to make it a row vector.

Generally, to characterize a linear classifier, it is required that $f(\vec{\mathbf{x}}) = 0$ on the linear discriminator (line, plane, or hyperplane), $f(\vec{\mathbf{x}}) > 0$ on one side (here, the +ve or rectangle side, or "above") and $f(\vec{\mathbf{x}}) < 0$ on the other side (the −ve side or circle side or "below"). The values of the discriminator function are shown in Fig. 6.12(b).

6.4.2 Formulating the separating hyperplane

The problem of finding a separating line or (hyper-)plane can be formulated as follows.

Given a set of N data points $\{\langle \vec{x}^{(i)}, y^{(i)} \rangle\}, i = 1 \cdots N$ with $y^{(i)} \in \{+, -\}$, find a hyper-plane

$$\vec{\theta} \cdot \vec{x} + \theta_0$$

such that for each data example in the $+$ class

$$\vec{\theta} \cdot \vec{x} + \theta_0 > 0$$

and for each data example in the $-$ class

$$\vec{\theta} \cdot \vec{x} + \theta_0 < 0.$$

It is possible to draw an infinite number of linear discriminators to separate examples of two classes and as a result, the formulation above has no unique solution. This is shown in Fig. 6.12(c). The question arises regarding which one should be picked as the best linear separator. One possibility is that it is a linear discriminator that maximizes some kind of error, similar to the discussion on linear regression in Chapter 5. In the 1980s, a Russian scientist named Vapnik came up with another idea—called the *Maximum-Margin Classifier* [7].

6.4.3 Maximum margin classifier

In particular, when building a linear classifier between two classes, some linear separators may be better than others. For example, a line too close to examples from one class and too far away from examples of the other class may not generalize well. This is because the purpose of drawing or learning the linear separator is to be able to generalize from seen examples, so that unseen examples can be classified well; and it is likely that there are examples in the empty area between the examples of the two classes that were not seen during training but may be encountered later during the testing or application of the trained classifier. A linear separator is situated in the middle of the separating region may be the best choice of a classifier because it will allow for the most amount of generalization. To be able to find such a classifier, it is necessary to identify the separating region between the two classes. If the region is drawn with a straight line (in 2D space) situated halfway from the two lines that separate the two classes, it may be the linear separator with potential for best generalization. In higher dimensions, the linear separator is a hyperplane, i.e., a plane in high dimensions. Fig. 6.13 shows such a classifier, clearly marking the upper and lower margins as well as the actual classifier situated halfway between the two. The margins are clearly parallel to each other as well as parallel to the classifier.

A SVM classifier is a maximum-margin classifier, one that was informally described in the previous paragraph. A Support-Vector Machine finds the region that separates the two

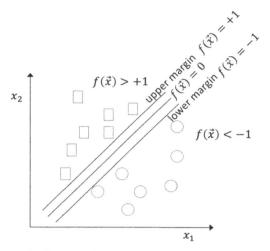

FIGURE 6.13 There is an empty region between the two classes. The linear separating plane is in the middle of this region. This is ensured by drawing two maximally separated planes margins that are equidistant from the separator.

classes in terms of the outer boundaries of the region or the margins. It posits the middle hyperplane as the separator. The classifier is characterized in terms of the separator and the margins. The Support-Vector Machine ensures that the margins are situated as far away from each other as possible or are maximally separated.

6.4.4 Formulating a maximum-margin classifier

How to find such a maximum-margin classifier? This requires setting up an optimization problem that involves the two margins on the two sides, making sure that the separation is maximal, i.e., solving a maximization problem. This maximization problem is not a direct maximization problem, but involves a number of constraints, just like the one set up earlier. Assuming there are examples from two classes, the separating hyperplane, which is situated in the middle of the two margins, must meet a constraint for each point in the dataset. The standard specification of SVM performs binary classification and requires one of the classes to be called the +1 class, and the other class to be called the -1 class. The margins lie at +1 and -1, and the middle of the margin lies at 0 in terms of the value for $\vec{\theta} \cdot \vec{x} + \theta_0$. The constraint associated with a point ensures that a point on the positive side of a margin has a value for the separator function ≥ 1, and a point that on the negative side of a margin has a value for the separator function ≤ -1. Hence, the optimization problem is set up in the following way:

Given a set of N data points $\left\{ \langle \vec{x}^{(i)}, y^{(i)} \rangle \right\}, i = 1 \cdots N$ with $y^{(i)} \in \{+1, -1\}$, find a hyperplane

$$\vec{\theta} \cdot \vec{x} + \theta_0$$

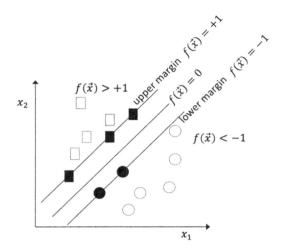

FIGURE 6.14 The filled black data examples or points are the support vectors. Note that we have moved points around compared to the previous figures to highlight the support vectors.

such that for each data example in the +1 *class*

$$\vec{\theta} \cdot \vec{x} + \theta_0 \geq +1$$

and for each data example in the −1 *class*

$$\vec{\theta} \cdot \vec{x} + \theta_0 \leq -1$$

and the distance between the margins is the widest.

Points that lie on the two margins are called support points or *support vectors* for classification. These are the ones that actually matter for classification. The number of support vectors is likely to be small. These points are crucial for classification. Points farther from the margins do not really matter for classification. In Fig. 6.14, there are three support vectors on the rectangle (or +ve) class, and two support vectors on the circle (or −ve) class.

Given the upper-margin hyperplane and the lower-margin hyperplane (with the separator hyperplane situated exactly in the middle), the distance between the two margins gives the width of the separation region between the two classes. This intermargin distance is optimized to be the widest in SVMs. To be able to formulate the optimization problem more clearly, it is necessary to obtain a formula for this intermargin distance first.

The optimization problem to solve is obtained step by step. The first step is to obtain the equation to the normal to a hyperplane since one deals with hyperplanes in discussing SVMs. Given the equation of a hyperplane, what is the equation of the normal to it? This equation requires to write a formula for the distance between the hyperplanes. Although the discussion is on 3D, the same idea applies to higher dimensions. Let $\vec{n}$ be a normal to a plane P. There are many normals, but they are all parallel to each other. Since P is a

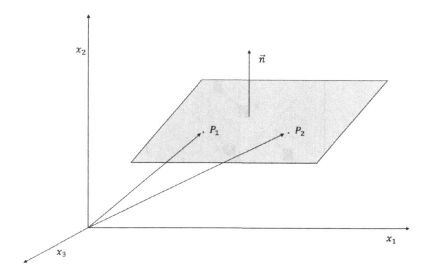

FIGURE 6.15 Equation of a normal to a plane.

hyperplane in n-dimensions, its equation can be expressed as a linear combination of the basis vectors as seen already. Let the hyperplane's equation be given as $\vec{\theta} \cdot \vec{x} + \theta_0 = c$, where c is a constant. This plane is shown in Fig. 6.15.

Consider two arbitrary points, P_1 and P_2 in the plane. Since both points are in the plane,

$$\vec{\theta} \cdot \vec{P}_1 + \theta_0 = c \tag{6.10}$$

$$\vec{\theta} \cdot \vec{P}_2 + \theta_0 = c. \tag{6.11}$$

Subtracting one from the other,

$$\vec{\theta} \cdot \left(\vec{P}_1 - \vec{P}_1 \right) = 0. \tag{6.12}$$

In other words, $\vec{\theta}$ is normal to the plane P. Thus, given a plane $\vec{\theta} \cdot \vec{x} + \theta_0 = c$, the normal to the plane is given by the vector $\vec{\theta}$. Thus, the vector $\vec{n}$ is in the same direction as the vector $\vec{\theta}$. If $\vec{n}$ is a unit vector normal to plane P, then

$$\vec{n} = \frac{\vec{\theta}}{|\vec{\theta}|} \tag{6.13}$$

$$= \frac{\vec{\theta}}{\sum_{i=1}^{n} \theta_i^2}.$$

In the characterization of SVM, there are three planes that are parallel to one another. In other words, once the expression for the unit normal to one of these planes is obtained, the unit normal for all three planes is the same.

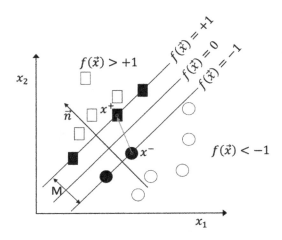

FIGURE 6.16 Obtaining the optimization formula for SVM.

The next step in the derivation of the optimization objective for SVM is to obtain the perpendicular distance between the two margin planes. To illustrate, see Fig. 6.16, which shows two support vectors, x^+ on the rectangle (+) side, and the other, x^- on the circle (−) side that are connected by a straight line. It also shows the direction of the normal $\vec{n}$ to the three planes, the two margins and the separator. The distance between the two margin planes is M, and the separator plane sits exactly in the middle. To compute M, it is necessary to project the line drawn from x^- to x^+ onto the unit normal in the direction of $\vec{n}$,

$$
\begin{aligned}
M &= \left(\vec{x}^+ - \vec{x}^-\right) \cdot \vec{n} && (6.14)\\
&= \left(\vec{x}^+ - \vec{x}^-\right) \cdot \frac{\vec{\theta}}{|\vec{\theta}|}\\
&= \frac{\vec{x}^+ \cdot \vec{\theta} - \vec{x}^- \cdot \vec{\theta}}{|\vec{\theta}|}\\
&= \frac{\left(\vec{x}^+ \cdot \vec{\theta} + \theta_0\right) - \left(\vec{x}^- \cdot \vec{\theta} - \theta_0\right)}{|\vec{\theta}|}\\
&= \frac{1 - (-1)}{|\vec{\theta}|}\\
&= \frac{2}{|\vec{\theta}|}\\
&= \frac{2}{\sqrt{\sum_{i=1}^{n} \theta_i^2}}.
\end{aligned}
$$

The *classification problem* now becomes an optimization problem given as below. Given a set of N data points $\{\langle \vec{x}^{(i)}, y^{(i)} \rangle\}, i = 1 \cdots N$ with $y^{(i)} \in \{+1, -1\}$,

Maximize $\frac{2}{|\theta|}$ such that for each data example in the $+1$ class

$$\vec{\theta} \cdot \vec{x} + \theta_0 \geq +1$$

and for each data example in the -1 class

$$\vec{\theta} \cdot \vec{x} + \theta_0 \leq -1.$$

It is a constrained optimization problem where the objective function is maximized based on a number of conditions, one condition arising from each labeled data point. Since the objective function that needs to be maximized is $\frac{2}{|\theta|}$, we can restate the problem as a minimization problem with the objective function as $\frac{|\vec{\theta}|}{2}$. As noted earlier, the numerator happens to be $\sqrt{\sum_{i=1}^{n} \theta_i^2}$. The objective function has a square root in it, which can be ignored for maximization. In addition, in the manner in which the SVM problem is formulated, the value of the label $y^{(i)}$ for the $+$ examples is $+1$ and is -1 for the $-$ examples. Therefore, the optimization problem can be further rewritten compactly as the following:

Maximize $\sum_{i=1}^{n} \theta_i^2$ such that for each data example

$$y^{(i)} \cdot \left(\vec{\theta} \cdot \vec{x} + \theta_0 \right) \geq +1.$$

6.4.5 Solving the maximum-margin optimization problem

The optimization process finds the values of the coefficients for the separator function. Thus, the optimization problem involves maximizing a quadratic function subject to a number of linear constraints. Such an optimization problem is called a *quadratic programming problem*. It is a well-studied problem in mathematics and operations research and there are many techniques to solve such problems. No matter how it is solved, the constraints are associated with each of the points in the training set. Hence, if the training set has a thousand or a million points, there are a thousand or a million constraints. This makes the problem slow to solve.

Another point to be noted is that with constrained optimization problems, often there are two ways of solving the same problem. The original problem is called the primal problem and the reformulated version that gives the same answer is called the dual problem. Sometimes, the primal problem is easier to solve than the dual and sometimes the reverse. In the case of the optimization problem for SVMs, the dual problem has been found easier to solve, although it is still slow. This book does not discuss how the optimization problem can be solved since the process is complex and is outside its scope. However, we have presented what the optimization problem looks like because it clearly demonstrates the complexity of the problem to be solved.

An aspect of SVM's setup and solution of the optimization problem is that although it solves a quadratic programming problem, the solution finds a linear separator between two classes such that the two classes are separated by the widest margin of a no-man's region where no training points may lie. Thus, classification problems that cannot be solved with linear separation cannot be solved by SVMs at a first glance. However, there is a clever approach called the *kernel trick* that has been invented by researchers. This trick takes the original points and transforms each point to a point in a higher dimension. For some problems, it can be shown that some carefully chosen kernel functions used to make the transformations by multiplication, can actually make the classification problems linear at a higher dimension. It also can be shown that the optimization problem does not become much more onerous because of this transformation. Once again, the kernel approach is not discussed in this text due to the mathematical complexity.

A linear separator seems like a very simple classifier since it cannot separate classes whose examples need a more sophisticated separator function to be learned. However, it has been shown that linear separators may actually be quite powerful, especially if the examples to be separated can be transformed into a higher dimension somehow, before training a classifier as just mentioned. Points that cannot be separated linearly in their original dimensions may be separable linearly in a higher dimension. This concept of transforming to a higher dimension is beyond the scope of this book, but is presented here to motivate the fact that enhanced linear classifiers are strong classifiers.

6.4.6 SVM in multiclass classification

There is another complication with SVMs. The basic formulation of SVMs prepares it to perform binary classification only. To perform multiclass classification, one has to adapt the binary classification algorithm so that it can classify among several classes instead of just two. This can be done in several ways, but a common approach is the *One-Against-One* approach. The most popular library for SVM across languages is libSVM [2]. This package uses a one-against-one multiclass classification approach. In this approach, one creates $\frac{1}{2}k(k-1)$ classifiers, creating one classifier for each possible pair of classes, with k being the number of classes. Thus, for the case of the Iris dataset, which has three classes: $\{setosa, versicolor, virginica\}$, one creates three classifiers pairwise: $\langle setosa, versicolor \rangle$, $\langle setosa, virginica \rangle$, and $\langle versicolor, virginica \rangle$. Each binary classification model predicts one class for an unlabeled example, and the class that is predicted a majority of times is considered the class of the unlabeled example.

SVMs are strong classifiers that produce excellent results, and can be trained well with small amounts of data. SVMs work well with large amounts of data as well, but are likely to become slow. When we run a linear SVM classifier (i.e., without a kernel) on the Iris dataset, by dividing the dataset randomly into three parts and training on two parts and testing on one, we obtain 98% accuracy. In the particular case, the SVM is trained on 100 randomly picked data examples. The test dataset has 50 examples and it is a multiclass classification problem. The confusion matrix created is given below.

```
                    Reference
Prediction   setosa versicolor virginica
   setosa       15          0         0
   versicolor    0         14         0
   virginica     0          1        20


Overall Statistics

          Accuracy : 0.98
```

6.5 Incremental classification

The classification algorithms we have discussed so far need the entire training dataset at the outset. Each algorithm builds the classification model based on patterns or properties of the entire training dataset. However, it is possible that all the data examples may not be available at one time. Data items may be generated one example (or a few examples) at a time and/or become available one (or a few) at a time. For example, in a social-media site like Twitter (renamed **X** in 2023), the posts on a certain topic are generated and become available one at a time. We may still want to classify the posts as they become available. Each of the classifiers discussed above can be modified to work in an incremental manner although the modified algorithms may not be simple. For example, the K-Nearest-Neighbor Algorithm can be easily converted to an incremental version [1]. Incremental versions of SVMs also have been created [3]. In this section, we discuss incremental versions of the Decision-Tree algorithm discussed earlier.

There are several issues of importance when dealing with incremental classifiers. First, the examples are generated one by one (or a few at a time). Thus, the classifier model that is built, i.e., trained, needs to be retrained or updated for the new example. That is, either the entire training computation has to be performed all over again or the model (e.g., the linear separator or hyperplane) needs to be updated slightly to adapt to the new examples, one at a time. Another important issue that some incremental algorithms consider is the issue of forgetting of examples.

6.5.1 Incremental decision trees

The Decision-Tree algorithm discussed earlier is the well-known ID3 algorithm, first discussed by Quinlan [4]. Several incremental versions of this algorithm were proposed by Schlimmer and Fisher [5]. The ID4 algorithm, introduced by Schlimmer and Fisher builds the tree incrementally, unlike ID3. The decision tree is kept in a global data structure and for each incoming data examples, the tree is updated. All information needed to compute entropy-gain values are available at the nodes. In other words, how many examples of which classes have various values of the attributes are recorded at the nodes. When a new example needs to be incorporated into the decision tree, the new example starts at the root and goes down as much as possible, but updating the counts stored at every node. At every node the new example reaches, one of the following happens. If the node is a leaf, i.e., had examples from one class only, but now becomes nonhomogeneous, it identifies the attribute that gives the highest amount of information gain at the node and build subtrees below the node as necessary. If the node where the new example is at happens to be an internal node (i.e., a nonhomogeneous node), it updates the counts at the node for the attribute values for the various classes, and computes the attribute that will give the highest information gain or entropy loss; if this attribute is not the one that is being used now at the node, it is necessary to change the attribute at the node to the one that gives the highest amount of information gain. Since all the necessary counts are there, this can be done easily. However, if the node at this level is changed, it is necessary to recursively construct the tree underneath. Since the attribute occurrence counts are available for each of the classes, it is possible to do these computations. The way the ID4 algorithm works is that after finding that the test attribute at the current node is not the one that produces the highest information gain after a new example is shown, it changes the attribute to the one that has the highest information gain, and performs a chi-squared test to ensure that the new attribute occurs by chance, and removes all subtrees below the node, before trying to reconstruct the subtrees again.

ID5R, proposed by Utgoff [6] works similarly to ID4 in keeping information at every node to enable recomputation of the attribute that produces the most information gain as an example moves through the nodes and branches of the tree. Unlike ID4, it actually keeps a list of all the examples associated with a node. At the root of the tree, the instances need to be kept with all their attribute values. This calls for a lot of space. The description in the original paper, it does not say how the instances are stored, although in the tiny example worked out, they are shown at every node. As one attribute is used to build a node of the tree, that attribute is no longer needed to represent examples at lower levels. Thus, the representation for instances becomes shorter by one attribute at every level of the tree. It is assumed that the attribute values are all categorical for this to work. However, in general a large amount of space is necessary to store examples associate with the nodes. However, unlike ID4, ID54 does not discard the subtrees below the old test attribute. ID5R restructures the tree so that the desired test attribute is at the root of the subtree with the current example. The restructuring process manipulates the tree in such as way as to preserve consistency with the observed training instances. The actual training instances do not need to be consulted during the restructuring process, although some consultation may be necessary since the paper claims that ID5R produces the exact tree that ID3 produces.

6.6 Summary

This chapter discussed several traditional supervised machine-learning algorithms. In particular, it discussed nearest-neighbor classifiers, decision trees, and support-vector machines. The chapter also presented a brief discussion on incremental learning. Understanding these algorithms will provide a strong foundation for how machine-learning algorithms can be used to classify featured datasets, which are commonplace in actual applications of data science.

References

[1] David W. Aha, Dennis Kibler, Marc K. Albert, Instance-based learning algorithms, Machine Learning 6 (1) (1991) 37–66.
[2] Chih-Chung Chang, Chih-Jen Lin, LIBSVM: a library for support vector machines, ACM Transactions on Intelligent Systems and Technology (TIST) 2 (3) (2011) 1–27.
[3] Pavel Laskov, Christian Gehl, Stefan Krüger, Klaus-Robert Müller, Kristin P. Bennett, Emilio Parrado-Hernández, Incremental support vector learning: analysis, implementation and applications, Journal of Machine Learning Research 7 (9) (2006).
[4] J. Ross Quinlan, Induction of decision trees, Machine Learning 1 (1) (1986) 81–106.
[5] Jeffrey C. Schlimmer, Douglas Fisher, A case study of incremental concept induction, in: AAAI, vol. 86, 1986, pp. 496–501.
[6] Paul E. Utgoff, ID5: an incremental ID3, in: Machine Learning Proceedings 1988, Elsevier, 1988, pp. 107–120.
[7] Vladimir Vapnik, The Nature of Statistical Learning Theory, Springer Science & Business Media, 1999.
[8] Ronald A. Fisher, The use of multiple measurements in taxonomic problems, Annals of Eugenics 7 (1936) 179–188, https://doi.org/10.1111/j.1469-1809.1936.tb02137.x.

Artificial neural networks

7.1 Introduction

The nominal objective in the efforts to create Artificial Neural Networks (ANNs) is to build effective and efficient learning machines that are inspired by the computational structures in the human brain, in order to perform tasks that most human beings seem to perform effortlessly and hence achieve a modicum of so-called artificial intelligence. A just-born human baby is endowed with an impressive and elaborate brain structure that humans have acquired through millions of years of evolution. However, this initial brain learns new things from infancy to adulthood and also grows in size to a degree with age as it learns. Following this remarkable engineering achievement of evolution, scientists and engineers have attempted to construct artificial intelligence by building structures or organizations (called architectures) of artificial neurons that can learn from data or experience. In the recent past, such artificial neuronal architectures that can learn have achieved excellent performance in many computational tasks; this can be thought of as endowing artificial intelligence on machines.

This chapter provides an introduction to the fundamentals of ANNs, including architectures and learning algorithms. We cover the basic concepts of feedforward networks, backpropagation, activation functions, and loss functions. We also explore a few popular types of ANNs, such as convolutional neural networks (CNNs), Autoencoders including Variational Autoencoders, and Transformers.

7.2 From biological to artificial neuron

Human beings are arguably the most evolved living forms on earth, endowed with superior intelligence, abilities for sophisticated reasoning, use of complex language, and introspection in addition to the felicity with use of fire, and manipulation of things with fingers and tools. Superior mental faculties and physical dexterity are hallmarks of a highly developed brain. Like any other organ in the body, the human brain is composed of cells. The predominant type of cells in the brain are the neurons (see Fig. 7.1), and scientists estimate that there are about ninety billion of them. Neurons are the basic units of the nervous system, and they receive information from other neurons through specialized structures called *dendrites*. The dendrites receive electrochemical signals from other neurons and transmit them to the cell body or soma of the neurons. The cell body integrates these signals, combining them and determining whether the resulting signal is strong enough to generate an

Fundamentals of Data Science. https://doi.org/10.1016/B978-0-32-391778-0.00014-4

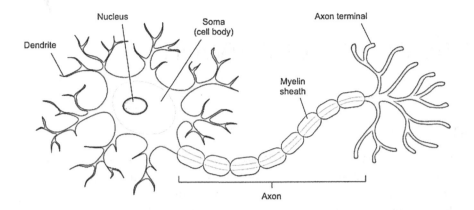

FIGURE 7.1 Cartoon of a biological neuron.

action potential, also known as a *spike* or a *nerve impulse*. If the signal exceeds a threshold level, an action potential is generated, which propagates down the *axons*, long thin fibers that extend from the cell body, and to the axon terminals, which form synapses or electrochemical connectors with other neurons. The action potential is a rapid and brief change in the electrical potential across the membrane of the neuron caused by the movement of ions across the membrane. This change in potential causes the release of neurotransmitters, which are chemicals that transmit the signal to the next neuron in the circuit. The process of communication between neurons through the transmission of electrical and chemical signals is the basis of all neural activity, including perception, cognition, and behavior.

Inspired by the biological neuron structure, a mathematical model, *Artificial Neural Networks* was developed to crudely mimic the learning procedure of humans. The development of ANNs was influenced by the work of many researchers over several decades, and thus, it is difficult to attribute the invention of ANNs to a single person or group.

7.2.1 Simple mathematical neuron

In the early 1940s, Warren McCulloch and Walter Pitts [5] proposed a mathematical model of neural networks, which they called "a logical calculus of ideas immanent in nervous activity." This work laid the foundation for the development of ANNs. Despite its simplicity, their model has demonstrated remarkable adaptability over more than eighty years. Modifications of such artificial neuronal models constitute the building blocks of most modern and advanced neural networks. The McCulloch–Pitts neuron model is made up of one neuron and n excitatory binary inputs $x_k \in \{0, 1\}$. The output of the McCulloch–Pitts neuron is also a binary output $y \in \{0, 1\}$, where $y = 1$ indicates the neuron is activated and its output is 1. The neuron can be visualized as in Fig. 7.2.

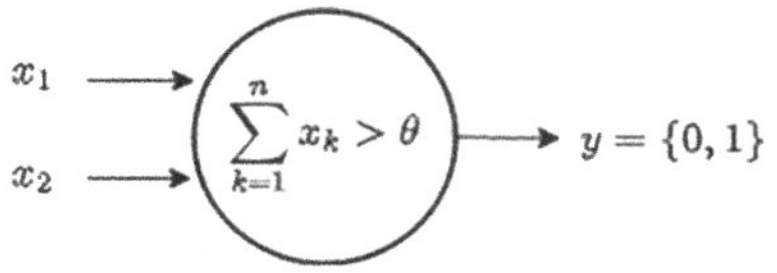

FIGURE 7.2 A two-input McCulloch–Pitts neuron with $k = 2$.

Given the input vector $\vec{x} = [x_1, x_2, \cdots, x_n]^T$,[1] and the threshold θ, the output y is simply computed using Eq. (7.1). Here, the neuron fires when the sum of its inputs is larger than θ.

$$y = \sigma(\vec{x}) = \begin{cases} 1 & \text{if } \sum_{k=1}^{n} x_k > \theta \\ 0 & \text{otherwise.} \end{cases} \qquad (7.1)$$

The McCulloch–Pitts model has limited functionality and cannot work with real data if the data are not linearly separable. Every synaptic weight is assigned a value of one, which means that each input has an equal impact on the output. In addition, there was no learning that was possible and a network to solve a specific problem had to be constructed by hand.

7.2.2 Perceptron model

In the late 1950s and early 1960s, Frank Rosenblatt invented the *Perceptron* [9], which was one of the earliest ANNs. The Perceptron was a type of artificial neuron that could learn to classify patterns in data. The perceptron model is a more generic computational model compared to the McCulloch–Pitts neuron. Unlike the McCulloch–Pitts neuron, it introduced synaptic weights (w_i) that act as learnable parameters. It operates by taking the input, computing the weighted sum of inputs, and outputting a 1 only when the aggregated sum exceeds a specific threshold; otherwise, it returns 0.

Eq. (7.1) can be extended to represent a perceptron with given input vector $\vec{x} = [x_1, x_2, \cdots, x_n]^T$ and weight vector $\vec{w} = [w_1, w_2, \cdots, w_n]^T$:

$$y = \sigma(\vec{x}, \vec{w}) = \begin{cases} 1 & \text{if } \sum_{k=1}^{n} x_k * w_k \geq \theta \\ 0 & \text{otherwise.} \end{cases} \qquad (7.2)$$

We can rewrite the above equation as:

$$y = \sigma(\vec{x}, \vec{w}) = \begin{cases} 1 & \text{if } \sum_{k=1}^{n} x_k * w_k - \theta \geq 0 \\ 0 & \text{otherwise.} \end{cases} \qquad (7.3)$$

In the above equation, the weighted sum of the $x_k * w_k$ values must be greater than θ for the perceptron to output a 1. In terms of the linear equation of the decision boundary, θ is

[1] The superscript T represents the transpose of a vector.

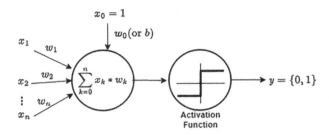

FIGURE 7.3 A perception with bias as input and sign activation function for calculating the outcome.

treated as an additional input to the perception, popularly called *bias* (*b*). The bias is an important parameter in the learning process of a perceptron. It allows the model to shift the decision boundary and make accurate predictions even when the input data is not centered around zero. In other words, the bias allows the perceptron to adjust the decision boundary to fit the data better. The above equation can be written as the perceptron rule for binary classification problems:

$$y = \sigma(\vec{x}, \vec{w}) = \begin{cases} 1 & \text{if } b + \sum_{k=1}^{n} x_k * w_k \geq 0 \\ -1 & \text{otherwise,} \end{cases} \tag{7.4}$$

where, $b = -\theta$ and +1 indicates the sample to be classified is on one side of the decision boundary (positive class) and -1 is the other side (negative class). In other words, the sign of the function $\sigma(\vec{x}, \vec{w})$ decides the class. Usually, the function, $\sigma(.)$, is an *activation function* called the *sign* or *signum* function. The refined pictorial representation of perception with an activation function is shown in Fig. 7.3.

For ease of implementation, the above rule can be represented in vector notation by replacing b with learnable weight w_0, multiplied with constant $x_0 = 1$. The modified form of the rule is as follows:

$$b + \sum_{k=1}^{n} x_k * w_k = \sum_{k=0}^{n} x_k * w_k = \vec{w}^T * \vec{x}, \tag{7.5}$$

where $\vec{w}^T$ is the transpose of a column vector and $\vec{w}^T * \vec{x}$ can be calculated as matrix multiplication.

7.2.3 Perceptron learning

The approach to learning a function by an ANN is loosely inspired by how (long-term) learning happens in the human brain. The human brain comes with weights of connections among neurons that have been learned through millions of years of evolution, unlike the random initialization of ANNs. The weights in a newly born baby's brain are modified slowly through learning as the individual is exposed to new experiences in life. Similarly, the weights in a neural network are the parameters that need to be learned from data. There may be hundreds of thousands or millions of such parameters.

The perceptron algorithm is a simple but powerful algorithm for binary classification tasks. While it can only classify linearly separable data, it can still be effective for many real-world problems. It involves the following steps:

1. **Initialize the weights**: The weights of the perceptron are initialized randomly within a range such as -1 to 1. The number of weights is equal to the number of input features.
2. **Calculate the weighted sum**: For each data point, calculate the weighted sum of the input features by multiplying each feature by its corresponding weight.
3. **Apply the activation function**: Apply an activation function (e.g., a step function or sigmoid function) to the weighted sum to produce the predicted output.
4. **Compare the predicted output to the actual output**: Compare the predicted output to the actual output of the data point to calculate the error.
5. **Update the weights**: Use the error to update the weights using a learning rate. The learning rate controls how much the weights are updated with each iteration of the training process.
6. **Repeat steps**: Iterate over all the data points and repeat steps 2–5 until the algorithm converges to a set of weights that accurately predict the output classes.

7.2.4 Updating perception weights

In the perceptron algorithm, the weights are updated based on the error between the predicted output (y) and the actual output ($\hat{y}$). The update rule for the weights in the perceptron algorithm can be expressed as:

$$w_i = w_i + \triangle w_i$$
$$\text{where, } \triangle w_i = \eta * (\hat{y} - y) * x_i,$$

(7.6)

where w_i is the weight of the x_i input. η is a positive constant, called the *learning rate*. This is a hyperparameter that controls the step size of the weight updates. It determines how much the weights are updated based on the error between the predicted output and the actual output. A higher learning rate can lead to faster convergence during training, but it can also cause the algorithm to overshoot the optimal solution and fail to converge. Conversely, a lower learning rate may require more iterations to converge, but it can result in more accurate weights. Usually, very small values are considered for the learning rate.

7.2.5 Limitations of the perceptron

While the perceptron has some advantages, such as simplicity and efficiency, it also has several limitations. The perceptron algorithm can only solve binary classification problems that are linearly separable, meaning that there must exist a line (or hyperplane) that can separate the examples of the two classes. The perceptron cannot converge to a solution if the problem is not linearly separable. The problem of learning the AND, OR, and

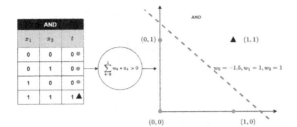
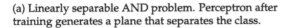
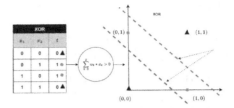

(a) Linearly separable AND problem. Perceptron after training generates a plane that separates the class.

(b) XOR is not separable by a single plane created by perceptron after training.

FIGURE 7.4 Limitation of a perceptron in handling nonlinearly separable problem.

XOR boolean functions when the truth tables are considered as data, are classic examples used to illustrate the limitations of a single-layer perceptron. The AND problem is a simple binary classification problem that involves two input variables (x_1 and x_2) and requires the perceptron to output a 1 when both inputs are 1, and a 0 otherwise. Similarly, in the case of the OR problem, the perceptron needs to output a 1 when either of the inputs is 1, and a 0 otherwise. The data, with the truth-table inputs considered as examples with the output as labels, for both problems are linearly separable and can be classified using a single-layer perceptron as it can find a single separating line between the 0 examples and the 1 examples. The XOR problem, on the other hand, is a more complex binary classification problem that also involves two input variables, but requires the perceptron to output a 1 when the inputs are different, and a 0 otherwise. A single-layer perceptron cannot solve the XOR problem, since the data points are not linearly separable. Both types of problems are illustrated in Fig. 7.4.

Further, the perceptron is sensitive to the input data, meaning that small changes in the data can lead to large changes in the decision boundary. This can make the model unstable and difficult to interpret. Sometimes, the perceptron algorithm may not converge to a solution, even if the problem is linearly separable. This can happen if the learning rate is too high or the data are noisy. It can only learn simple linear patterns, making it unsuitable for problems requiring more complex decision boundaries. These limitations have led to the development of more advanced neural-network models, such as multilayer perceptrons, which can overcome some limitations using more complex architectures and training algorithms.

7.3 Multilayer perceptron

A multilayer perceptron (MLP) is an extension of the single-unit (or the single-layer) perception, consisting of one or more hidden layers (Fig. 7.5) between the input and output layers. An MLP is a *feedforward neural network*, meaning that information flows through the network only in one direction, from input to output, when performing classification.

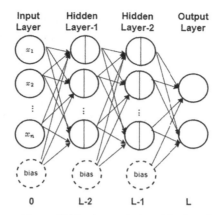

FIGURE 7.5 A Multilayer Perceptron with two hidden layers designed for handling two-class classification problems. It can be thought of a cascaded layers of perception functions that are composed together, $f^L(f^{L-1}(f^{L-2}(\vec{x})))$. f^i denotes the function that corresponds to the application of all the activation functions in layer i on all the inputs coming to layer i.

Each layer of an MLP comprises multiple nodes, or neurons, connected to nodes in the previous and following layers by weighted connections. Each neuron applies a mathematical function to the weighted sum of its inputs to produce its output, which is then passed to the next layer.

It is customary that every node in an ANN layers uses the same activation function. Often, all nodes in all layers, except possibly the top layer in a deep network, use the same activation function.

An MLP is a dense or fully connected neural network. In a dense neural network, every node in a certain layer is connected to every node in the layer above. Thus if a layer has l_1 nodes and the layer above has l_2 nodes, the number of edges or weights required to connect the two layers is $l_1 \times l_2$. This often becomes expensive in terms of the number of weights to learn. For example, assume the input is a picture with 25×25 pixels, which are entered in a linearized manner with 625 inputs. If the hidden layer above has 250 nodes, just connecting the input layer to this layer requires $625 \times 250 = 156250$ weights or parameters to be learned.

7.4 Learning by backpropagation

Backpropagation is the primary algorithm used to train a multilayer perceptron (MLP). The goal of backpropagation is to adjust the weights of the neurons in the network so that the output of the network more closely matches the desired output.

Like the single-layer perception, the weights of an MLP network are initialized with random values. This initial network is asked to classify the first example from the training dataset. Note that the example has a label or class associated with it in the dataset, giving the expected output. It is quite likely that the network, which has random knowledge (the set of all weights is considered knowledge), will make an error in classification when producing the predicted output. Note that even though the network is performing classification, it is generating a number at the output, usually an estimate of the probability that the data example belongs to a certain class. For example, if the example being shown should have an output of, say 1, saying it belongs to a certain class with 100% probability, the network may wrongly produce another value, say 0.2. Thus, there is likely to be a numeric difference, which can be computed in different ways, between what the network should produce as output, and what it actually produces. This is called *loss* for the specific data example (more details are given in Section 7.5). This individual data example's loss is computed at an output node of the network. The loss at the output node can be used to compute updates to the weights on the edges coming to the output node. In other words, the loss at the output node or layer can be backpropagated to the layer directly beneath it. Next, it is possible to perform computations so that the errors committed at the nodes in the penultimate layer (due to the error at the output, of course) are backpropagated further to the layer just before the penultimate layer. In a similar way, layer by layer, the error at the output can be backpropagated all the way to the layer just above the input layer. At each layer, based on the error at a node on this layer, the weights on the edges coming up to the node can be changed a little so that the next time around the network may make a slightly smaller error. Thus the training of a network with respect to a training example happens in a bidirectional manner. Training is performed with each example in the dataset. The process is summarized below.

Forward Propagation: During forward propagation, the input signal is passed and modified through the network, layer by layer, until it reaches the output layer. The output of each neuron in the network is calculated similarly to that in a single-layer perception. The weighted sum is the sum of the input signals multiplied by their corresponding weights, plus a bias term with its own weight. The activation function is typically a nonlinear function, such as the sigmoid function (discussed in Section 7.6), that introduces nonlinearity into the network, starting at the layer above the input layer and moving forward to the output layer. Each neuron in a layer performs the required computation and passes the output value to the next layer. This happens until the network's output(s) is (are) produced.

Backpropagation: During backpropagation, the error between the network's actual output and the expected output is calculated, and then the output neurons' weights are adjusted to reduce this error. The error is propagated backwards through the network, layer by layer. At each node, an update to a weight on an edge arriving at the node from a lower layer is made considering the rate of change (gradient) of the loss at the node with respect to the weight.

7.4.1 Loss propagation

To update a neural network's weights, it is necessary to compute the difference between what the output of the network should be and what the actual output is, for each example in the training dataset. As mentioned before, this is called *loss* or error. If there are several output nodes, the total loss is a combination of the losses at each output node. The

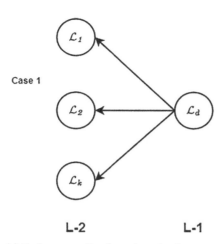

Case 1

L-2   L-1

(a) Backpropagating loss along *k* edges connected to lower layer.

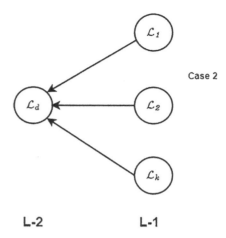

Case 2

L-2   L-1

(b) Backpropagating losses from all incoming edges from the higher layer to calculate the loss of the node.

FIGURE 7.6 Loss propagation in backward learning of MLP for nodes at intermediate levels. The figures show the edges in the backward direction to emphasize that backpropagation takes place in the direction opposite to the feedforward direction.

(total) loss can be computed in many different ways. We discuss a few loss functions in Section 7.5.

Given the loss function $\mathcal{L}$ at the output node, we can use its first derivative $\frac{d\mathcal{L}}{dw}$ to compute the change that must be made to the weight w of an edge connected to the output node. Updates or corrections are made to all weights connected to the output nodes from the layer below. The algorithm called *Gradient descent* (discussed next) requires the computation of the gradient of the loss function at the output node with respect to the weights on edges connected to this node. The gradient contains the derivatives of the loss with respect to each of the weights connected to a node. The derivative w.r.t. a weight is partial, assuming all other weights are constant. The gradient has all the partial derivatives at a node, written as a column vector. This gradient is used to update all weights connected to this node from below. To make updates to weights between the last hidden layer (the penultimate layer) and the hidden layer before it, the loss must be computed at the end of these particular weights. This requires the loss at the output node(s) to be propagated backwards to the end of the edges under consideration. Such backpropagation of loss can be done by considering two cases.

Case 1: When there are several edges from nodes at a lower layer to *one* node at an immediately higher layer, as shown in Fig. 7.6(a). In such a case, during the backpropagation of loss, the loss at the higher-level node is propagated backwards. The weights of the edges also play a role at this computation, although details are not discussed here to keep things simple.

Case 2: This case applies when there are several edges from *one* node at a lower layer to several nodes at a higher layer, as in Fig. 7.6(b). In such a case, the loss at the lower node *d* is the sum of the losses at the higher nodes. The weights of the edges also should be taken into consideration, although mathematical details are not presented here.

Thus, we see that using these two cases, loss at an output node or a number of output nodes can be backpropagated to any node in any layer in a neural network, even if there are many layers of nodes with many nodes at each layer, and with millions of edges in the entire network. Once the loss at a node is known, the gradient of the loss w.r.t. weights connected to it can be computed. The gradient can be used to update all these weights.

7.4.2 Gradient descent for loss optimization

When the training examples are linearly separable, the perceptron rule works well. However, it fails to converge if the classes in the dataset are not linearly separable. In such a case, an ANN uses *gradient descent* to find weights that best fit the training examples. Gradient descent is an iterative algorithm for the optimization (minimization) of a function. In the case of neural networks, gradient descent is used to optimize the weights given any loss function. During gradient descent, the weights of the network are adjusted in small steps to minimize the value of the loss function over the entire dataset. This process is repeated until the value of the loss function is minimized or until a stopping criterion is met.

Given a neural architecture and a loss function, we can plot the loss as a function of parameters, which are the learnable weights in the network. Since it is easy to visualize with just one parameter (maybe, two; beyond two, visualization is challenging), a plot with just one parameter is shown in Fig. 7.7. The loss function is assumed to be convex to make the discussion easy to follow. Although gradient descent is designed to work with convex functions, in practice it is used with nonconvex functions also.

The objective of gradient descent is to find the value of w_{min}, (assuming there is a single weight or parameters) where the least value of the loss function occurs. Finding w_{min} can potentially be performed using concepts of minimization using first and second derivatives in basic calculus, but the problem becomes difficult when the number of parameters or weights is large—possibly millions or even billions in a modern artificial neural network.

Based on the discussion in the previous section, given any edge *i* with weight w_i, we are able to compute an expression for the loss at the top end of it. Once the loss at the higher end of an edge is known, the change in weight that needs to be made to it can be computed. Iteratively, we can compute loss at the output of every edge in the network during backpropagation and make an update to the weights. Next, we discuss how a single weight can be updated when the loss at its top end or the output is known. The basic algorithm is called *Stochastic Gradient Descent*. This algorithm considers only one data example at a time, although examples can be put together in small batches for efficiency. The algorithm needs to be applied to every weight in the network for every example in the dataset to complete an epoch of training. Usually, several epochs of training are necessary.

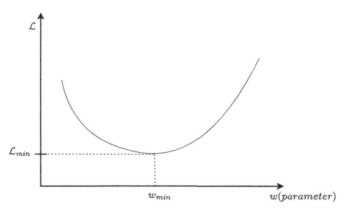

FIGURE 7.7 Loss-function plot assuming there is only one parameter or weight in the entire neural network. In a modern ANN, there are millions, if not billions of weights or parameters.

Stochastic Gradient Descent (SGD) takes an iterative approach to optimization (here, minimization) of a loss function. It picks a random initial value for the parameter w and calls it estimate w_0 for the value of w_{min}. We do not use the superscript for an edge or weight to keep the notations simple. Since it is random (it may also be a somewhat informed guess), it is likely to be far from the actual value of w_{min}. Stochastic Gradient Descent attempts to perform a simple computation to improve the initial estimate to obtain a second, and possibly improved estimate for the value of w_{min}. The approach continues to attempt to iteratively improve the estimate for a number of times until the estimate does not improve any further. The last estimate is assumed to be the value of w_{min} that the algorithm can find. This high-level algorithm for an iterative improvement algorithm for SGD is given in Algorithm 7.1.

Algorithm 7.1: SGD Algorithm, First Attempt.

Input: Loss function $\mathcal{L}$ that is defined in terms of one parameter w
Input: Dataset **D** of example points
Output: Estimated value of w at which the loss function is minimum.
$w^{(0)} \leftarrow$ A randomly picked initial value
$k \leftarrow 0$
foreach *point in the dataset D* **do**
  repeat
    $w^{(k+1)} \leftarrow$ slightly improved version of estimate than $w^{(k)}$
    $k \leftarrow k + 1$
  until *no further improvements are possible, i.e.,* $\left| \left(w^{k+1} - w^{(k)} \right) \right| \leq \epsilon$;
return *the last value of w as w_{min}*

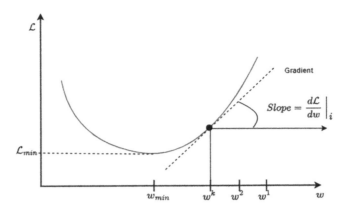

FIGURE 7.8 Stochastic Gradient Descent with one parameter.

Let us now consider how Stochastic Gradient Descent (SGD) improves the current estimate of w_{min} in each iteration as a data point is being examined. Let $w^{(k)}$ be the kth estimate for the value of w_{min}, initially starting from $w^{(0)}$ and having updated it k times. Note that the tangent line to the loss curve at $w = w^{(k)}$ gives the direction of the fastest increase in the value of the loss function in a small neighborhood around $w = w^{(k)}$. The slope of the tangent line to the loss function at $w = w^{(k)}$ is given by $\frac{d\mathcal{L}}{dw}$ at this point. Since the goal is to minimize the loss $\mathcal{L}$, instead of moving along the direction of the tangent, we move the estimate against it because we want to move it closer to w_{min}. This discussion assumes that we are to the right of w_{min}, as shown in Fig. 7.8. The approach SGD takes is to move a very small distance α along the direction of the negative tangent to obtain the next estimate. α is called the learning rate. The algorithm for SGD is given in Algorithm 7.2, assuming the neural network has single parameter.

Algorithm 7.2: SGD Algorithm, Second Attempt.

Input: Loss function $\mathcal{L}$ that is defined in terms of one parameter w
Input: Dataset **D** of example points
Output: Estimated value of w at which the loss function is minimum.
$w^{(0)} \leftarrow$ A randomly picked initial value
$k \leftarrow 0$
foreach *point in the dataset D* **do**
   repeat
      $w^{(k+1)} \leftarrow w^{(k)} - \alpha \frac{d\mathcal{L}}{dw}\big|_{w^{(k)}}$
      $k \leftarrow k + 1$
   until *no further improvements are possible, i.e.,* $|(w^{k+1} - w^{(k)})| \leq \epsilon$;
return *the last value of w as w_{min}*

Although our discussion has pertained to points to the right of the point given as w_{min}, the formula works even if the points are to its left. This is because to the left of w_{min}, the value of the first derivative $\frac{d\mathcal{L}}{dw}$ is negative, and the presence of a negative sign in front of it causes the new estimate to move toward w_{min}. Thus whether an estimate is to the left or the right of w_{min}, the update moves it toward w_{min}.

The above discussion has also assumed that there is only one parameter w on which the loss function depends. However, in an artificial neural network, each weight that can be learned is a parameter, and potentially there may be hundreds of thousands, or millions of parameters, if not more. When there is more than one parameter, we write all the parameters as a parameter vector, with $n + 1$ components, corresponding to the n weights in the neural network, and the bias w_0:

$$\vec{w} = \begin{bmatrix} w_0 \\ w_1 \\ \vdots \\ w_n \end{bmatrix}. \tag{7.7}$$

In such a case, we initialize all parameters randomly to obtain the initial value $\vec{w}^{(0)}$. For iterative improvement, each component of $\vec{w}$ is updated in each iteration. The derivative taken for each parameter is partial, assuming all other parameters are held constant. For example, to improve the kth parameter in the jth iteration, the formula used is given below:

$$\vec{w}_k^{(j+1)} \leftarrow \vec{w}_k^{(j)} - \alpha \frac{d\mathcal{L}}{dw}\bigg|_{\vec{w}^{(j)}}. \tag{7.8}$$

The modified algorithm is given in Algorithm 7.3.

Algorithm 7.3: SGD Algorithm with Parameter Vector.

Input: Loss function $\mathcal{L}$ that is defined in terms of parameter vector $\vec{w}$
Input: Dataset **D** of example points
Output: Estimated value of $\vec{w}$ at which the loss function is minimum.
$\vec{w}^{(0)} \leftarrow$ Randomly initialized parameter vector
$j \leftarrow 0$
foreach *point in the dataset D* **do**
    repeat
        $\vec{w}^{(j+1)} \leftarrow \vec{w}^{(j)} - \alpha \nabla\mathcal{L}\big|_{\vec{w}^{(j)}}$
        $j \leftarrow j + 1$
    until *no further improvements are possible, i.e.,* $|(\vec{w}^{j+1} - \vec{w}^{(j)})| \leq \epsilon$;
return *the last value of $\vec{w}$ as $\vec{w}_{min}$*

Here,

$$\nabla \mathcal{L}(w_0, w_1, w_2, \ldots, w_n) = \begin{bmatrix} \dfrac{\partial \mathcal{L}}{\partial w_0}(w_0, w_1, w_2, \ldots, w_n) \\[2mm] \dfrac{\partial \mathcal{L}}{\partial w_1}(w_0, w_1, w_2, \ldots, w_n) \\[2mm] \dfrac{\partial \mathcal{L}}{\partial w_2}(w_0, w_1, w_2, \ldots, w_n) \\[2mm] \vdots \\[2mm] \dfrac{\partial \mathcal{L}}{\partial w_n}(w_0, w_1, w_2, \ldots, w_n) \end{bmatrix} \qquad (7.9)$$

is the gradient of $\mathcal{L}$ with respect to the vector $\vec{w}$.

A point to be noted concerns the termination condition of the **repeat** loop. The algorithm obtains the difference between two vectors. The usual way to obtain this difference is by computing the Euclidean distance between the two vectors:

$$\vec{w}^{(j+1)} - \vec{w}^{(j)} = \sqrt{\left\{w_0^{(j+1)} - w_0^{(j)}\right\}^2 + \cdots + \left\{w_n^{(j+1)} - w_n^{(j)}\right\}^2}, \qquad (7.10)$$

assuming there are n weights in the neural network.

7.4.3 Epoch of training

This process of forward propagation followed by backpropagation is repeated for multiple epochs until the network's cumulative loss becomes low and stabilizes. Learning from all the examples once is called an *epoch* of learning. Thus an epoch of learning in a neural network is a very computation-intensive task. In ANNs, one epoch of learning is not enough. Often, the ANN model needs to go through tens of epochs to learn properly. Usually, the model can run for a certain number of epochs till the cumulative loss for an epoch no longer decreases, compared to the loss in the immediately previous epoch. Often, the cumulative loss is reduced dramatically for a few of the first epochs and then the rate of reduction slows, and finally levels off, or even becomes a little worse.

We should note that at the output layer, we may have not just one node, but several. For example, in a network that classifies the MNIST dataset of hand-written digits, there are ten output nodes corresponding to the ten digits 0–9. For a single training example, each output node computes a loss. The losses at the output nodes are used to make corrections to edge weights connecting to the output layer.

7.4.4 Training by batch

The previous discussion of backpropagation assumes that one example is input to the neural network at a time, the loss is computed at the output, and updates to weights are made in a backward fashion. However, performing backpropagation for each training example, one at a time, is expensive, considering that a large number of parameters or weight values need to be updated for each example. Thus, in practice, a small number of training examples, say 32, 64, or 128, are used to make updates to weights. Such a subset of examples is called a *batch*. For each example in the batch, the weight updates are computed for each edge in the neural network, but the updates are not actually made. The updates for a weight are averaged over all examples in the batch. All weight updates are made from these averages at the end, when processing a batch of examples.

7.5 Loss functions

In a neural network, a loss function ($\mathcal{L}$) is a mathematical function that measures the difference between the predicted output and the actual output for a given set of input data. The loss function is used to calculate the error in the output of the network, which is then used to adjust the weights of the network during the process of backpropagation.

The choice of the loss function in an ANN depends on the nature of the problem being solved and the type of relationship between the input and the output being predicted. For example, for a binary classification problem, a *crossentropy loss* function may be used, while for a regression problem, a *mean-squared error* loss function may be appropriate. In addition to these common loss functions, there are several other loss functions that can be used depending on the specific problem being solved. Many published papers create their own loss functions best suited to the problems at hand. In this section, we discuss a few of the popular loss functions in brief.

7.5.1 Mean-squared loss

As we train a neural network, we can compute the loss at an output node in various ways. A simple way to compute the loss at an output node for a training example i is to simply compute the difference between y_i, the expected or target value from the dataset for this example, and $\hat{y}_i$, the actual output produced by the neural network with the weights it has at this time. Such a loss at the output node can be written as $(y_i - \hat{y}_i)$. We can obtain a cumulative or total raw loss over all the examples in the training set by adding up all the errors. The total loss at this output node for the entire dataset can be written as

$$\mathcal{L}_{sumraw} = \sum_{i=1}^{N}(y_i - \hat{y}_i) \tag{7.11}$$

and the mean raw loss can be obtained as

$$\mathcal{L}_{meanraw} = \frac{1}{N}\sum_{i=1}^{N}(y_i - \hat{y}_i), \tag{7.12}$$

where N is the total number of examples in the training set. However, the loss at an output node can be positive or negative, and if we compute the total (or mean) of such losses over all the training examples, some positive losses may cancel negative losses, making the total (or mean) smaller than it really is. Therefore instead of computing the sum (or average) of raw losses, we can compute the sum of squared raw losses or the mean of squared losses as the loss that we want to reduce by changing the weights of the network:

$$\mathcal{L}_{sumsquared} = \sum_{i=1}^{N}(y_i - \hat{y}_i)^2 \tag{7.13}$$

and the mean-squared loss can be obtained as

$$\mathcal{L}_{meansquared} = \frac{1}{N}\sum_{i=1}^{N}(y_i - \hat{y}_i)^2. \tag{7.14}$$

Often, the loss is not computed over all the examples in a dataset, but a smaller *batch* of training examples. In such a case:

$$\mathcal{L}_{meansquared} = \frac{1}{B}\sum_{i=1}^{B}(y_i - \hat{y}_i)^2, \tag{7.15}$$

where B is the number of examples in a batch. The entire dataset of N examples is processed in $\left\lceil \frac{N}{B} \right\rceil$ batches.

The backpropagation algorithm discussed earlier, used to make neural networks learn from a training set, attempts to reduce a loss such as the mean-squared loss between observed outputs and expected outputs (or labels given to the examples) during training.

7.5.2 Crossentropy loss

Another commonly used loss function is the crossentropy loss. It assumes that we have more than one output node. In the case of classification, each output node corresponds to one class, and therefore, if there are n classes, there are n output nodes. Given a data example from the training set, the last layer of a modern classification neural network produces a probability that the example belongs to each of the n classes. The predicted probability values for a class can be anywhere between 0 and 1; all probabilities for one specific example belonging to various classes must add up to 1. The dataset also has a label or class associated with each data example. Thus, for an example, the label gives a probability of 1 for the class it belongs to, and a probability of 0 for all other classes.

In Fig. 7.9, the X-axis labels correspond to all the training examples. There are N such examples. On the Y-axis, we show the values for the two discrete probability distributions. In particular, $p_1^{(i)}$ is the probability that the ith example $\vec{x}^{(i)}$ belongs to a certain class C, obtained the first way, and $p_2^{(i)}$ is the probability that $\vec{x}^{(i)}$ belongs to class C obtained the second way. The assumption is that there are only two classes.

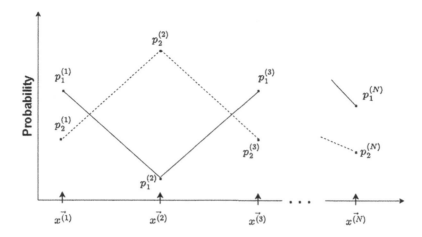

FIGURE 7.9 Two Discrete Probability Distributions.

To measure how different the two (discrete) distributions are, we can compute what is called the crossentropy between the two discrete distributions, defined as:

$$\mathcal{L}_{cross\text{-}entropy} = -\frac{1}{N} \sum_{i=1}^{N} \left[p_1^{(i)} \ln p_2^{(i)} + \left(1 - p_1^{(i)}\right) \ln \left(1 - p_1^{(i)}\right) \right],$$ (7.16)

where N is the total number of examples in the training dataset. The negative sign in front of the equation makes the value of crossentropy positive since the ln values of probabilities (which are between 0 and 1, inclusive at both ends) are negative. Note that ln 0 is undefined, but the way the algebra works out below, we do not really have to use a value for it.

In particular, assume the first discrete distribution comes from the probability of the labels of a training dataset with two classes. Since there are two classes, if the probability for class C for one example is known, the probability for the other class can be computed, because the two probabilities for an example add up to 1. The $p_1^{(i)}$ value is simply derived from the label $y^{(i)}$ on example $x^{(i)}$,

$$p_1^{(i)} = 1 \text{ if } y^{(i)} = 1, \text{ otherwise } p_1^{(i)} = 0; \ i = 1 \cdots N.$$ (7.17)

This is because the label, which now corresponds to the probability of belonging to class C, is either 1 or 0, depending on whether the example belongs to the class or not. The $p_2^{(i)}$ value for element $x^{(i)}$ is the probability obtained by the ANN classifier (such probabilities can be obtained using the so-called softmax layer—which we have not discussed yet—at the top of a classification neural network) that $x^{(i)}$ belongs to the class C. In other words,

$$p_2^{(i)} = \hat{y}^{(i)}, \ i = 1 \cdots N,$$ (7.18)

where $\hat{y}^{(i)}$ is the output of the neural network after the softmax layer, for class C. Note that $\hat{y}^{(i)}$ values are real in the range $[0, 1]$. Considering these particulars, the crossentropy formula can be written as

$$\mathcal{L}_{cross\text{-}entropy} = -\frac{1}{N}\sum_{i=1}^{N}\left[y^{(i)}\,ln\,\hat{y}^{(i)} + \left(1 - y^{(i)}\right)\,ln\left(1 - \hat{y}^{(i)}\right)\right], \qquad (7.19)$$

where the $y^{(i)}$ values are either 0 or 1, depending upon if $x^{(i)}$ belongs to the class or not.

For a training example for which $y^{(i)} = 0$, i.e., the example does not belong to the class as per the label in the dataset, the contribution of the example to crossentropy is simply

$$0\,ln\,\hat{y}^{(i)} + (1 - 0)\,ln\left(1 - \hat{y}^{(i)}\right) = ln\left(1 - \hat{y}^{(i)}\right).$$

For a training example for which $y^{(i)} = 1$, i.e., that the example belongs to the class as per the dataset label, the contribution of the element to crossentropy is

$$1\,ln\,\hat{y}^{(i)} + (1 - 1)\,ln\left(1 - \hat{y}^{(i)}\right) = ln\,\hat{y}^{(i)}.$$

Thus the crossentropy loss for the set S can also be written as

$$\mathcal{L}_{cross\text{-}entropy} = -\sum_{i|y^{(i)}=1}^{N}ln\,\hat{y}^{(i)} - \sum_{i|y^{(i)}=0}^{N}ln\left(1 - \hat{y}^{(i)}\right). \qquad (7.20)$$

Crossentropy for a single example is simply $ln\,\hat{y}^{(i)}$ or $ln\left(1 - \hat{y}^{(i)}\right)$, depending on whether $y^{(i)}$ is equal to 1 or 0, respectively. Eq. (7.20) works for the case where the two classes and the probabilities are given for one of the two classes.

If there are n classes, $n > 2$, the approach to computing crossentropy can be generalized. For each example i, where the classifier produces n probabilities corresponding to the n classes; these probabilities add up to 1 for each example. Thus, there are n discrete probability distributions corresponding to the n classes; of these, if we know $n - 1$ probabilities, the last one can be obtained easily. In addition, there are n probability distributions corresponding to the labels; among the labels for example, only one value is 1 and the rest are 0 each. The crossentropy loss can be computed for each class and then these can be added to obtain the total loss.

7.6 Activation functions

An activation function transforms the weighted sum of input signals (from the previous layer or input data) into an output signal. The role of an activation function is to make nonlinear the mapping from input to output of a neural-network model, allowing it to learn more complex patterns and relationships in the data. Without the use of a nonlinear activation function, a neural network would simply be a linear regression model, unable to

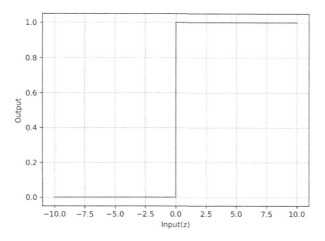

FIGURE 7.10 *Binary step* function for discrete decision making within 0 and 1.

model complex, nonlinear relationships between the input and the output. Many different activation functions have been used in constructing artificial neural networks.

7.6.1 Binary step function

A binary step function uses a positive numeric threshold value δ to decide whether a neuron becomes activated. If the input is greater than δ, the neuron is activated with its output at 1; else it is not, meaning that its output is 0. The step function given in Fig. 7.10 has a δ value of 0. There is a variation of the step function where the steady values are -1 and +1 instead of 0 and 1.

Mathematically, it is represented as Eq. (7.21):

$$f(z) = \begin{cases} 0 & \text{if } z \leq 0 \\ 1 & \text{if } z > 0 \end{cases}, \tag{7.21}$$

where z is the weighted sum of the inputs to the neural unit.

7.6.2 Sigmoid activation

A neuron that uses the binary step function for computing the output suffers from several issues. A small change in the weight of any single input can potentially cause the output of a neuron to flip, say from 0 to 1. This may then cause the behavior of a network that this neuron belongs to, to change in complex ways. To overcome this, another activation function called the *sigmoid* activation, has been traditionally used. The sigmoid activation is such that small changes in the weighted sum of inputs and the bias cause only a small change in its output, which is likely to allow a network containing such a sigmoid neuron to behave consistently as well as learn effectively. The output of a sigmoid unit maps any

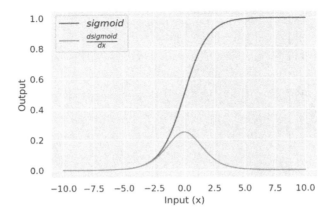

FIGURE 7.11 *Sigmoid* activation function for smooth transition within the range [0,1].

input value to a value between 0 and 1. The output can be interpreted as the probability of an example belonging to a particular class. The sigmoid activation is a continuous nonlinear function. It is also differentiable.

Mathematically, the sigmoid function is defined as

$$\sigma(z) = \frac{1}{1 + e^{-z}}, \tag{7.22}$$

where z is the weighted sum of the inputs and the bias. The sigmoid function is illustrated in Fig. 7.11.

As seen in this figure, the value of the function is close to 0 when the value of the input weighted sum is negative and is below -4 or so, and its value is close to 1 when the value of the input is positive and above +4 or so. Between -4 and 4, there is a smooth transition from negative values to positive values. On the negative side, the value of the function saturates to 0, and the value of the function saturates to 1 on the right side. Saturation means the value does not change much at all at the two ends.

7.6.3 *tanh* activation

The *tanh* function is similar to the sigmoid activation function and even has the same shape with a different output range of -1 to +1. It can be shown that *tanh* is a scaled and shifted version of the sigmoid function. The *tanh* function's value saturates to -1 and +1 at the two ends, and around the weight-summed input value of 0, it transitions from negative to positive. Mathematically, it is defined in Eq. (7.23):

$$f(z) \equiv \frac{e^z - e^{-z}}{e^z + e^{-z}}, \tag{7.23}$$

where $z = \sum_j w_j x_j + b$. The output of the *tanh* activation function is zerocentered. Hence, it can easily produce output values that are negative, neutral, or positive, whereas the sig-

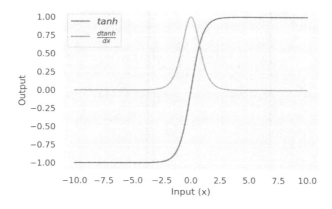

FIGURE 7.12 Visualizing hyperbolic transitions with a *tanh* activation function.

moid function's value is never negative. The *tanh* function is a continuously differentiable function (see Fig. 7.12). It is used in certain modern neural networks.

7.6.4 ReLU activation

ReLU stands for Rectified Linear Unit. It is a nonlinear function with two linear pieces. The ReLU function is the identity function if the input is positive and the zero function otherwise. Overall, as a single function, it is considered nonlinear. The neuron is inactive if the weighted sum of the inputs is less than 0; otherwise, it transmits the weight-summed input to the output. It is usually used in the hidden layers of modern neural networks. Although a simple-looking function [7], it is currently the most commonly used activation function in high-functioning neural networks of all kinds. Mathematically, it is defined as in Eq. (7.24):

$$r(z) = max(0, z),$$

(7.24)

where $z = \sum_j w_j x_j + b$. The function is illustrated in Fig. 7.13.

We list a few more activation functions with their derivatives in Table 7.1.

7.7 Deep neural networks

Many advanced neural-network architectures beyond the simple multilayer perceptron are commonly used in practice [3]. Deep Neural Networks are a class of artificial neural networks that have several layers of nodes. There is no clear definition of what the word "deep" means. In practice, any network that has more than 3 or 4 layers is usually considered deep. An illustration of a deep network is shown in Fig. 7.14, although real deep networks are usually not dense networks.

Table 7.1 Activation functions with their derivatives and plots. x can be any input value. α is a continuous value (usually 0.001). e is the exponent. The plot shows the trend of the actual activation function and its derivative.

S.No	Activation Function	Formula	Derivative	Plot
1.	Sigmoid	$S(x) = \dfrac{1}{1+e^{-x}}$	$S'(x) = S(x)(1 - S(x))$	
2.	Hyperbolic Tangent (tanh)	$tanh(x) = \dfrac{(e^x - e^{-x})}{(e^x + e^{-x})}$	$tanh'(x) = 1 - tanh(x)^2$	
3.	Rectified Linear Unit (ReLU)	$ReLU(x) = max(0, x)$	$ReLU'(x) = \begin{cases} 1 & \text{if } x > 0 \\ 0 & \text{if } x < 0 \end{cases}$	
4.	LeakyReLU	$LReLU(x) = max(\alpha . x, x)$	$LReLU'(x) = \begin{cases} 1 & \text{if } x > 0 \\ \alpha & \text{if } x < 0 \end{cases}$	

continued on next page

Table 7.1 (continued)

S.No	Activation Function	Formula	Derivative	Plot
5.	SoftPlus	$SP(x) = log(1 + e^x)$	$SP'(x) = \dfrac{1}{1+e^{-x}}$	
6.	SigNum	$SigNum(x) = \begin{cases} 1 & \text{if } x > 0 \\ 0 & \text{if } x = 0 \\ -1 & \text{if } x < 0 \end{cases}$	*Nondifferentiable*	
7.	Binary Step (BS)	$BS(x) = \begin{cases} 1 & \text{if } x >= 0 \\ 0 & \text{if } x < 0 \end{cases}$	*Nondifferentiable*	

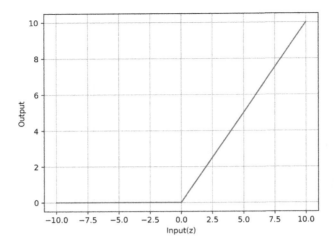

FIGURE 7.13 Exploring rectifying nonlinearity with the *ReLU* activation function outputting in the range [0, *z*].

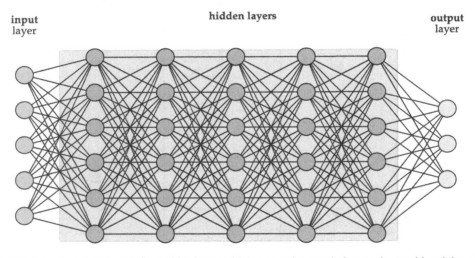

FIGURE 7.14 A neural network with five hidden layers. This is a neural network that can be considered deep. Deep neural networks usually have specialized architectures such as Convolutional Neural Networks and Transformers.

Deep neural networks have been implemented with hundreds of layers, some even more than a thousand layers. Given an input, the layer immediately above the input layer in a (deep) neural network finds low-level patterns or features in the input. The network layer immediately above takes these patterns or features as input and extracts slightly higher-level features or patterns. In such a fashion, the network's multiple layers progressively extract higher-level features from the raw input. The highest-level features are used to perform tasks such as classification or regression in the case of supervised learning, and clustering in the case of unsupervised learning.

Many different neural-network architectures have been developed for deep learning. These include Convolutional Neural Networks for image processing, Recurrent Neural Networks, and Attention Mechanisms for text and sequence processing.

7.7.1 Convolutional neural networks

In the year 1980, Yann LeCun initially presented convolutional neural networks, or ConvNets [4]. LeCun had expanded upon the research of Japanese scientist Kunihiko Fukushima, who had developed the neocognitron [11], a primitive neural network for image recognition.

A Convolutional Neural Network (CNN) is a type of artificial neural network that uses a special architecture that is particularly well adapted to classify images. A CNN is different from a traditional neural network, which is "dense" in nature in the sense that every node in a certain layer is connected to every node in the layer immediately above. One characteristic of dense neural networks is that the number of edges or weights becomes quite high quite quickly. For example, consider a dense neural network that has 100 input nodes, 256 hidden layer nodes and 10 output nodes. Such a network has $100 \times 256 + 256 \times 10 = 28160$ weights whose optimal values need to be learned as parameters. If a dense neural network is deeper with, say 10 layers, and takes a realistic image of size 256×256 say, it is likely to have millions of parameter values that need to be learned from data. Thus, a dense neural network is usually quite inefficient. The inefficiency arises because every node at a hidden (intermediate) layer is examining every node at the layer below and is performing similar computations, without any differentiation in labor.

In a CNN, a node in a higher layer looks only at a portion of the lower layer on which it is focused. This region is called the upper node's *receptive field*. Unlike a dense neural network, the responsibility of finding patterns or features in the entire lower layer is not blindly replicated for all nodes in the upper layer, but divided more equitably among the nodes in the higher layer. It turns out that such a division of responsibility for finding patterns is also true in the human visual system. Due to its construction, a CNN is more efficient in terms of parameters that need to be learned from data. It has also been shown to produce better results in terms of performance metrics such as accuracy.

7.7.1.1 Layers of CNN

A Convolutional Neural Network (CNN) typically consists of several layers (Fig. 7.16) with different functions. A few of the common layers are introduced below.

1. *Input Layer:* The input layer receives the raw input data, such as text, images or audio signals, and prepares the input for processing by the next layer. Unlike an MLP discussed earlier, the input to a CNN may be laid out in 1D space for text, 2D for a black and white image, and 3D for an image with three colors.

2. *Convolutional Layer:* A convolutional layer applies a set of filters to its input, with each filter extracting a specific feature from the input. The output of this layer is a set of feature maps, where each map contains the activations of a filter over the entire input.

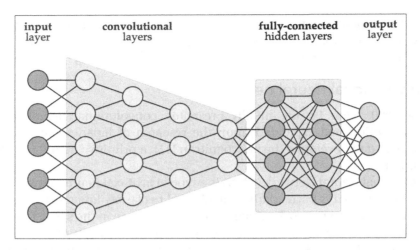

FIGURE 7.15 A simple CNN with a lower convolutional part, and an upper dense part that performs classification. In this diagram, the assumption is that the neural-network layers are linearly organized.

CNNs are named because of the use of convolutional layers, which are the main ingredients, although other layers play crucial roles in making CNNs effective and efficient.

3. *ReLU Layer:* A Rectified Linear Unit (ReLU), which computes a nonlinear activation function, is often applied to the output of each placement of convolutional filters.

4. *Pooling Layer:* A pooling layer reduces the size of the feature maps by performing a form of downsampling operation, such as maxpooling or average pooling. This helps make the feature maps compact and processing efficient.

5. *Fully Connected Layers:* The high-level features extracted by the highest convolutional layer are linearized and fed into one or more fully connected layers. The outputs of the fully connected layers correspond to classification or regression results.

6. *Output Layer:* The output layer produces the final output of the CNN. Depending on the task, the output layer may have the softmax activation function for classification or a linear function for regression.

As shown in Fig. 7.15 (also in Fig. 7.16), a CNN architecture has two parts: a convolutional part and a fully connected dense neural network. The convolutional part of a CNN consists of one or more convolutional (and other) layers, which extract "features" from the input, with higher layers capturing more complex features. Unlike the input to a dense neural network, the input to a CNN is usually not linear in nature but is a higher-order matrix. For example, if the input is a black-and-white image, the input is a two-dimensional array, with the dimensions depending on the size of the input. If the input is an RGB color image, the input to a CNN is a three-dimensional matrix, which is composed of three two-dimensional matrices each corresponding to one of the red (R), green (G), and blue (B) colors. The convolutional part of the CNN produces a three-dimensional matrix of features. The final processing in a CNN is performed by the dense neural network. This last

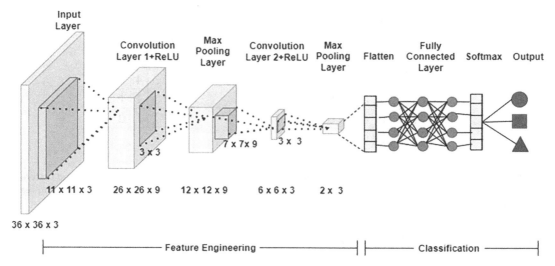

FIGURE 7.16 Typical layers in a Convolutional Neural Network. The convolutional part of the CNN extracts features and is said to perform feature engineering. In the feature-engineering section of the network, each layer is convolved or pooled using a filter shown as a smaller block inside a bigger block. Numericals indicate dimension of the matrix. 36x36x3 means input image is of size 36x36 with 3 channels, i.e., R-G-B. It is convolved through a kernal of size 11x11x3. It passes through four rounds of convolution, ReLU and Pooling with varying kernal sizes. Finally, the matrix is downsized to 2x3 matrix, which is then converted to a vector form, called Flattened vector. The vector (as a feature vector) is then passed to fully connected neural network for classification.

layer of features obtained by the convolutional part is linearized or written out as a vector and is input to a small dense neural network. The dense neural network produces the output desired by the entire CNN. For a classification CNN, the output of the fully connected layer is passed through the *softmax function* to obtain a probability distribution over the classes. The softmax layer is typically used as the final layer of the network for multiclass classification tasks. Softmax is a mathematical function that takes a vector of real numbers as input and normalizes it into a probability distribution (where each output number is between 0 and 1, and all the output numbers add up to 1), with each output element representing the probability of the corresponding class. The softmax function is defined as:

$$softmax(z_i) = \frac{e^{z_i}}{\sum_{j=1}^{K} e^{z_j}},$$ (7.25)

where z_i is the ith element of the vector input to the softmax function, K is the number of classes, and e is Euler's number. The output of the softmax function is a vector of the same dimensionality as the input vector, where each element represents the probability of the corresponding class. During training, the entire CNN is often trained using crossentropy loss.

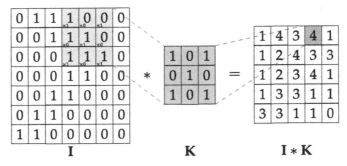

FIGURE 7.17 Convolution filter **K** of size 3 × 3 is applied on a specific input area (**I**) to obtain an extracted feature value. The filter is dragged up and down the input, and applied at each placement to obtain a corresponding feature value.

7.7.1.2 Local receptive fields and convolutions

A simple convolutional layer consists of a single convolution (also called a filter or a kernel) that transforms the content of a lower layer of neurons to values that represent extracted features to be processed further. In practice, a convolutional layer is likely to have several, if not many filters that are looking for various kinds of features in the input layer's values.

A 2D convolution focuses on a small region of the input, with the region representing pixel values at certain contiguous locations of the input layer of an image (here, two-dimensional). The convolution, which is also a matrix (of the same size as that of the focused input region) of numbers, is used to transform the input values in the region into a single number that becomes available as a feature value in a neuron at the next higher level. A 3 × 3 convolution is illustrated in Fig. 7.17. A convolution **K** is a small square matrix of numbers—usual sizes are $k \times k$, where k is a small integer like 3 or 5. The transformation performed is simple, and consists of multiplying a number in the convolution by the number in the corresponding position in the input region of the same size. Such pairwise multiplications are performed at each position and the individual products are added. The integers given in the filter are for illustration purposes only. In a CNN, the values in a convolutional filter are learned, and thus, are real numbers.

In this case, the input neurons are the pixel intensities of an input image, and on the right is one hidden neuron in the first hidden layer. Each neuron is connected to only a region of the input layer; this region in the input image is called the *local receptive field* for the hidden neuron. As shown in Fig. 7.18, the focused rectangular region of the input space is connected to a hidden neuron in the higher convolutional layer. The edges connecting the input to the output have weights on them; these weights are parameters that are learned during the training of the neural network. In order to cover the entire input space, kernal representing the receptive field is slided over the entire input image. For placement of the kernal, there is a different neuron in the convolutional layer, although the weights on the edges from input to the output are always the same for a specific convolution. The kernal is dragged across from left to right as well as up and down the image by sliding it over the input so that the entire input space is covered, and a feature value is generated for each

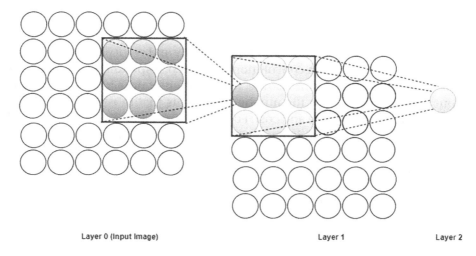

Layer 0 (Input Image)   Layer 1   Layer 2

FIGURE 7.18 Receptive field for a convolutional filter with two-dimensional input.

position of the kernal at a corresponding output node. The positions of the kernal may or may not overlap. The amount of dragging of the kernal over the input is called *stride*. If the horizontal stride length is one, then the kernal is moved by one pixel at a time from left to right. If the vertical stride is also one, the kernal is moved by one pixel down the input space. The convolutional filter is dragged first from its initial position on the top left of the input by the horizontal stride to go across the entire input space's breadth. It is then placed on the top left again, but dragged down by the vertical stride and moved across the input again. In this manner, the entire input space is covered with the (overlapped) placement of the convolution, and features are generated at the next layer for further processing. The values of horizontal and vertical strides are configurations or hyperparameters of the neural network. In practice, a number of different convolutional filters of the same size or even different sizes are used to transform one layer into another layer. The complete application of one convolution over the entire input produces a two-dimensional matrix of numbers at the next level. Thus when a number of convolutions are used to transform an input layer to an output layer, the output layer becomes a three-dimensional matrix of numbers. If the convolutions used are of different sizes, the strides need to be properly chosen so that each of the two-dimensional components of the output layer are of the same size. In addition, depending on the size of the input, the size of the filter and the values of the two strides, it may be necessary to pad the output to produce an output matrix of proper size.

Note that if the input image has three channels, corresponding to three RGB colors, the convolutional kernel has a depth of three so that the sum of products covers all channels. Convolutions are used not just in the input layer, but can be used in higher layers also. The process of applying a convolution at a higher layer is exactly the same as in the input layer. The only difference is that at a higher layer, the depth of the convolution (or the number of "channels") in it depends on the depth of the particular layer.

7.7.1.3 Efficiency of CNNs

As mentioned earlier, in a CNN, the same weights are used for each placement, possibly overlapping regions, as a convolution is slid horizontally and vertically over the input. This means that all the neurons in the convolutional layer, corresponding to a single convolutional filter, detect exactly the same feature, just at different possibly overlapping regions in the input image. This reduces the number of weights that must be learned, which in turn reduces model-training time and cost. For a 3×3 convolution, when used on a black-and-white image input, only 9 weights need to be learned. If the input image has three channels, the number of weights learned increases to 27. If there are k kernels, each of size 3×3 used on a depth of d, the total number of parameters to learn between the two layers is $3 \times 3 \times k \times d$ or $9kd$. For example, even if there are 30 different filters each of size 3×3 applied to an input RGB image, the total number of parameters to learn is $3 \times 3 \times 3 \times 30 = 270$, which is a dramatically smaller number than if the two layers were connected in a dense manner. Thus CNNs are much more cost effective than dense neural networks. Since the number of parameters to learn from one layer to another is small, CNN architectures can be built with many more layers, tens or even hundreds. It has also been found that although much lighter in terms of the number of parameters, CNN architectures are very effective in classification problems, usually much better than dense networks with a much larger number of parameters to learn.

7.7.1.4 Pooling layers

Pooling layers are usually used immediately after convolutional layers to reduce the amount of computation in a CNN. Pooling aggregates the feature information in the output from the convolutional layer. A pooling layer takes the output from the convolutional layer and prepares a condensed version of the captured features. At the same time, the use of pooling produces some generality by capturing the largest feature value within a small region ignoring smaller noisy values or when features may not be aligned properly, e.g., some straight lines may become a little crooked in an image. One common procedure for pooling is known as *maxpooling*. In maxpooling, a pooling unit focuses on a small square region of the input and simply outputs the maximum value in the region. In Fig. 7.19, a pooling unit is shown to output the maximum activation in the input region. Maxpooling picks the maximum out of all the values in the region and transfers the maximum to the output, ignoring all other values. Another commonly used pooling approach is average pooling. In average pooling, the numbers in the region are replaced by their average. As a result, the output number is likely to reflect the general contents of the region well and in some cases this may be important.

7.7.2 Encoder–decoder architectures

An encoder–decoder architecture learns to create a compact representation or encoding of an input such as an image, and uses this encoding to generate a related version of the input. For example, an encoder–decoder architecture can be used to produce a superresolution version of an input image, a version of the original image with enhanced resolution

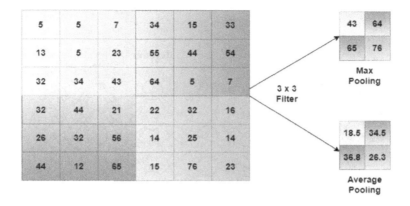

FIGURE 7.19 Max- and Average Pooling using 3 x 3 filter.

and details. An encoder–decoder architecture can also be used to clean up images that have issues to improve clarity and definition, or remove occlusions.

An encoder–decoder architecture has two neural networks, one called encoder and the other decoder, that are trained together. The two neural networks may or may not be of the same type, although often they are. For example, both encoder and decoder can be dense neural networks, or both CNNs or Transformers or a combination of two different kinds of neural networks. The training of an encoder–decoder network can be supervised as in machine translation; or can be unsupervised as in learning to clean up images that were taken badly with blurriness or have other kinds of noise.

The encoder module compresses the input data into a condensed latent representation space, and the decoder module reconstructs the latent representation back to a form similar to the original, which may or may not be of the same size as that of the input data. Fig. 7.20 represents a typical encoder–decoder architecture. During training, the reconstructed data is brought as close to the input data by minimizing the reconstruction error between the input data and the reconstructed data. Various parameters govern the modeling performed by an encoder–decoder architecture, including the number of layers, size of latent representation space, activation functions, and the optimization algorithm. The commonly used notations while modeling using an encoder–decoder architecture are discussed below.

Input Data: The input data $\vec{x}$, can be an image, text document, or a set of features. If it is a text document, it is converted to a numerical form first, using a method like Word2Vec [6] or Glove [8], not discussed in this book. The numeric word vectors are usually 300 long. The word vectors can be concatenated to make a longer document vector or each word vector can be written in a row by itself in sequence to create an image-like 2D matrix with 1 channel. The input can be of any dimension, e.g., 1-D, 2-D, or 3-D.

Encoder Function: The encoder function, $Enc(\vec{x})$, maps the input data to the encoded latent representation $\vec{z}$. If the encoder is a CNN, no softmax layer is used and the extracted

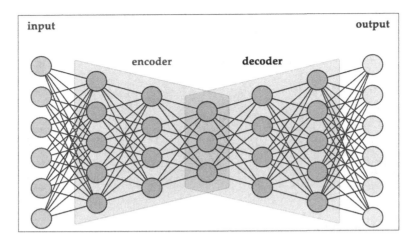

FIGURE 7.20 Encoder–Decoder Architecture. In an Autoencoder, the output is a variant of the input.

features from the last convolutional layer of the CNN are linearized to form the encoding of the input. The encoding is also referred to as a bottleneck in encoder–decoder networks. In general, the encoding is linear and its size is a hyperparameter in the encoder–decoder setup.

Decoder Function: The decoder function constructs the input data from the latent representation: $Dec(Enc(\vec{x}))$. The encoder and decoder architectures are usually similar, but this is not necessary. The encoder learns to produce a compact representation of the input that can be expanded by the decoder to produce a version of the input as needed.

Output: The reconstructed output $\hat{\vec{x}}$ is usually of the same size as that of the input data.

Loss Function: The loss function $\mathcal{L}(\vec{x}, \hat{\vec{x}})$ measures the difference between the input and the reconstructed data, which is minimized during training. A simple loss function can compute the squared error loss between pixel values of the image to be generated and the image produced by the neural network. Such a loss function can be used to update the weights of the neural network during training using backpropagation. Other loss functions can also be used.

7.7.3 Autoencoders

The success of deep neural networks largely depends on the availability of large, labeled datasets for supervised learning. In many applications, obtaining such labeled data is not feasible due to either the cost or the complexity of the data. Unsupervised learning methods offer an alternative approach, where the network learns to extract useful features from the data without any explicit labels. Autoencoders are a type of unsupervised neural network that can learn to compress the input data into a lower-dimensional representation while preserving its essential features. This makes autoencoders an attractive option for tasks such as data compression, image cleaning, and data visualization, where the goal is to extract the underlying structure of the data without relying on explicit labels.

The idea of autoencoders (AE) has been around for a while. In the mid-2000s, Geoffrey Hinton and his colleagues [1] explored the notion of utilizing neural networks to construct autoencoders and gave them their current shape. The autoencoder neural-network model tries to learn an encoding of the input data and then uses the encoded representation to reconstruct the original input. It is a variant of the encoder–decoder architecture where the output is a "simple" variant of the input, whereas in a general encoder–decoder network, the relationship between the input and the output can be more complex.

Autoencoders represent an unsupervised class of deep-learning architectures that work without labeled data. They use a nonprobabilistic autoencoding approach toward modeling the input data in an end-to-end learning framework.

In practice, an autoencoder is just like a regular encoder–decoder architecture except that there is no labeled data for training. For example, to learn to clean and denoise images, given an original clean image, we can introduce random noises using a certain probability distribution or using a certain algorithm, and create a number of noisy images from the original good image. The autoencoder can be trained using the noisy pictures as input and the original clean picture as the desired output. An autoencoder trained in this fashion can be used to denoise or clean previously unseen noisy images to obtain clean images.

7.7.3.1 *Variational autoencoders*

The main issue with traditional autoencoders is that they are purely deterministic models, meaning that they map an input to a fixed output, and do not provide any measure of uncertainty or probability. For example, a denoising autoencoder takes many different noisy versions of an image and attempts to produce the same clean image. In other words, the output is always the same or is deterministic. A traditional autoencoder is unable to produce variations of the input, e.g., after being trained on the image of faces, a traditional autoencoder cannot produce a smiling version or a crying version of an input face. This inability can make autoencoders less effective in scenarios where we want to generate new data points or perform tasks such as anomaly detection, where we need to be able to generate a distribution over possible outputs.

Variational Autoencoders (VAEs) are *generative models* that aim to overcome this limitation by introducing a probabilistic interpretation into the autoencoder architecture. VAEs are *probabilistic models* also, and therefore they are probabilistic-generative models. Generative models can learn to produce creative outputs based on a given input. Generative models have recently ushered in a new period of excitement and interest in artificial intelligence and machine learning.

VAEs use a probabilistic latent variable model to represent the input data, where the encoder maps the input to a distribution over latent variables, and the decoder maps latent variables to a distribution over the output data. This allows VAEs to model the uncertainty in the input data and generate new data points by sampling from the latent variable distribution.

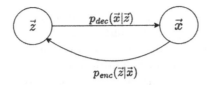

FIGURE 7.21 Underlying probability distribution assumptions for a Variational Autoencoder. $\vec{x}$ is the observed multidimensional random variable generated by the hidden multidimensional random variable $\vec{z}$. Note that it may have been more "natural" to draw $\vec{z}$ to the right of $\vec{x}$ with the arrows reversed.

A VAE makes a basic probabilistic assumption about the inputs on which it is trained (see Fig. 7.21). The assumption is that there is an unobserved or hidden random variable responsible for generating the observed training examples. If the training examples were one-dimensional, this unobserved random variable would also be one-dimensional. However, since the training examples have multiple dimensions or features, the unobserved random variable is likely to be multidimensional as well. Let us call the hidden random variable $\vec{z}$ with components or features $\langle z_1 \cdots z_m \rangle$ assuming there are m features in the hidden generating variable. Each feature $z_i, i = 1 \cdots m$ takes values that come from a multivariate probability distribution, i.e., there is a mean and standard deviation for each dimension. Such a probability distribution is the *prior probability distribution*. The $z_i, i = 1 \cdots m$, features may be dependent on each other in various ways, but we can assume that they are independent of each other to keep discussions simple. Thus corresponding to each z_i, there is a mean μ_i and standard deviation σ_i as the parameters for its assumed Gaussian distribution. All covariances for σ_{ij} values are assumed to be 0 and this allows us to write $\vec{\sigma}$ as a vector instead of a matrix. This is written as $\vec{z} \sim N(\vec{\mu}, \vec{\sigma})$ where $\vec{\mu} = \langle \mu_1 \cdots \mu_m \rangle$ and $\vec{\sigma} = \langle \sigma_i \cdots \sigma_m \rangle$.

Given the random variable $\vec{z}$, it generates using a certain process the multidimensional input variable $\vec{x} = \langle x_1 \cdots x_n \rangle$ with n visible features. For example, $\vec{z}$ could be a random variable (about which we do not know much) whose components describe the inner workings of a car, and the random variable $\vec{x}$ contains features like velocity and acceleration, which are easily observed.

If $p(\vec{x} \mid \vec{z})$—the probability (distribution) for the input variable $\vec{x}$ given the unobserved internal variable $\vec{z}$—were known, it is likely to be given in terms of a certain formula using exponentials, and it would be possible to compute the scalar probability of generating a specific value $\vec{x}$ given a specific value of $\vec{z}$.

A VAE is structured like a normal encoder–decoder network (see Fig. 7.22). However, the assumption is that the bottleneck or the encoding does not contain one simple value for each encoded high-level feature f_i, but contains two corresponding values μ_i and σ_i, the parameters for the probability distribution for f_i. As a whole, the assumption is that a VAE, when trained on values of $\vec{x}$, learns to encode the parameters, $\vec{\mu}$ and $\vec{\sigma}$, of the probability distribution for $\vec{z}$. To be even more particular, the VAE learns the weights $\vec{w}_{enc}$ of the encoder neural network (which is usually a simple network such as a dense network or CNN) that helps learn $\vec{\mu}$ and $\vec{\sigma}$. A VAE also learns using another neural network that is trained

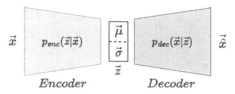

FIGURE 7.22 First attempt at building a VAE. $\vec{x}$ represents the input to the encoder; the encoded representation has two parts: $\vec{\mu}$ and $\vec{\sigma}$, the means and standard deviations for all the components of the hidden variable $\vec{z}$. $\hat{\vec{x}}$ is the output of the decoder and is of the same dimensions as $\vec{x}$. $p_{enc}(\vec{z} \mid \vec{x})$ represents the encoder network and $p_{enc}(\vec{z}) \mid \vec{x}$ represents the decoder network.

together with the encoder neural network, the weights $\vec{w}_{dec}$ that take the parameters of the probability distribution for $\vec{z}$ and produce an output that is similar to the input $\vec{x}$. The decoder is responsible for learning weights that correspond to the probabilistic mapping $p(\vec{z} \mid \vec{x})$. Since the encoder neural network encodes the inputs in a compact manner and the decoder decodes the encoding, this encoder–decoder setup is still similar to the ordinary autoencoders even though the encoding is in terms of probabilistic parameters. A network like this is likely to produce outputs $\hat{\vec{x}}$ that do not vary and are deterministic.

The goal of VAE is not to produce the same output every time, but a different one when a trained VAE is used. For this to happen, the network in Fig. 7.22 needs to be modified. This is achieved by sampling values of the hidden variable $\vec{z}$ and producing an output $\hat{\vec{x}}$ that corresponds to this sampled value. Sampling a value for $\vec{z}$ means sampling a value for each of its components $z_i, i = 1 \cdots m$, given the corresponding means and standard deviations μ_i and σ_i, respectively.

7.7.3.1.1 Loss function for variational autoencoders

To train a neural network, a loss function is necessary. The loss function has two components:

(i) a reconstruction loss that ensures that the output $\hat{\vec{x}}$ produced is not very far from the image desired; and

(ii) a loss that ensures that the parameters of the probability distribution, $\vec{\mu}$ and $\vec{\sigma}$ learned for components of the hidden variable $\vec{z}$ conform to a Gaussian distribution with covariance values equal to 0.

The reconstruction loss used is the same as that of a regular encoder–decoder neural network:

$$\mathcal{L}_{reconst}(\vec{x}, \hat{\vec{x}}) = \frac{1}{N} \sum_{i=1}^{n} (x_i - \hat{x}_i)^2. \tag{7.26}$$

This is the average squared per-pixel loss, assuming N is the total number of pixels. Some authors multiply it by $\frac{1}{2}$.

The loss from probability-distribution considerations is computed at the end of the encoder, *not* at the end of the decoder like the reconstruction loss. However, since the decoder does not change the value of the loss component, it can be added to the total loss

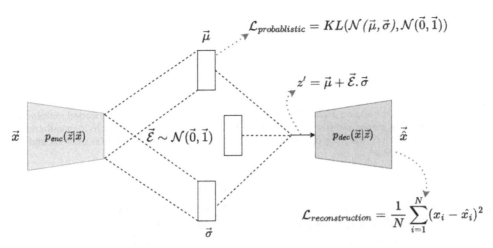

FIGURE 7.23 Training a VAE to produce a creative output. A new value $\vec{z}'$ of $\vec{z}$ is created to send through the decoder using a random number generator through which backpropagation does not flow.

computed at the end of the decoder. The probabilistic parameter values learned for the unobserved variable $\vec{z}$ are Gaussian, i.e., $\vec{z} \sim N(\vec{\mu}, \vec{\sigma})$. We want this learned distribution to be "well structured" with two properties:

(i) Continuity: Samples of $\vec{z}$ that are close to each other produce similar outputs, once decoded; and

(ii) Completeness: Any sample $\vec{z}$ produces a reasonable output when decoded.

Work on VAE [2] recommends that this be achieved by making the probability distribution $N(\vec{\mu}, \vec{\sigma})$ as close to a reduced Gaussian distribution $N(\vec{0}, \vec{1})$, where $\vec{0}$ means of each the m components corresponding to the means is 0, and $\vec{1}$ means that the standard deviations are all 1. To make $N(\vec{\mu}, \vec{\sigma})$ as close to $N(\vec{0}, \vec{1})$, we need to compute the distance between the two probability distributions using Kulbeck–Liebler (KL) Divergence. This component of the loss is written as $KL(N(\vec{\mu}, \vec{\sigma}), N(\vec{0}, \vec{1}))$. The training setup for a VAE is given in Fig. 7.23.

7.7.3.1.2 Using a trained VAE

Once a VAE has been trained, in each instance of application, one needs to sample a value of the hidden variable $\vec{z}$, and produce an output corresponding to this sampled value. To sample a value of $\vec{z}$, each component of it has to be sampled. Thus to obtain a sample $\vec{z}'$ that is sent through the decoder, we need an additional mechanism that is placed outside the encoder–decoder framework. This mechanism, shown in Fig. 7.24 produces a random number ξ between 0 and 1 outside of the neural network, and creates $\vec{z}'$ by adding to the mean $\vec{\mu}$ a random multiple of the standard deviation. With this mechanism in place, researchers have shown that the VAE can be trained using gradient descent. Note that during training, backpropagation flows through the $\vec{\mu}$ and $\vec{\sigma}$ portions of the bottleneck representation but not through the $\vec{\xi}$ computation.

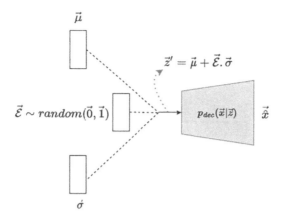

FIGURE 7.24 Using a trained VAE decoder to produce a creative output. A new value $\vec{z}'$ of $\vec{z}$ is created to send through the decoder using a random-number generator through which backpropagation does not flow.

7.7.4 Transformer

A deep neural-network architecture, Transformer, has gained significant attention and popularity in recent years. It also uses a modified form of encoder–decoder like VAE, and what is called a self-attention mechanism in addition. The Transformer was originally introduced in 2017 by Vaswani et al. [10], for sequence-based natural language processing (NLP) tasks for capturing long-range dependencies in text. It shows exceptional performance in a wide range of NLP tasks such as language translation, text summarization, and question answering. Variants of it have also become dominant in computer vision, geometric deep learning, speech recognition, and other areas. There are multiple components that contribute to the success of Transformer architecture.

Encoder: The encoder takes an input sequence and converts it into a series of vectors ready for capturing the information in the sequence. The encoder consists of a series of identical layers that are stacked on top of each other, each of which has two sublayers: a so-called self-attention mechanism and a feedforward neural network.

Self-Attention: Self-Attention is a key innovation inside the Transformer model, allowing it to pay attention to or weight different parts of the input sequence while processing the input itself. For example, certain words in an input sentence may be highly related to certain other words in the input sequence, and these relationships need to be learned to understand the input well. When generating the encoded representation of the input, Transformer computes attention or interrelation scores that represent the relevance of an element in a certain position in the input sequence to other elements in other positions. Attention scores are computed at various levels of abstraction.

Multihead Attention: One attention head or mechanism is used to compute the interrelation scores between one element of the input and all other elements in one way. Multihead Attention is an enhancement to this self-attention process that enables the model to focus on several elements of the input sequence in different ways when processing a single

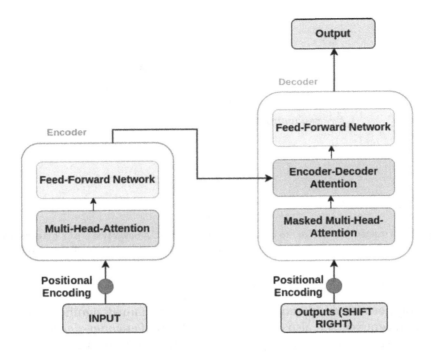

FIGURE 7.25 A basic architecture of a Transformer network.

element. This improves the model's ability to capture complex relationships between the elements of the input.

Feedforward Neural Network: The feedforward neural network is straightforward and is fully connected and is used in every encoder block. It transforms the representation of the sequence into a higher-dimensional space before passing it to the next layer.

Positional Encoding: Positional encoding is a technique that helps the model record the location of elements in the input sequence since the same element (e.g., a word) may occur in several locations (in a sentence). It is achieved by adding a set of fixed vectors to the input embeddings that encode each element's position in the sequence.

Decoder: Similar to the encoder, the decoder consists of a stack of identical layers, each of which has three sublayers: a self-attention mechanism, an encoder–decoder attention mechanism, and a feedforward neural network. Each decoder layer's self-attention component enables the output production to focus on various elements of the output when processing one element of the same. The output of the highest layer of the encoder is fed to each layer of the decoder.

A typical Transformer framework is shown in Fig. 7.25. It works with sequence data. The sequence is given as input to the encoder part. To enable the order of sequence during encoding, the positional encoding technique helps remember each element's position. The encoder part channels the input sequence by passing it through the first sublayer, i.e., the multihead attention unit and further passes it towards the feedforward sublayer. This diagram shows only one layer of multihead attention and feedforward network; usually, there are several. The output of the encoder part is then directly passed to the decoder module's encoder–decoder attention unit, which allows the decoder to attend to specific parts of the encoder's hidden representations that are relevant to generate the current token in the output sequence. The figure shows only one attention and feedforward block in the decoder; usually, there are several. If there are several such blocks, the encoder output is fed to each such block. It is worth noting that the decoding module adds a few preprocessing units before it addresses the encoder–decoder attention unit. Similar to the encoder part, the decoder part adds the output's (shift right) operation, which is another type of input sequence processing activity. The outputs (shift right) aim to enable the model to predict the next word in a sentence by attending to the previous token or element. This helps the model to participate in the earlier words when predicting the next word. After passing it through position encoding, the outputs (shift right) operation is given as input to the masked multihead attention unit. The masked multihead attention unit is a variant of the multihead attention mechanism. The purpose of masked multihead attention is to apply a mask to the attention weights to prevent the model from attending future tokens in the input sequence. Further, this information is passed to the encoder–decoder attention unit, allowing the decoder to attend to specific parts of the encoder's hidden representations relevant to generating the current token in the output sequence. Then, merging masked multihead attention output and the encoder's hidden representation gives the merged information to the feedforward network layer to generate the final output.

7.8 Summary

This chapter started with a brief discussion of how real neural networks have inspired the design and implementation of artificial neural networks. It started with a basic perceptron model followed by the multilayer perception model. It introduced various loss functions, functions popularly used in neural networks during training. The introduction of deep Convolutional Neural Networks to computer-vision problems and the high successes they have achieved have ushered in the age of Deep Learning impacting the world in expansive ways, allowing for effective and efficient machine learning and the mining of vast amounts of data of various modalities for beneficial purposes. It also introduced two recent deep network models, Autoencoder and Transformer that have gained popularity in recent years as generative learning architectures. This chapter barely scratches the surface of the study of artificial neural networks and deep learning, but has laid the foundation on which the reader can build by future reading and research.

References

[1] Geoffrey E. Hinton, Ruslan R. Salakhutdinov, Reducing the dimensionality of data with neural networks, in: Science, vol. 313, American Association for the Advancement of Science, 2006, pp. 504–507.
[2] Diederik P. Kingma, Max Welling, et al., An introduction to variational autoencoders, Foundations and Trends® in Machine Learning 12 (4) (2019) 307–392.

[3] Yann LeCun, Yoshua Bengio, Geoffrey Hinton, Deep learning, Nature 521 (7553) (2015) 436–444.

[4] Yann LeCun, Léon Bottou, Yoshua Bengio, Patrick Haffner, Gradient-based learning applied to document recognition, Proceedings of the IEEE 86 (11) (1998) 2278–2324.

[5] Warren S. McCulloch, Walter Pitts, A logical calculus of the ideas immanent in nervous activity, The Bulletin of Mathematical Biophysics 5 (1943) 115–133.

[6] Tomas Mikolov, Ilya Sutskever, Kai Chen, Greg S. Corrado, Jeff Dean, Distributed representations of words and phrases and their compositionality, Advances in Neural Information Processing Systems 26 (2013).

[7] Vinod Nair, Geoffrey E. Hinton, Rectified linear units improve restricted Boltzmann machines, in: International Conference on Machine Learning, 2010.

[8] Jeffrey Pennington, Richard Socher, Christopher D. Manning, Glove: global vectors for word representation, in: Proceedings of the 2014 Conference on Empirical Methods in Natural Language Processing (EMNLP), 2014, pp. 1532–1543.

[9] Frank Rosenblatt, The perceptron: a probabilistic model for information storage and organization in the brain, Psychological Review 65 (6) (1958) 386.

[10] Ashish Vaswani, Noam Shazeer, Niki Parmar, Jakob Uszkoreit, Llion Jones, Aidan N. Gomez, Łukasz Kaiser, Illia Polosukhin, Attention is all you need, Advances in Neural Information Processing Systems 30 (2017).

[11] Kunihiko Fukushima, Neocognitron: a self-organizing neural network model for a mechanism of pattern recognition unaffected by shift in position, Biological Cybernetics 36 (4) (1980) 193–202.

8

Feature selection

8.1 Introduction

A feature or an attribute is an individual measurable property or characteristic of a phenomenon. A measurable property can be quantified in many ways, resulting in different types of values. For example, a quantity can be represented as nominal, ordinal, ratio, or interval. For any task of interest, we can usually identify a number of features to describe a data example suitable for the application domain. Ideally, as a whole, the features used to describe data examples should give us valuable insights into the phenomenon at hand. For example, in classification, data examples are separated into different categories or classes based on distinguishing feature values they contain. Similarly, network traffic examples can be classified either as an attack or normal using values of network-traffic features or attributes. However, if there are redundant and/or irrelevant features, they complicate the data-analysis process. If several features convey the same essential information, the analysis algorithms may become confused regarding how much information each similar feature should contribute to the process. An irrelevant feature also muddies the process by sending the analysis on the wrong path, unless we can determine that it is irrelevant and ignore it from consideration.

Reducing the number of features to consider is useful in Data Science. Often, when data are collected for a phenomenon or a dataset is given for analysis, the number of features is high. Using every feature may make data analysis slower, especially when dealing with datasets with many examples. If the number of features can be reduced, the analysis will likely be more efficient. Feature selection and feature extraction are used as dimensionality-reduction methods. Feature selection, also known as an attribute, or variable subset selection, is used in machine learning or statistics for the selection of a subset of relevant features so that efficient models can be constructed to describe the data [15–19]. Two important issues to keep in mind when performing feature selection are (i) minimum redundancy, which requires that the selected features are as independent of one another as possible, and (ii) maximum relevance, which requires that each selected feature plays a role in the classification task [20]. In addition, data scientists use feature selection for dimensionality reduction and data minimization for learning, improving predictive accuracy and increasing the comprehensibility of models.

Artificial Neural Networks (ANNs), in particular, Deep-Learning methods such as Convolutional Neural Networks (CNNs), are excellent feature extractors on their own. ANNs accept raw-data examples, such as pixels representing images, and extract features of increasing complexity from lower to higher layers. In a CNN, the highest layer in the convolutional part of the network extracts the highest-level features. These features are used by the

Fundamentals of Data Science. https://doi.org/10.1016/B978-0-32-391778-0.00015-6

dense part of the network (also called a multilayer perceptron or MLP) above that to perform classification. When we discuss feature selection or feature extraction in this chapter, our focus is not on ANNs or CNNs that perform feature extraction as an integral part of the learning process. This chapter deals with situations where data examples are given regarding several features. We use specialized feature-selection (extraction) algorithms to extract or select features that are used to send the "reduced" examples to downstream algorithms for tasks such as classification and clustering. These algorithms are, in some sense, akin to the convolutional part of a CNN. In particular, this chapter discusses feature selection in depth. Data scientists who develop cost-effective solutions for real-life problems often need to use such algorithms.

8.1.1 Feature extraction vs. feature selection

A feature-selection method selects a subset of relevant features from the original feature set. In contrast, a feature-extraction method creates or extracts features based on combinations or transformations of the original feature set. Both approaches are used to overcome the curse of dimensionality in machine learning, and statistical data analysis [5]. The curse of dimensionality refers to the phenomenon that when the number of dimensions or features is large in a data-analysis problem, it is difficult to distinguish among the data examples since the space of examples becomes large and sparse. Thus to solve a data-analysis problem effectively, at a basic level, we must be able to perform computations to find which examples are similar and which ones are dissimilar. The main objective of feature selection is to identify m most informative features out of the n original features, where $m < n$. A feature-selection algorithm may be simple and greedy, using a statistical measure to find correlations among the features themselves (unsupervised approach), or the features or the response variable (supervised approach), or maybe more elaborate combining search techniques to discover a new subset of features from the original feature set, followed by evaluation of the selected subset of features using a score computed by an evaluation measure. A simple algorithm may test each possible subset of features finding the one that minimizes the error rate or maximizes the classification accuracy or any other downstream task.

Feature extraction differs from feature selection. Feature extraction attempts to reduce the dimensionality of the original feature set by creating and computing new features to ease subsequent processing. In contrast, a feature-selection method reduces the dimensionality by selecting a most relevant and nonredundant subset of features to achieve the best possible accuracy. In addition, feature extraction attempts to create new features by applying various transformation techniques to the original raw features. In contrast, a feature-selection technique attempts to select a subset of relevant and nonredundant features by applying a suitable statistical or information-theoretic measure with reference to a user-defined threshold to filter out unwanted features or following a more complex algorithm.

Fig. 8.1 demonstrates both the dimensionality-reduction processes, which can be linear and nonlinear projections. The selection of an optimal subset of relevant features can

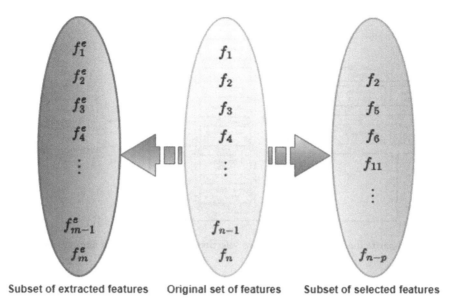

Subset of extracted features   Original set of features   Subset of selected features

FIGURE 8.1 Feature Extraction vs. Selection. Feature selection extracts $m = n - p$ features from the original set of n features. Feature extraction creates m new features by combining the original features.

help improve classification accuracy as well as classification efficiency. Feature-selection algorithms have been applied to many Data Science problems in many application domains.

8.2 Steps in feature selection

This section discusses steps in a generic feature-selection approach, whose components can be instantiated in various ways, simple to complex. As already mentioned, the objective of feature selection is to select a subset of features that are relevant to the task at hand. Different algorithms vary in the specifics of the steps. Some algorithms may short-circuit the steps by combining or eliminating some feature. Usually, the downstream task is classification. In general, feature selection involves four major steps:

   i) Generation of feature subsets;
  ii) Evaluation of the feature subsets that have been generated using various criteria;
 iii) Determination of the goodness of a feature subset based on the values of evaluation criteria; and
  iv) Finally, validation of the selected feature subset on a downstream task.

Fig. 8.2 shows these steps with additional details. We discuss each of these steps next.

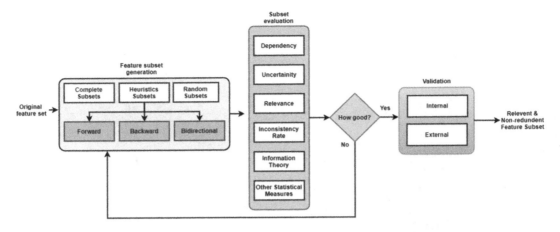

FIGURE 8.2 The summarize the major steps in Feature Selection along with different alternatives for each step.

8.2.1 Generation of feature subsets

This step is responsible for generating many candidate subsets of features from the raw feature set. One can take several approaches to generate candidate subsets from a dataset that has n features. These include an exhaustive search over all possible subsets that can be generated, performing a heuristic search over the set of all possible subsets, or performing a random search. A simple approach may create the desired subset of features by choosing features to add to or subtract from a working set of features using a statistical measure. Such an approach consists of only one candidate subset at a time.

8.2.1.1 Feature-subset generation using exhaustive search

Given a set of features, this approach generates all possible subsets of features. Thus for a dataset with n features, it will generate 2^n unique subsets. Obviously, for a high-dimensional dataset, the number of possible subsets is exponentially large and consequently, the cost of the evaluation will also be very high. As a result, this approach is impractical for higher-dimensional datasets. However, this is the only approach that guarantees the selection of the absolute best possible subset of features. In practice, it is likely that Feature Selection takes shortcuts, making greedy choices, and adding or subtracting one feature at a time to a working set, instead of generating subsets in a more elaborate manner.

8.2.1.2 Feature-subset generation using heuristic search

An exhaustive search of feature subsets is exponential in computational complexity and therefore, infeasible if the number featured in the original dataset is larger than a relatively small number, say 15, since $2^{15} = 32\,768$, resulting in the generation of a large number of features followed by subsequent evaluation. In practice, datasets with tens or even hundreds of features or more are common in real life.

A heuristic is a rule of thumb. It usually involves a quick and dirty computation to come to an informed but greedy or "hasty" relevant decision regarding a problem at hand, usually producing good results, while not guaranteed to do so. For example, assume that we have a heuristic that first computes the correlation between each individual feature and the label, which is the dependent variable. The heuristic then chooses the feature that shows the most positive correlation with the label. The assumption is that the features and the label are all numeric, and hence it is a simple regression problem. This is a heuristic because the entire set of numeric features may relate to the numeric label in various ways as a whole, but the heuristic looks at each feature in isolation assuming independence among features, and scores it. It makes a decision regarding the current most influential feature quickly, without looking at complex relationships that may exist between the features and the label. In addition, once a decision is made, often it cannot be undone.

Depending on how the heuristic is used as features are selected or deselected to produce the "optimal" subsect results in three commonly used heuristic approaches: (i) forward feature selection, (ii) backward feature selection, and (iii) bidirectional feature selection. In the *forward* heuristic approach, the approach starts with an empty feature set and successively adds highest-scoring features (typically, one by one) using the heuristic, whereas in the *backward* heuristic approach, it starts with the entire feature set and successively, eliminates lowest scoring features (typically one by one) from the whole set using the heuristic criterion. Each of these heuristic approaches has its own advantages and limitations. The *bidirectional* approach searches for the "optimal" subset of features using a combination of forward and backward searches to produce the result more efficiently. There are different ways to implement a bidirectional search. One simple way is to start with, say the empty set of features and add one or more iteratively, but as a feature is added, the current subset of features is analyzed to see if any of the current features in it should be eliminated. It can be performed in the reverse fashion also, by starting with the entire set of features and eliminating features one or more at a time, but every time a feature is eliminated, the approach looks to see if there are features that can be added. There are other more complex ways to achieve bidirectional feature selection. However, like the forward and backward heuristic approaches, the bidirectional approach also does not give a guarantee to select the best possible feature subset. The quality of a feature subset given by this approach depends on the (i) effectiveness of the heuristic used for selection and (ii) criterion used for its evaluation.

In practice, there may be no explicit subset generation at all. Instead, the algorithm may start from an empty set of features and add a new feature at a time, growing the subsect incrementally.

8.2.1.3 Random feature-subset generation

This approach, unlike the approaches discussed already, attempts to select feature subsets at random and evaluate their effectiveness. On average, this approach may be more time efficient, but it also does not give a guaranteed selection of the best possible feature subset. Random feature selection is used in building an ensemble of decision trees known as Random Forests, where a number of trees are built by sampling features and examples from a featured dataset. It has also been seen that the use of what is called stratified random sampling involves the division of the feature space into smaller subgroups (known as *strata*), and then sampling from the subgroups, may be helpful in finding better feature subsets in less time. However, additional computation such as clustering is needed to find the subgroups.

8.2.2 Feature-subset evaluation

This step deals with the evaluation of the feature subsets generated in the previous step using an appropriate measure or metric. Depending on the nature of the computation, evaluation metrics are categorized into two types: *independent metrics* and *dependent metrics*.

An *independent* metric evaluates the goodness of a feature subset based on the intrinsic properties of the selected features. On the other hand, to evaluate the quality of a feature subset, a *dependent* metric analyzes the results generated by the feature subset when used on a downstream task. Based on the approach used for the evaluation of the results, a dependent criterion can be *external* or *internal* in nature. In *external* dependent metric-based evaluation, the quality of results is evaluated using prior knowledge, whereas, in *internal* dependent metric-based evaluation, prior knowledge is either not used or is not available. Rather, it considers the intrinsic properties such as cohesiveness or homogeneity in the results or separability among the instances.

Often, a simple approach to feature selection may not evaluate feature subsets in a formal way. Instead, the approach may start from an empty set, evaluate the individual features in a greedy fashion, and decide which feature to add to the subset to enhance it in a greedy fashion, without evaluating the entire feature set at the time.

8.2.3 Feature-selection methods

Based on the approaches used for the selection of the features, the methods are categorized into four types: (a) filter methods, (b) wrapper methods, (c) embedded methods, and (d) hybrid methods. In addition, there are two other approaches, namely incremental and ensemble feature selection, that are used to handle dynamic datasets, and to obtain the best possible feature subset.

8.2.3.1 Filter approach

The filter approach [2] selects a subset of features with the help of an evaluation function, without using a machine-learning algorithm. Most filtering algorithms use statistical measures such as correlation and information gain. A complex filter approach may even use particle-swarm optimization, ant-colony optimization, simulated annealing, and genetic algorithms.

Depending on the data types of the feature values, it is necessary to use different statistical methods to compute how the features are similar. For example, if the original feature values are all numeric, Pearson's or Spearman's correlation coefficients can be used. In case of categorical features, either χ^2 (Chi) or mutual information can be used to rank the features. While feature set is a combination of both numeric and categorical features, there are other appropriate methods that can be used. Below, we discuss Pearson correlation and mutual information based feature section methods.

8.2.3.1.1 Pearson correlation-based feature selection

Pearson's correlation coefficient takes two vectors and produces a number between -1 to 1 to indicate the degree of correlation between the two. A value close to 0, whether positive or negative, implies a weak correlation between the two vectors, with a value of 0 implying no correlation. A value close to 1 indicates a very strong positive correlation, indicating that when one feature's value changes in a certain way, the other feature's value changes in a very similar way. Values above 0.5 indicate strong positive correlations, and a value between 0 and 0.3 is considered a weak positive correlation. A value close to -1 indicates a negative correlation, indicating the values of the two features change in opposite ways. Often, a high positive correlation is used to determine if one feature can be dropped, although some approaches use a negative correlation as well. If we want to use either a positive or negative correlation, we can use the absolute value of the correlation. Low correlation can also be used to find a set of nonredundant features.

Suppose we want to determine if two features f_1 and f_2 are correlated, given values of the two features for the examples in a dataset. To start, we create two vectors, $\vec{f_1}$ containing the values of feature f_1 for all examples in the dataset one by one, and $\vec{f_2}$ containing values of the feature f_2 for the entire set of examples in the same order. Assuming there are N examples in the dataset, the two vectors of feature values are given by the following:

$$\vec{f_1}^T = [f_{11}, f_{21}, \cdots f_{N1}]$$
$$\vec{f_1}^T = [f_{11}, f_{21}, \cdots f_{N1}].$$

Pearson's correlation coefficient between the two vectors is given by

$$r_{Pearson} = \frac{N \sum_{i=1}^{N} (f_{i1} f_{i2}) - \left(\sum_{i=1}^{N} f_{i1}\right)\left(\sum_{i=1}^{N} f_{i2}\right)}{\left[\sqrt{N \sum_{i=1}^{N} f_{i1}^2 - \left(\sum_{i=1}^{N} f_{i1}\right)^2}\right]\left[\sqrt{N \sum_{i=1}^{N} f_{i2}^2 - \left(\sum_{i=1}^{N} f_{i2}\right)^2}\right]}. \tag{8.1}$$

It assumes that both feature vectors are quantitative, the values of the features are normally distributed, there are no outliers, and the relationship between the values of the two feature vectors is linear.

Supervised feature selection can be performed when the dataset is labeled, as if for classification or regression. For Pearson correlation to be used, the label must be numeric, i.e., the downstream task is regression.

Table 8.1 Comparing features against one another. The entry in the cell $\langle i, j \rangle$ gives the Pearson-correlation value between features i and j. The table's diagonal elements are 1. The table is symmetric, i.e., $\langle i, j \rangle$ values are the same as the $\langle j, i \rangle$ values.

	f_1	f_2	$\cdots\cdots$	f_j	$\cdots\cdots$	f_{n-1}	f_n
f_1	1		$\cdots\cdots$		$\cdots\cdots$		
f_2		1	$\cdots\cdots$		$\cdots\cdots$		
$\vdots$			$\cdots\cdots$		$\cdots\cdots$		
f_i			$\cdots\cdots$		$\cdots\cdots$		
$\vdots$			$\cdots\cdots$		$\cdots\cdots$		
f_{n-1}			$\cdots\cdots$		$\cdots\cdots$	1	
f_n			$\cdots\cdots$		$\cdots\cdots$		1

If there are n features in the dataset, we can compute pairwise n^2 Pearson's correlation values. These can be displayed in tabular form as shown in Table 8.1. Each cell holds the Pearson-correlation value between two features f_i and f_j, $1 \le i, j \le n$. The main diagonal values are all 1 since they represent the correlation between a feature and itself, and this correlation is always perfect.

In addition to correlation between pairs of features, we compute correlation between each feature and the label y. We obtain the vector of feature values $\vec{f_i}$ for feature f_i, and the vector of label values $\vec{y}$ for the label y.

A simple feature-selection algorithm may be described at high level in the following way; details have to be filled in to create the actual algorithm. The algorithm starts with an empty selected-working-set of features. It also has an empty excluded-working-set of features. It also has a yet-to-be-considered-set of features; this set's initial value is the entire set of features. From the yet-to-be-considered-set, features are chosen to be included in either the selected working set or the excluded working set. The algorithm finds a new feature from the yet-to-be-considered-set to add to the selected working set; this feature is most highly correlated to the label among the features in the yet-to-be-considered-set. Next, the algorithm looks for features in the yet-to-be-considered-set that are most highly correlated to the features of the selected working set, on average; one of these features is added to the excluded working set. The algorithm repeats the process of selecting a feature from the yet-to-be-considered-set to include in the selected working set and putting one in the excluded working set as many times as necessary. The algorithm stops when the yet-to-be-considered set becomes empty.

8.2.3.1.2 Mutual information in feature selection

Knowing the value of one feature for a data example may sometimes not give any clue about what the value of the label (or, another feature) may be, but sometimes it may to a small extent, or even to a large extent. This idea is quite similar to the correlation between the values of a feature and the label, or between the values of two features, as we have discussed above. Information Theory deals with a similar idea in terms of what is called information content. The problem can be cast in terms of information contained in the following way: How much information about the label y (or feature f_j) can be gleaned if we have information about feature f_i? This idea of the amount of information obtained about the value of one random variable (here, label y or feature f_j) from the value of another random variable (here, feature f_i) is quantified in terms of *mutual information* between the two random variables.

The amount of mutual information between a feature f_i and the label y, assuming both are discrete is given as

$$MI(f_i, y) = \sum_{f_{ij}} \sum_{y_k} p(f_{ij}, y_k) \, log \frac{p(f_{ij}, y_k)}{p(f_{ij}), p(y_k)}, \tag{8.2}$$

where f_{ij} is the jth possible value for feature f_i and y_k is the kth possible value of the label y. In the formula above, in the *log* computation, the numerator is the probability of a feature value f_{ij} occurring together with the label value y_k or the joint probability distribution, whereas the denominator contains the product of the probability of the feature value f_{ij} and the label value y_k occurring independently. Thus, the ratio gives us an estimate of how much more likely the two values occur together if there is some dependency between the two compared to if they were independent of each other. It is a measure of the dependency between f_{ij} and y_k. Taking *log* of this ratio still gives us a measure of the same dependency; it only changes the scale of the ratio. The formula above adds the values for all f_{ij} and y_k pairs to give us a measure of how much information the feature contains about the label across all values of both.

We can also compute $MI(f_i, f_j)$ to find the mutual information between two features f_i and f_j. Similar to our discussion on Pearson-correlation-based feature selection earlier, mutual information-based feature selection can be either supervised or unsupervised. Feature–feature and feature–label mutual information values can be combined in various ways to develop more complex algorithms for feature selection. Mutual information and correlation values may also be combined in various ways in sophisticated algorithms.

Mutual information can be computed for continuous values as well. If both feature f_i and label y are continuous numerical values, the equation for mutual information given above can be changed as follows:

$$MI(f_i, y) = \int_{f_{ij}} \int_{y_k} p(f_{ij}, y_k) \, log \frac{p(f_{ij}, y_k)}{p(f_{ij}), p(y_k)} \, df_i \, dy. \tag{8.3}$$

In this case, we see that we have to perform double integration over all values of the feature f_i and all values of label y. Since the values are continuous, we have to assume probability-density functions for the feature as well as the label. For example, we can assume that both have values that are distributed normally, and hence, use the formula for a Gaussian probability distribution. To specify a Gaussian probability distribution, we need to know the mean and standard deviation for it. The two distributions will have different means and standard deviations.

Mutual information can also be computed for all pairs of feature pairs, and can be used to develop complex algorithms along with the mutual information values between individual features and the label [4]. Note that mutual information can be computed between a discrete variable and a continuous-valued variable also. For the discrete variable, we perform summation over all possible values, whereas for a continuous-valued variable, we perform the integration. Thus it is possible to develop complex feature-selection algorithms using mutual information to cover datasets with a variety of features.

8.2.3.1.3 Other methods

Many advanced filter feature-selection algorithms [30], [29], [32] have been proposed for use with classification as the downstream task. The objective of these algorithms is to search for an optimal set of features that are used later to obtain good classification results. An unsupervised feature subset-selection method using feature similarity was proposed by Mitra et al. [28] to remove redundancy among features. They use a new measure called a maximal information compression index to calculate the similarity between two random variables for feature selection. Bhatt et al. [26] use fuzzy rough set theory for feature selection based on natural properties of fuzzy t-norms and t-conforms. A mutual information-based feature-selection algorithm called MIFS was introduced by Battiti [31] to select a subset of features. This algorithm considers both feature–feature and feature–class mutual information for feature selection. It uses a greedy technique to select a feature subset that maximizes information about the class label.

8.2.3.2 Wrapper approach

Unlike filter feature selection which relies on statistical measures to assess the relevance of features, wrapper feature selection uses machine learning algorithms to evaluate feature subsets based on their impact on model performance [3]. The wrapper approach [24] uses a learning algorithm to evaluate the accuracy produced by the use of the selected features in classification. Wrapper methods can give high classification accuracy for particular classifiers, but generally, they have high computational complexity. Fig. 8.3 illustrates the working mechanism of filter and wrapper approaches. Kwak and Cho [25] developed an algorithm called MIFS-U to overcome the limitations of MIFS to obtain better mutual information between input features and output classes than MIFS. Peng et al. [27] introduced a mutual information-based feature-selection method called mRMR (Max-Relevance and Min-Redundancy) that minimizes redundancy among features and maximizes dependency between a feature subset and a class label. The method consists of two stages. In the first stage, the method incrementally locates a range of successive feature subsets where a consistently low classification rate is obtained. The subset with the minimum error rate is used as a candidate feature subset in stage 2 to compact further using a forward selection and backward elimination-based wrapper method. Estevez et al. [23] proposed a mutual information-based feature-selection method as a measure of relevance

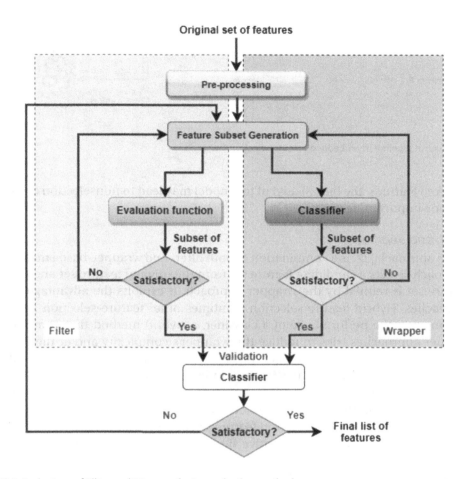

FIGURE 8.3 Basic steps of Filter and Wrapper feature selection methods.

and redundancy among features. Vignolo et al. [14] introduced a novel feature-selection method based on multiobjective evolutionary wrappers using a genetic algorithm.

8.2.3.3 Embedded approach

Embedded feature selection is a technique that integrates feature selection directly into the training process of a machine learning model [2]. Instead of judging feature significance separately, embedded approaches learn the most essential features automatically during model training. By selecting the most useful features for the specific job at hand, this technique can increase both model performance and computational efficiency while reducing overfitting. Lasso regression and tree-based algorithms such as Random Forests are examples of embedded approaches. A conceptual framework of this approach is presented in Fig. 8.4. Although the embedding of a prediction model is helpful in selecting

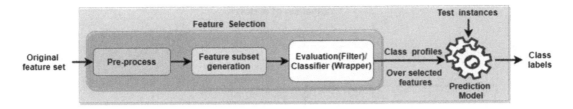

FIGURE 8.4 Typical pipeline of embedded feature selection.

highly ranked features, the bias (if any) of the model may lead to nonselection or the dropping of some important feature(s).

8.2.3.4 Hybrid approach

The hybrid approach [22] is a combination of both filter- and wrapper-based methods. The filter approach selects a candidate feature set from the original feature set and the candidate feature set is refined by the wrapper approach. It exploits the advantages of these two approaches. Hybrid feature selection combines other feature-selection approaches that can improve the performance of a classifier. A hybrid method that combines filter and wrapper approaches tries to reduce the wrapper's complexity and at the same time improves the filter's predictive accuracy. It is a two-step feature-selection method where the output of the filter method is used as input to the wrapper method to boost the overall performance of a classifier. The filter step chooses the high-ranked features using an objective function, and the wrapper method reevaluates all possible subsets of features using a machine-learning model. Hybrid feature selection is useful for the classification of high-dimensional data with an informative subset of features. The most important characteristic of the hybrid method is that it can overcome the common issues of the filter and wrapper methods. Hybrid methods can also handle data instances that have diverse distribution patterns. Most hybrid FS methods have been developed to select subsets of informative features from high-dimensional voluminous data. Alejandro et al. [11] developed a hybrid FS method for the classification of botnet-based IoT network traffic. The method uses the Pearson Correlation Coefficient in the filter step and the k-NN classifier in the wrapper step. Their method gives high detection accuracy in classifying benign and malignant instances. Liu et al. [12] proposed an effective rule-based hybrid feature-selection method. The method known as HGAWE uses a hybrid Genetic Algorithm (GA) with wrapper-embedded approaches for feature selection. The main objective of HGAWE is to improve learning performance and enhance the searching performance to identify the relevant feature subsets. The population of GA solutions was fine tuned by selecting some signature features using an embedded method and constructing the learning model based on efficient gradient-regularization approaches. The population of GA solutions is induced by a wrapper method using heuristic search strategies. Many hybrid evolutionary algorithms have also been developed for effective feature-subset selection.

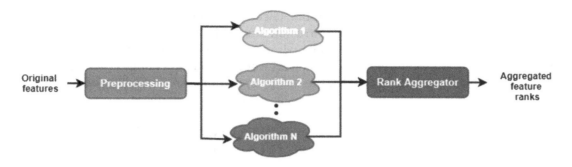

FIGURE 8.5 Ensemble Feature Selection.

In addition to these four basic approaches of feature selection, a data scientist prefers to use two more effective approaches, i.e., ensemble approach and incremental approach, especially for handling dynamic datasets and achieving the best possible feature subset.

8.2.3.5 Ensemble approach

The ensemble feature-selection approach uses a diverse set of feature-selection algorithms and combines them to achieve the best possible accuracy. The basic concept behind an ensemble approach is that a team's decision is usually better than the decision of an individual. Ideally, an ensemble approach should be designed to overcome the poor performance of any individual algorithm. A high-quality ensemble is often one in which the base algorithms are from diverse families and are independent of each other [10], [1]. This is necessary because the errors made by base algorithms should be uncorrelated so that any possible miscalculation by any one or a few of them is outweighed by the better performance of the other algorithms. The goal of constructing the ensemble is to obtain a composite model from several simpler submodels to obtain high overall accuracy. A combination function is used to aggregate the decisions made by the individual algorithms to obtain an aggregated list of ranked features. Fig. 8.5 illustrates the process of ensemble feature selection. For rank aggregation, a good combination function should try to eliminate the bias of individual ranker algorithms and generate the best possible ranked list of features.

8.2.3.6 Incremental feature selection

An incremental feature-selection algorithm computes an optimal subset of features as new data examples become available dynamically [6]. In other words, the data examples are not available at the same time as we have assumed in the prior discussions on feature selection. Incremental feature selection computes an optimal subset of features that is relevant and nonredundant based on information obtained from the currently available instances, and when a new instance or a number of new instances arrive, it updates the feature set dynamically without processing the earlier instances again. The purpose of incremental feature selection is to avoid repetitive feature selection as new examples come onboard. For example, incremental feature selection is important in applications dealing with gene expression [8], network intrusion detection [21], and text categorization [9] where real-time classification needs to be performed on the data instances. Incremental feature selection is useful for dynamic datasets whose character changes over a period of time. For such datasets, traditional feature-selection methods are inefficient for knowledge discovery as the same learning algorithm needs to be run repeatedly.

An incremental feature-selection algorithm was proposed by Liu & Setiono. [13] using a probabilistic approach. The algorithm is an incremental version of a filter algorithm called Las Vegas Filter (LVF) [7]. The main feature of a Las Vegas algorithm is that it must eventually produce the proper solution. In the context of feature selection, this means that the algorithm must create a minimal subset of features that optimizes some parameter, such as classification accuracy, which was designed to ensure that features selected from the reduced data should not generate inconsistencies more than those generated optimally from the whole dataset. However, in the original LVF algorithm, if the dataset size becomes small, the number of inconsistency checks becomes small as well. As a result, the set of features selected from a reduced dataset may not be suitable for the whole dataset. The incremental version of LVF overcomes this problem without sacrificing the quality of feature subsets in terms of the number of features and their relevance.

8.3 Principal-component analysis for feature reduction

Feature Selection attempts to find an optimal subset of the entire feature set whereas Feature Reduction or Dimensionality Reduction obtains a set of newly created features where each feature is a linear combination of the original features, and the total number of new features is smaller than the number of original features.

One of the most common feature or dimensionality reduction techniques is Principal-Component Analysis (PCA). Given a dataset with n features, PCA can be used to obtain k features with $k < n$ or even $k \ll n$ such that the new set of features expresses the variance in the original dataset well. In other words, the new features are almost as expressive as the original features, although some loss in expressiveness is expected. Another way to look at PCA is that it is able to transform the original features that may be correlated, expressing similar information, to obtain a smaller set of uncorrelated features. PCA is also able to rank the newly transformed features in terms of expressiveness, where a higher-ranked feature explains the variations in the dataset better than a lower-ranked feature. Thus of the k resulting transformed features, the highest ranked one explains most of the variance (how the data values are spread out) in the dataset; the next ranked feature explains the next highest amount of variance in the dataset, and so on. Therefore it is possible to use PCA to obtain a small set of k ranked features, often much smaller, that explain $p\%$ of variation in the data, with p being a high number like 90, 95, or 99.

To motivate what PCA accomplishes, let us look at a two-dimensional dataset given in Fig. 8.6(a). The dataset is given in terms of an original feature set $\{f_1, f_2\}$. To reduce the number of features, the first step is to extract a principal component, somewhat similar

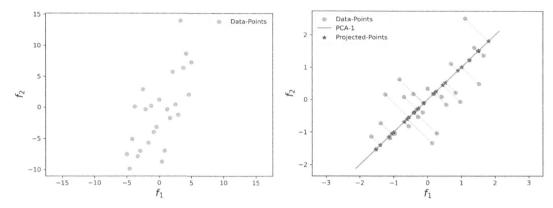

FIGURE 8.6 (a) A scatter plot of the original points in 2D. (b) A principal-component line is obtained by minimizing the variance in the set of values of the projected points. The principal-component line can also be used to reduce the two-dimensional data points into one dimension. The projection of the data points onto the principal-component line are shown as blue solid circle (mid gray in print version).

to a least-squares linear regression, but a little different. In the least-squares linear regression, we obtain the line that gives the trend in the dataset with the smallest amount of sum of squared errors, whereas, PC line, also known as the first principal component, is the line that represents the direction of the maximum variance in the data. A PC is a linear combination of the original variables in a dataset. This line is obtained by finding the direction in which the projected points have the maximum variance, and then rotating the coordinate system so that this direction becomes the first coordinate axis.

To perform PCA, we project the points to the PC line we draw. In other words, we drop perpendiculars to the line we are trying to find (see Fig. 8.6(b)); this line is also called a principal-component line. The projected points are also shown on the PC line in Fig. 8.6(b). Once the points have been projected to the PC line, each point can be thought of simply as its magnitude along the PC line, which becomes a new axis. Among all possible lines we can draw, the one chosen as the PC line minimizes the variance of the projected set of points, considering the magnitude of the points along the new PC axis. The new feature, given by PCA-1, can be used as the single (only) transformed feature for the dataset. The PC line can be thought of as an axis representing a new composite feature or dimension, and the original points can now be redescribed by their values on this new axis. The PC is also called the first principal component for the original dataset. This principal component accounts for most of the variations in the dataset.

Since the original dataset is two dimensional, it is possible to obtain a second principal component. The second principal component is orthogonal or perpendicular to the first principal component and is shown in Fig. 8.7 as PCA-2. By looking at the projections of the original points on the line representing the direction of the second principal component, we see that the distances from the new origin, which is the intersection of the two principal components PCA-1 and PCA-2, are small, and thus the second principal component does not explain as much variation in the original dataset as the first principal component. As

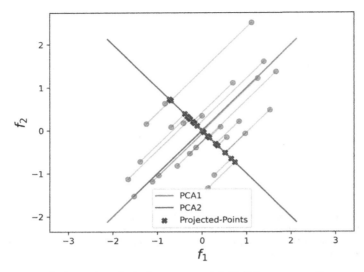

FIGURE 8.7 The line perpendicular to the principal-component 1 (PCA 1) line serves as the second principal component (PCA 2) for the dataset. It is necessary to drop perpendiculars to project the examples to the second principal component. We see that the spread of the projected points (shown in blue; mid gray in print version) is much tighter, and variances in values of the projected points are much smaller than the spread and variance along the first principal component in Fig. 8.6(b).

a result, if we choose to, we can ignore the second principal component as the second transformed feature without losing a significant amount of explanatory information.

In the example in Figs. 8.6 and 8.7, we start with a dataset in two dimensions, obtain two principal components, and decide to ignore the second, thus effectively reducing the number of dimensions in the dataset. However, in a realistic dataset, the number of features is likely to be higher, possibly in the tens, hundreds or more. In such cases, we can use PCA to obtain a sequence of ranked principal components in descending order of informativeness or explaining capacity for variances in the data. Unless there are features in the original dataset that are completely collinear (i.e., two features whose values are multiples of each other), we can obtain n ranked principal components from the original n features. However, it is likely that the first k features, $k \ll n$, explain most of the variances in the dataset, and we do not have to produce any principal components beyond the first k. For example, it is possible that given an original feature set with 100 distinct features, only 5 or so features may be able to explain 99% of the variance in the dataset, thus achieving feature reduction from 100 to 5, a reduction of 95% in the feature-space size with loss of only 1% in explainability.

The prime task of PCA is to compute a principal components. A number of such PCs can be extracted (possibly up to the number of independent variables) from a dataset. We first need to standardize the data by subtracting the mean from each variable and dividing by the standard deviation. This ensures that each variable has equal weight in calculating the principal components. Next, we need to calculate the covariance matrix of the standardized data. The covariance matrix describes the relationships between the variables in the data set and is used to determine the directions of the principal components. The eigenvectors of the covariance matrix represent the directions of the principal components, and the eigenvalues represent the amount of variance explained by each principal component. To select the principal components with the highest variance, we need to sort the eigenvectors in descending order based on their corresponding eigenvalues. The top k eigenvectors are selected to explain the desired amount of variance in the data. Finally, we project the data onto the new coordinate system defined by the selected principal components. This transforms the original dataset into a new space with reduced dimensionality.

As new principal components are computed one by one, it must be noted that each principal component is orthogonal to every prior principal component. Thus if we have n, $n = 100$, say, original features, we need to imagine an n-dimensional space, which of course, cannot be visualized, and imagine constructing a new principal component, which is a linear combination of the original features, that is orthogonal to all other principal components constructed so far.

We have described the concept of principal components conceptually, outlining the process of computation in sequence. The efficient mathematical process of computing the principal components for a given dataset involves the use of linear algebra, primarily the concepts of eigenvalues and eigenvectors. These are not discussed here, and are left for the readers to pursue on their own.

After performing PCA, data examples need to be described in terms of the smaller number of transformed features. It can be observed that in the reduced and transformed feature space, originally similar data examples become more similar, and originally unlike examples become more unlike. It helps to remember that the new features are ordered from most informative to least. When features that do not show meaningful variations are removed, it is quite likely that it removes noise in the examples. Since data examples, expressed in terms of the new features, are shorter vectors, the new feature values make downstream computation faster.

8.3.1 Summary

Feature selection plays a crucial role in Data Science. The use of an appropriate feature-selection method can reduce the cost of computation and resource requirements significantly. This chapter has discussed the basics of feature selection. It has also presented the steps in feature selection, various approaches for feature-subset generation and their evaluation, and finally, the metrics used to validate the selected features. The chapter has also briefly presented a few recent feature-selection approaches such as incremental feature selection and ensemble feature selection. The chapter ends by introducing Principal-Component Analysis (PCA), the most common method used for feature or dimensionality reduction.

References

[1] Ren Diao, Fei Chao, Taoxin Peng, Neal Snooke, Qiang Shen, Feature selection inspired classifier ensemble reduction, IEEE Transactions on Cybernetics 44 (8) (2014) 1259–1268.

[2] Isabelle Guyon, André Elisseeff, An introduction to variable and feature selection, Journal of Machine Learning Research 3 (2003) 1157–1182.

[3] Mark A. Hall, Lloyd A. Smith, Feature selection for machine learning: comparing a correlation-based filter approach to the wrapper, in: FLAIRS Conference, 1999, pp. 235–239.

[4] Nazrul Hoque, Dhruba K. Bhattacharyya, Jugal K. Kalita, MIFS-ND: a mutual information-based feature selection method, Expert Systems with Applications 41 (14) (2014) 6371–6385.

[5] G. Hughes, On the mean accuracy of statistical pattern recognizers, IEEE Transactions on Information Theory 14 (1) (1968) 55–63.

[6] Ioannis Katakis, Grigorios Tsoumakas, Ioannis Vlahavas, Dynamic feature space and incremental feature selection for the classification of textual data streams, in: Knowledge Discovery from Data Streams, 2006, pp. 107–116.

[7] Huan Liu, Rudy Setiono, A probabilistic approach to feature selection:a filter solution, in: Proc.of the 13th Int.Conf. on Machine Learning, 1996, pp. 319–327.

[8] Roberto Ruiz, José C. Riquelme, Jesús S. Aguilar-Ruiz, Incremental wrapper-based gene selection from microarray data for cancer classification, Pattern Recognition 39 (12) (2006) 2383–2392.

[9] Yiming Yang, Jan O. Pedersen, A comparative study on feature selection in text categorization, in: ICML, vol. 97, 1997, pp. 412–420.

[10] Robi Polikar, Ensemble learning, in: Ensemble Machine Learning, Springer, 2012, pp. 1–34.

[11] Alejandro Guerra-Manzanares, Hayretdin Bahsi, Sven Nõmm, Hybrid feature selection models for machine learning based botnet detection in IoT networks, in: International Conference on Cyberworlds (CW), IEEE, 2019, pp. 324–327.

[12] Xiao-Ying Liu, Yong Liang, Sai Wang, Zi-Yi Yang, Han-Shuo Ye, A hybrid genetic algorithm with wrapper-embedded approaches for feature selection, IEEE Access 6 (2018) 22863–22874, IEEE.

[13] Huan Liu, Rudy Setiono, Incremental feature selection, Applied Intelligence 9 (3) (1998) 217–230, Springer.

[14] L. Vignolo, D. Milone, J. Scharcanski, Feature selection for face recognition based on multi-objective evolutionary wrappers, Expert Systems with Applications 40 (13) (2013) 5077–5084.

[15] A. Arauzo-Azofra, J. Aznarte, J. Benítez, Empirical study of feature selection methods based on individual feature evaluation for classification problems, Expert Systems with Applications 38 (2011) 8170–8177.

[16] J. Cadenas, M. Garrido, R. Martínez-España, Feature subset selection Filter–Wrapper based on low quality data, Expert Systems with Applications 40 (2013) 6241–6252.

[17] K. Polat, S. Güneş, A new feature selection method on classification of medical datasets: kernel F-score feature selection, Expert Systems with Applications 36 (2009) 10367–10373.

[18] H. Liu, L. Yu, Toward integrating feature selection algorithms for classification and clustering, IEEE Transactions on Knowledge and Data Engineering 17 (2005) 491–502.

[19] M. Dash, H. Liu, Feature selection for classification, Intelligent Data Analysis 1 (1997) 131–156.

[20] A. Unler, A. Murat, R. Chinnam, Mr2PSO: a maximum relevance minimum redundancy feature selection method based on swarm intelligence for support vector machine classification, Information Sciences 181 (2011) 4625–4641.

[21] K. Khor, T. Yee, S. Phon-Amnuaisuk, A feature selection approach for network intrusion detection, in: Information Management and Engineering, International Conference on, 2009, pp. 133–137.

[22] H. Hsu, C. Hsieh, M. Lu, Hybrid feature selection by combining filters and wrappers, Expert Systems with Applications 38 (2011) 8144–8150.

[23] P. Estevez, M. Tesmer, C. Perez, J. Zurada, Normalized mutual information feature selection, IEEE Transactions On Neural Networks / A Publication Of The IEEE Neural Networks Council 20 (2009) 189–201.

[24] A. Blum, P. Langley, Selection of relevant features and examples in machine learning, Artificial Intelligence 97 (1997) 245–271.

[25] N. Kwak, C. Choi, Input feature selection for classification problems, IEEE Transactions on Neural Networks 13 (2002) 143–159.

[26] R. Bhatt, M. Gopal, On fuzzy-rough set approach to feature selection, Pattern Recognition Letters 26 (2005) 965–975.

[27] H. Peng, F. Long, C. Ding, Feature selection based on mutual information criteria of max-dependency, max-relevance, and min-redundancy, IEEE Transactions on Pattern Analysis and Machine Intelligence 27 (2005) 1226–1238.

[28] P. Mitra, C. Murthy, S. Pal, Unsupervised feature selection using feature similarity, IEEE Transactions on Pattern Analysis and Machine Intelligence 24 (2002) 301–312.

[29] L. Yu, H. Liu, Redundancy based feature selection for microarray data, in: Proc. 10th ACM SIGKDD Conf. Knowledge Discovery and Data Mining, 2004, pp. 737–742.

[30] R. Caruana, D. Freitag, Greedy attribute selection, in: International Conference on Machine Learning, 1970.

[31] R. Battiti, Using mutual information for selecting features in supervised neural net learning, IEEE Transactions on Neural Networks 5 (1994) 537–550.

[32] D. Bhattacharyya, J. Kalita, Network Anomaly Detection: A Machine Learning Perspective, CRC Press, 2013.

Cluster analysis

9.1 Introduction

Classification and regression perform the task of prediction based on prior experience. In the real world of data-driven decision making, discovering underlying data patterns and interrelationships among the data elements is a crucial analytical task besides prediction. For instance, a marketing strategist of a retail corporation might like to understand the customers' purchasing patterns for individualized marketing. Accordingly, an organization may offer attractive price discounts or recommend specific products to a group of customers based on an analysis of their buying habits. More specifically, cluster analysis may help separate groups of customers based on purchase histories, addresses, and locations of favorite stores. Google uses cluster analysis for generalization, data compression, and privacy preservation of YouTube videos.[1] YouTube-video metadata are generalized through clustering to promote and recommend less popular videos with hit videos. Data compression helps reduce YouTube data storage by grouping data for similar videos with a single cluster identifier. YouTube-user clustering can hide the user identity by grouping and tagging the search-history details with cluster identifiers corresponding to similar search histories.

Besides organizing the data into meaningful groups, cluster analysis may also help label extensive unlabeled datasets when few representative labeled samples are available for predictive learning, popularly known as semisupervised learning.

9.2 What is cluster analysis?

Let us distinguish the task of clustering from that of classification. When we group labeled data instances into prespecified classes (or groups), it is called classification. In the absence of labels, if we try to group the instances based on a notion of similarity or proximity, it is called clustering. It is a data-driven approach, where data items themselves decide their groupings based on certain features or attributes. Unlike other predictive machine-learning activities, no parameter learning is involved in clustering. It is also known as *data segmentation.*

Cluster analysis is an *unsupervised learning* method for exploring interrelationships among a collection of data instances by organizing them into similar clusters. The objec-

---

[1] https://developers.google.com/machine-learning/clustering/overview.

Fundamentals of Data Science. https://doi.org/10.1016/B978-0-32-391778-0.00016-8

181

tive of clustering is to group the elements so that elements within a group have strong intragroup similarity and weak intergroup similarity with other groups; i.e., individual clusters are largely independent. In turn, a good cluster is well separated from other clusters. The intragroup similarity is measured as an average of pairwise similarity between the instances within a single cluster. A high intragroup similarity among the instances in the same cluster indicates strong dependency among the instances. The formal definition of clustering can be given as follows.

Definition 9.2.1 (Exclusive Clustering). Given a database $D = \{x_1, x_2, \cdots, x_n\}$ of n objects, the clustering problem is to define a mapping $f : D \rightarrow \{C_1, C_2, \cdots C_k\}$, where each x_i is assigned to only one cluster C_j. A cluster C_j contains precisely those objects mapped to it; i.e., $C_j = \{x_i | f(x_i) = C_j, 1 \leq i \leq n$ and $x_i \in D\}$ and $\forall C_i, C_j, C_i \cap C_j = \phi$.

Consider a set of playing cards. We need to group them into similar clusters. The cards can be grouped according to color, symbol, or shape. Hence, clustering may give rise to multiple outcomes depending on even when one attribute, here color, is considered. Whether we consider single or multiple attributes, the similarity or proximity measure used also determines the outcome. Hence, the merit of good clustering depends on attribute selection, effective proximity measures and the quality of clustering functions used.

Often, a database contains foreign elements that do not go well with the other data elements. Such elements do not share common properties with the members of any of the clusters. These elements are class *outliers* or *noise*.

Definition 9.2.2 (Outliers). Given a database $D = \{x_1, x_2, \cdots, x_n\}$ of n objects and a set of clusters $C = \{C_1, C_2, \cdots C_k\}$, an object x_j is said to be an outlier with respect to C if $x_j \notin C_i$, $\forall C_i \in C$.

Outlier detection (also known as anomaly detection) is an important activity when solving real-life application problems. For example, a piece of malware or an instance of fraud can be treated as an outlier as its characteristics deviate from usual software or user activity, respectively.

The concept of an exclusive cluster is not the last word in real-life examples. For instance, social-media users may be interested in cricket and film [39]. In molecular biology, it has been observed that certain groups of genes participate in several DNA-metabolism processes such as replication, repair, and transcription [2]. Out of 1628 proteins in a hand-curated yeast-complex dataset [44], 207 proteins participate in more than one complex. It may not be possible to describe all these complexes using disjoint or exclusive clustering [40]. Hence, it is not always justified to group the entities exclusively into only one particular group following the classical definition of a cluster. The above definition should be referred to as *exclusive clustering*. To describe the overlapping or shared participation in multiple groups, as discussed above, we need a new definition of clustering, which can be called *nonexclusive* or *overlapping* [37].

Definition 9.2.3 (Nonexclusive Clustering). Given a database D of objects, the nonexclusive clustering problem is to define a mapping $f : D \rightarrow \{C_1, C_2, \cdots C_k\}$, where each $x_i \in D$ may be assigned simultaneously to one or more clusters from a set of clusters $C = \{C_1, C_2, \cdots, C_k\}$. Cluster C_i contains those objects that are mapped to it exclusively and also those objects that may belong to other clusters C_j of C. (for $\forall j = 1, \cdots, k \wedge j \neq i$), i.e., $C_i \cap C_j \neq \phi$.

There may be some situations when a subset of elements in a cluster share more closeness than the rest of the elements in the group, forming a different type of overlapping cluster, called *embedded* or *intrinsic* clusters [47].

Definition 9.2.4 (Intrinsic Community). A cluster C_i is intrinsic or embedded inside C_j if $C_i \subset C_j$ and intracluster closeness of C_i is significantly higher than that of C_j.

For example, a few of the members of a music group may possess an additional interest in drama. Hence, unlike the rest of the members, they have equal interests in music as well as drama, forming a more compact group within the parent group. Formation of intrinsic clusters is common in genetic networks [37], protein–protein interactions [33], and social networks [25].

To understand with the help of examples, let us consider a set of geometric shapes of different colors as shown in Fig. 9.1. The placement of the shapes indicates their relative physical locations in two-dimensional space. Considering physical closeness as the only attribute parameter for grouping, the outcome may be the three clusters, as shown in Fig. 9.1B. There are outliers (yellow and blue shapes) that belong to none of the clusters. However, the exclusive clustering outcomes will be different if we consider color or shape as the attribute. When we consider both color and shape as attributes for clustering, it would be difficult and inappropriate to form exclusive clusters. In such a case, overlapping clusters (Fig. 9.1C) may be a correct solution to describe the situation. When we try to group both types of circles, blue circles will form another group due to their physical closeness and similar color, forming an embedded cluster (Fig. 9.1D).

To perform effective cluster analysis, it is important to compute the closeness or similarity of data elements; it is the prerequisite for clustering. The computation of true closeness or proximity among the elements determines the quality of the outcome. Next, we discuss a few proximity measures used for cluster analysis.

9.3 Proximity measures

Proximity (similarity) or the distance measure is the key mathematical concept required for quantifying the degree of likeness between two data points or a set of data points or between two probability distributions. The proximity measure used determines how data samples are related to each other. The specific measure used depends on the downstream activity following clustering. Generally, it is measured in a particular range of continuous values where a smaller distance score represents a higher degree of similarity and

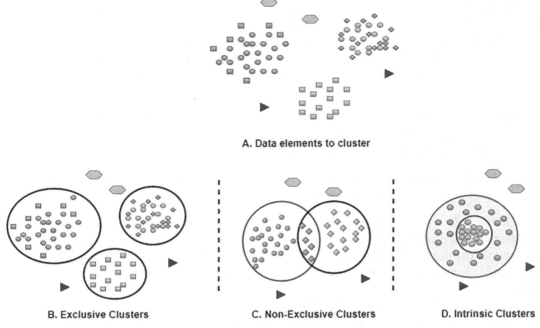

FIGURE 9.1 Different types of clusters that can be formed on the sample data elements (A) depend on varying grouping criteria and clustering techniques. The triangle and hexagone shapes are noises. They usually ignored by any good clustering algorithm.

vice versa. In other words, $Similarity = 1 - Distance$. Mathematically, it is a real-valued function that determines to what extent two data objects or samples vary from each other. Formally, it can be defined as follows.

Definition 9.3.1 (Proximity Measure). Given two vectors or probability distributions, namely $\vec{x}$ and $\vec{y}$, the proximity S, between the two vectors return a score as $S(\vec{x}, \vec{x}) \in R$ (or, $dist(\vec{x}, \vec{y})$), which is normally a continuous value representing the degree of similarity (or distance) between them.

In particular, a given distance (proximity) score for two observations $\vec{x}$ and $\vec{y}$ satisfies the following conditions:

- **Symmetry:** $S(\vec{x}, \vec{y}) = S(\vec{y}, \vec{x})$, $\forall \vec{x}, \vec{y}$;
- **Positivity:** $S(\vec{x}, \vec{y}) \geq 0$, $\forall \vec{x}, \vec{y}$;
- **Reflexivity:** $S(\vec{x}, \vec{y}) = 0$, $\iff \vec{x} = \vec{y}$.

Various types of proximity measures are used to compute the similarity between data objects. Based on the type of measure, similarity measures can be placed in four categories: 1) *Standard Measures*, 2) *Statistical Measures*, 3) *Divergence Measures*, and 4) *Kernel Measures*. Next, we present a brief discussion for each of them.

9.3.1 Standard measures

A measure for calculating the distance between two data points indicates how far these points are from each other. It is a positive value except when calculating the distance from a point to itself, which must be zero. If $\vec{x}$ and $\vec{y}$ are the two features (input vectors), the distance or the similarity represents a score in the range $\mathbb{R}^{[0,1]}$.

1. **Euclidean distance (E_d):** The Euclidean distance represents the distance between two data points in the geometric space. For two scalar values, the distance between two points can be computed by obtaining the absolute value of their difference. To compute the distance between two two-dimensional data points, the underlying intuition involves the application of Pythagoras's theorem in the Cartesian plane:

$$E_d(x, y) = \sqrt{\sum_{i=1}^{n}(x_i - y_i)^2}.$$

2. **Manhattan distance (M_d):** The Manhattan distance, often called the City-Block Distance or Taxi-cab distance, obtains the sum of the absolute differences between the respective dimensions when the input vector points are placed in such a way that one can only move in right-angled directions on a uniform grid. n is the number of dimensions or attributes:

$$M_d(\vec{x}, \vec{y}) = \sum_{i=1}^{n}|\vec{x}_i - \vec{y}_i|.$$

3. **Cosine similarity (C_s):** Cosine similarity refers to the cosine of the angle between two vectors. The measure is more concerned with the orientation of the two points in the space than it is with the exact distance from one to the other. It measures the level of similarity between two vectors via an inner product if they were normalized to both have length one:

$$C_s(\vec{x}, \vec{y}) = \frac{\vec{x} . \vec{y}}{\|\vec{x}\| \, \|\vec{y}\|}.$$

4. **Dot-Product similarity (DP_s):** In contrast to cosine similarity, the dot-product-based similarity incorporates the magnitude of the input vectors. The dot-product score increases as the input vector lengths increase. It indicates how two vectors are aligned in the space considering the directions they point:

$$DP_s(\vec{x}, \vec{y}) = |\vec{x}| \, |\vec{y}| \, cos(\theta).$$

9.3.2 Statistical measures

These measures employ statistical correlation techniques to evaluate the similarity between two vectors. If $\vec{x}$ and $\vec{y}$ are the two data instances (input vectors), then the correlation represents the measure of similarity where a value greater than 0 (positive correlation) represents the high similarity, and a value less than 0 represents the dissimilarity (negative correlation).

1. **Pearson correlation coefficient (PCC):** The Pearson correlation coefficient is a measure that quantifies the strength of the linear correlation between two vectors or data instances. It does so by computing the ratio between the covariance of the two sets of data and the product of their standard deviations:

$$PCC\,(\vec{x}, \vec{y}) = \frac{\sum_{i=1}^{n}(\vec{x}_i - \bar{x})\,(\vec{y}_i - \bar{y})}{\sqrt{\sum_{i=1}^{n}(\vec{x}_i - \bar{x})^2\,(\vec{y}_i - \bar{y})^2}},$$

where $\bar{x}$ and $\bar{y}$ are the mean of $\vec{x}$ and $\vec{y}$, respectively.

2. **Spearman rank correlation coefficient (SRCC):** The Spearman rank correlation measures the rank correlation between two data instances or vectors by assessing their relationship using a monotonic function regardless of their linear relationship. It is high when observations have a similar rank:

$$SRCC(\vec{x}, \vec{y}) = \frac{n\sum \vec{x}_i \vec{y}_i - \sum \vec{x}_i \sum \vec{y}_i}{\sqrt{n\sum \vec{x}_i^2 - (\sum \vec{x}_i^2)}\sqrt{n\sum \vec{y}_i^2 - (\sum \vec{y}_i^2)}}.$$

3. **Kendall rank correlation coefficient (KRCC):** The Kendall rank correlation coefficient computes the correlation between two vectors or data samples. Intuitively, it measures the similarity of sampled data when ordered by each vector. The association is based on the number of concordant (match the rank order) and discordant (mismatch the rank order) pairs, adjusted by the number of ties in ranks. The Kendall rank correlation coefficient, commonly known as the Kendall τ_b coefficient can be defined as:

$$KRCC(\vec{x}, \vec{y}) = \frac{n_c - n_d}{\sqrt{(n_0 - n_1)(n_0 - n_2)}},$$

where $n_0 = n(n - 1)/2$, n the length of the vector $\vec{x}$ or $\vec{y}$.
$n_1 = \sum_i t_i(t_i - 1)/2$,
$n_2 = \sum_j u_j(u_j - 1)/2$,
n_c = number of concordant pairs,
n_d = number of discordant pairs,
t_i = number of tied values in the ith group of ties for 1st vector,
u_j = number of tied values in the jth group of ties for 2nd vector.

9.3.3 Divergence measures

A divergence is a statistical function that measures the distance (dissimilarity, proximity, closeness) between two probability distributions. If $\vec{x}$ and $\vec{y}$ are the two input probability distributions, then the divergence $Div \in \mathbb{R}^{[0,\infty]}$ between them follows the two reasonable properties of positivity and reflexivity as mentioned above.

1. **Kullback–Liebler Divergence (KL$_{Div}$) [43]:** The Kullback–Liebler divergence has its roots in information theory for measuring information lost when comparing two probability distributions over a chosen approximation. If $\vec{x}$ is the probability distribution and $\vec{y}$ is the approximating distribution on the same probability space d, it can be mathematically described as:

$$KL_{Div}(x \| y) = \sum_{i=1}^{n} \vec{x}(d_i) \cdot log \frac{\vec{x}(d_i)}{\vec{y}(d_i)}.$$

   When used in clustering, the values of features for the two examples are thought of as two discrete probability distributions so that the KL divergence can be computed. KL divergence is not symmetric because

$$KL_{Div}(\vec{x} \| \vec{y}) \neq KL_{Div}(\vec{y} \| \vec{x}).$$

2. **Jensen–Shannon Divergence (JS$_{Div}$):** The Jensen–Shannon divergence also quantifies the similarity or difference between two probability distributions. It is a derived version of the Kullback–Liebler divergence that provides the symmetric capability and a normalized way to represent the scores between 0 (identical) and 1 (different) range. The Jensen–Shannon divergence between $\vec{x}$ and $\vec{y}$ probability distributions can be computed as follows:

$$JS_{Div}(\vec{x} \| \vec{y}) = \frac{1}{2} * KL_{Div}(\vec{x} \| m) + \frac{1}{2} * KL_{Div}(\vec{y} \| m),$$

   where $m = \frac{1}{2} * (\vec{x} + \vec{y})$ is a mixed distribution.

9.3.4 Kernel similarity measures

In machine learning, a kernel is a mathematical trick that solves a nonlinear problem using linear modeling strategies. It simplifies the execution process and makes the calculation faster that would otherwise be much more complex, if even possible. Mathematically, for two vectors $\vec{x}$ and $\vec{y} \in \mathbb{R}^n$, the kernel k is a transforming function $f : \mathbb{R}^n \to \mathbb{R}^m$, thereby mapping $\mathbb{R}^n$ to some feature space $\mathbb{R}^m$, such that $k(\vec{x}, \vec{y}) = < f(\vec{x}), f(\vec{y}) >$ represents the dot product in the transformed space.

1. **Linear Kernel (L$_k$):** The Linear Kernel is the simplest kernel that can be applied on linearly separable data, i.e., data where instances of two classes can be separated by a straight line or a linear function in general. The linear kernel can be defined as the dot product between two observed vectors:

$$L_k(\vec{x}, \vec{y}) = \vec{x}^T \vec{y}.$$

2. **Polynomial Kernel (P$_k$):** The Polynomial Kernel deals with a d-degree similarity between two data instances or feature vectors. It computes the relationship between the

188 Fundamentals of Data Science

feature vectors in a nonlinear fashion, which otherwise may not be possible for other variants of input data samples. The polynomial kernel can be defined as:

$$P_k(\vec{x}, \vec{y}) = (\vec{x}^T \vec{y} + c)^d,$$

where c is a constant, used as a tradeoff to fit the training data and the adjustment of margin, and d represents the degree or the order.

3. **Radial Basis Function Kernel (RBF$_k$):** The Radial Basis Function Kernel is the popularly used kernel in machine learning that measures the similarity between two sets of input features or vectors. It can be mathematically represented as:

$$RBF_k(\vec{x}, \vec{y}) = exp(-\frac{\|\vec{x} - \vec{y}\|^2}{2\sigma^2}),$$

where $\|\vec{x} - \vec{y}\|^2$ is the squared Euclidean distance between two vectors and σ is the variance.

4. **Laplacian Kernel (L$_k$):** The Laplacian Kernel is a variant of the RBF kernel, which is more robust in handling data with noisy labels. Therefore it fits the random data well:

$$k(\vec{x}, \vec{y}) = exp(-\frac{\|\vec{x} - \vec{y}\|_1}{\sigma}),$$

where $\|\vec{x} - \vec{y}\|_1$ is the Manhattan distance between two vectors.

5. **Sigmoid Kernel (S$_k$):** The Sigmoid Kernel is generally called a hyperbolic tangent function, which is used as the activation function in artificial neurons. It computes the input feature vectors just as similar to a sigmoid function in logistic regression. It can mathematically be described as:

$$S_k(\vec{x}, \vec{y}) = tanh(\gamma \vec{x}^T \vec{y} + c),$$

where $\vec{x}$ and $\vec{y}$ are input feature vectors, γ represents the slope and C is known as the intercept.

9.4 Exclusive clustering techniques

There are numerous exclusive clustering algorithms. The type of data being clustered, the size of the dataset, and the specific goal and application all influence the choice of algorithm to use.

Real-world databases vary in dimension, with some being low dimensional (like the iris-flower dataset[2] consisting of four features) and others being high dimensional (like gene expression of thousands of genes or sensor data from IoT devices). Most clustering techniques that are currently in use cannot handle high-dimensional datasets. Some have been specifically designed for larger dimensions, while certain algorithms can only be used

---

[2] https://archive.ics.uci.edu/ml/datasets/iris.

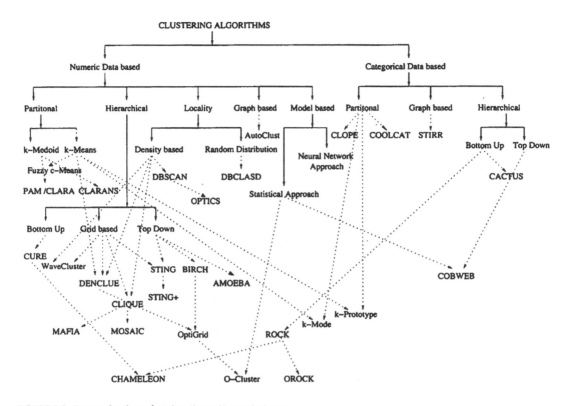

FIGURE 9.2 Categorization of Major Clustering Techniques.

with lower dimensions. The classification of clustering algorithms as lower dimensional or higher dimensional can be made on the basis of their design.

Clustering algorithms can be categorized in various ways. One way is based on the type of data for which they were originally designed. The three forms of data are numerical, categorical, and mixed. Partitional, hierarchical, graph-based, locality-based, and model-based algorithms are also some common categories used to describe clustering algorithms. Ensembles of several techniques have become popular recently. A majority of algorithms were designed with low-dimensional data clustering in mind at first. Top-down, bottom-up, and grid-based algorithms all fall into the category of hierarchical algorithms. There are two categories of locality-based algorithms: density-based and random distribution-based. Similarly, model-based algorithms can be divided into two groups: neural-network-based and statistical-based.

The categorization presented in Fig. 9.2 shows some of the methods for handling categorical and numerical data. As stated above, most low-dimensional data-clustering techniques are covered by Partitional, Hierarchical, and Density-based techniques. Therefore we arrange our discussion focused on these three basic techniques. The following sections

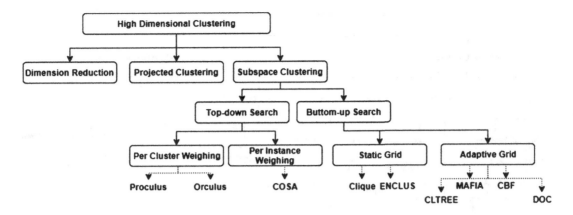

FIGURE 9.3 Hierarchy of High-Dimensional Clustering Algorithms.

introduce some algorithms that can handle high-dimensional data. Techniques for higher-dimensional clustering are categorized into dimensionality reduction or feature selection, projected clustering, and subspace clustering. A considerable amount of work has been done on higher dimensions, especially on subspace clustering. A categorization of higher-dimensional clustering approaches is given in Fig. 9.3.

9.4.1 Partitional clustering

Partitional clustering is one of the most fundamental forms of exclusive clustering techniques that partition or segment data elements into k clusters. If we consider Definition 9.2.1 of exclusive clustering above, the parameter k is usually a user-given value when performing partitional clustering. The primary objective of partitional clustering is to minimize the intracluster similarity and maximize intercluster similarity. In other words, elements within a cluster should have high similarity and low similarity with elements of other clusters. Therefore in a nutshell, partitional-clustering approaches perform an iterative refinement process to optimize a cost function. Approaches primarily vary on the cost function used and the nature of the data type handled.

9.4.1.1 K-means

The K-means algorithm is a typical partition-based clustering method. It is simple and fast, and easy to implement and was considered a top 10 Data Mining algorithm in the year 2006 [54]. The term *k-means* was introduced by James MacQueen in the year 1967 [36]. However, the algorithm was first published by Edward W. Forgy in 1965, and therefore it is popularly also called the Lloyd–Forgy algorithm [22]. It is based on the concept of a cluster centroid. A centroid is a representative element or the center of a cluster. Given the desired number of clusters k, the algorithm partitions the dataset into k disjoint subsets that optimize the following objective function:

$$\mathcal{E} = \sum_{i=1}^{k} \sum_{O_j \in C_i} \| O_j - \mu_i \|^2.$$

O_j is a data object in cluster C_i, and μ_i is the centroid (mean of objects) of C_i. Thus the objective function $\mathcal{E}$ tries to minimize the sum of the squared distances of objects from their cluster centers.

The algorithm starts with k randomly selected initial seed cluster centers or centroids, and iteratively refines the clusters formed based on the current centroids. In every iteration, new cluster centers are computed and clusters are adjusted with the goal to optimize the above cost function. The algorithm converges or is stopped when no changes in centroids happen between two consecutive iterations of refinement.

Algorithm 9.1: K Means algorithm.

Data: D: Database; k: Desired number of clusters
Result: C: Cluster Set

1   Select k initial means, $\mu_1, \mu_2, \cdots \mu_k$
2   **while** *Not Converged* **do**
       `// Assign elements to closest cluster`
3   **for** $O_i \in$ D **do**
4   **if** dist (O_i, μ_m) *is minimum* **then**
5            $C_m \leftarrow O_i$
6   **end**
7   **end**
       `// Update all means`
8   **for** $\mu_{p=1\cdots k}$ **do**
9         $\mu_p = \frac{1}{|C_p|} \sum_{x=1}^{|C_p|} t_x \in C_p$
10   **end**
11 **end**

The time complexity of K-means is $O(l * k * n)$, where l is the number of iterations, and k and n are the number of clusters and data elements in D, respectively. It has been observed that the K-means typically converges in a small number of iterations.

K-means is not equipped to handle noise or outliers. It inclusively assigns each element to one of the clusters. Since it is a centroid-based approach, it works well with well-separated clusters that have spherical distributions and uniform cluster densities. Deciding on an appropriate value for k is challenging without understanding the underlying data distribution. To detect the optimal number of clusters, one can run the algorithm repeatedly with different values of k and compare the clustering results. However, it is an unrealistic approach for large datasets.

9.4.1.2 PAM

A calculated mean in K-means clustering may be an arbitrary value, not necessarily a member of the input data. This limitation reduces the interpretability of the cluster center. A solution has been provided by Leonard Kaufman and Peter J. Rousseeuw as the Partitioning Around Medoids (PAM) [31] algorithm. PAM belongs to the k-medoid family of clustering algorithms that use medoids as realistic centers in place of means. Another advantage of using medoids is their nondependency solely on Euclidean distance to calculate the closest cluster center. Instead, any arbitrary distance or proximity measure can be used. PAM minimizes a sum of pairwise dissimilarities between centers or medoids (M_i) and data elements instead of a sum of squared Euclidean distances. The cost function used by PAM is given as:

$$\mathcal{E}_{ih} = \sum_{j=1}^{n} \|O_j - O_h\|.$$

PAM proceeds in the same way as K-means except for the way the centers of the clusters are calculated based on optimization of the above cost function. The algorithm starts with arbitrarily selected k-medoids and iteratively improves upon this selection. In each step, a swap between a selected object O_i and a nonselected object O_h is made as long as it results in an improvement of the quality of clustering in terms of the cost function. To calculate the effect of such a swap between O_i and O_h, a cost $\mathcal{E}_{ih}$ is computed, which is related to the quality of partitioning the nonselected objects to k clusters. While swapping nonmedoid element O_h with medoid M_i, it calculates the overall cost (if swapped) in terms of distance change with O_h (as new medoid) and all other elements, O_j. The cost function can be given as follows. If the overall cost improves with respect to all the medoids, the algorithm converges and the selected medoids are considered for assigning the elements to the nearest medoid and the corresponding cluster.

The original PAM algorithm has a run time complexity of $O(k(n-k)^2)$ in each iteration of cost calculation. $O(n^2k^2)$ time is needed for a simple implementation that recalculates the whole cost function each time. A fast version of PAM [51] that takes a better time of $O(n^2)$ by improving the cost calculation step is also available.

9.4.1.3 Improvement over PAM

Although PAM suffers inherent limitations of partitional clustering, it is more effective than K-means and other partitional approaches in terms of faster convergence and less sensitivity towards noise. A few more k-medoid-based clustering algorithms have also been proposed. The main computational challenge for PAM is to use iterative optimization to determine the k-medoids. *CLARA* (Clustering LARge Applications) [32] was an attempt to reduce the number of iterations. Although CLARA adheres to the same premise, it makes an effort to minimize the computational demands. Instead of searching for representative items for the full dataset, CLARA selects a sample of the data instances and runs PAM on it. The remaining objects are then categorized using the partitioning principles. The medoids

---

Algorithm 9.2: PAM algorithm.

Data: D: Database; k: Desired number of clusters
Result: C: Cluster Set

1 Select k initial medoids, $M_1, M_2, \cdots M_k$
2 **while** *No change in cost* **do**
   // Update all medoids with nonmedoids
3   **for** $O_{h=1\cdots(n-k)}$ **do**
4   **for** $M_{i=1\cdots k}$ **do**
5   Calculate $\mathcal{E}_{ih}$
6   **end**
7   **end**
8   Replace M_i with O_h for which $\mathcal{E}_{ih}$ is minimum.
9 **end**
   // Assign elements to closest cluster
10 **for** $O_j \in$ D **do**
11   **if** dist (O_j, M_h) *is minimum* **then**
12          $C_h \leftarrow O_j$
13   **end**
14 **end**

---

of the sample roughly match the medoids of the complete dataset if the sample were drawn in a sufficiently random manner.

Clustering Large Applications based on RANdomized Search (CLARANS) [42] is based on a graph-theoretic framework, inspired by PAM and CLARA. The method uses a randomized search of a graph to locate the medoids that serve as the cluster representative. The algorithm requires *numlocal* and *maxneighbor* as inputs. *Maxneighbor* specifies the highest number of neighbors of a node that can be looked at. The number of local minimums that can be gathered is *numlocal*. Starting off, CLARANS chooses a random node. A sample of the node's neighbors is then checked, and if a superior neighbor can be located based on the "cost difference of the two nodes," processing shifts to the superior neighbor and continues until the *maxneighbor* condition is satisfied. If not, a new pass is begun to look for further local minima and the current node is declared a local minimum. The technique gives the best of these local values as the medoid of the cluster after a predetermined number of local minima (*numlocal*) have been gathered.

9.4.2 Hierarchical clustering

A series of partitioning processes is carried out during hierarchical clustering. A set of nested clusters is formed and organized as a tree. Often, the hierarchical structures are represented diagrammatically as a *dendrogram* (Fig. 9.4). Clustering can be carried out top-down by repeatedly splitting the data until a certain threshold is met, or bottom-up by repeatedly amalgamating sets of data.

9.4.2.1 Agglomerative and divisive approaches

Hierarchical clustering techniques vary depending on how individual clusters are obtained. *Agglomerative* clustering is a bottom-up approach where initially n data elements form n clusters. In every iteration, a pair of closest clusters are merged. At the end, it merges all the clusters into one big cluster. *Divisive* clustering is the reverse approach compared to agglomerative clustering. It starts with a single cluster containing all n data elements. Iteratively, it splits a cluster into subclusters that are relatively dissimilar from each other. It finishes with n individual clusters. Let us consider six (6) data elements (A, B, C, D, E) given in Fig. 9.4 that need to be clustered. In the case of agglomerative clustering, A and B (D and E too) are clubbed together according to a proximity measure. Cluster (AB) is grouped with C, and in the next step (ABC) is clustered with C to form the cluster (ABC). Finally, (ABCDE) and (F) are grouped to obtain the final single cluster (ABCDEF). The divisive approach starts with (ABCDEF) as a single cluster and iteratively splits it into 6 different clusters. There are various methods for determining the steps in clustering, and they are frequently referred to as Linkage Methods. Common linkage techniques include the following:

- **Complete-linkage**: The distance between any two clusters, A and B, is determined by calculating the maximum distance between any two elements within the clusters:

$$max\{dist(a, b), \forall a \in A \wedge \forall b \in B\}.$$

- **Single-linkage**: The distance between any two clusters, A and B, is determined by calculating the minimum distance between any two elements within the clusters. Single-linkage helps isolate outliers that are large values and merge at the end (say, F in Fig. 9.4):

$$min\{dist(a, b), \forall a \in A \wedge \forall b \in B\}.$$

- **Average-linkage**: The distance between any two clusters, A and B, is determined by calculating the average distance between any two elements within the clusters. A popular average-linkage clustering method is *UPGMA* (Unweighted Pair Group Method with Arithmetic mean) [52]:

$$\frac{1}{|A| \cdot |B|} \sum_{a \in A} \sum_{b \in B} dist(a, b).$$

- **Centroid-linkage**: This method identifies the centroid of clusters A and B, C_A and C_B, respectively, and then measures the distance $(dist(C_A, C_B))$ between them before merging.

The advantage of hierarchical clustering is its inherent capability to detect intrinsic cluster structures. However, deciding upon the number of clusters is not straightforward, like any other clustering approach. One commonly used solution is to use a cut line in

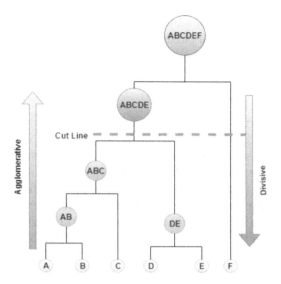

FIGURE 9.4 Example of agglomerative and divisive hierarchical clustering. The cut line decides the number of clusters.

the dendrogram. As shown in Fig. 9.4, the cut line intersects two vertical lines in the dendrogram that finally produces two clusters, (ABCDE) and (F). Although it depends on how many clusters are ultimately needed, usually, cut lines are set at the maximum height from the base of the dendrogram without intersecting the merging points.

Below, we discuss a few popular hierarchical clustering techniques that have been proposed considering issues such as data types to handle or faster execution.

9.4.2.2 BIRCH
Balanced Iterative Reducing and Clustering using Hierarchies (BIRCH) [55] is an integrated hierarchical clustering technique. In order to produce cluster representations, it proposes two new ideas: *clustering feature* (CF) and *clustering-feature tree*. In huge databases, these structures aid the clustering method in achieving good speed and scalability. Additionally, BIRCH is useful for the dynamic and gradual grouping of incoming objects. A single scan of the dataset produces a basic, decent clustering in BIRCH, and one or more further scans may be utilized (optionally) to further enhance the quality.

9.4.2.3 CHAMELEON
CHAMELEON combines a network-partitioning technique with a dynamic clustering-based hierarchical clustering scheme. The algorithm's initial phase divides the data into sections using a technique based on the k-nearest-neighbor method for graph partitioning. In a graph, a region's density is represented by the weight of the connecting edge. Numerous small subclusters of data are created first. An algorithm for multilevel graph partitioning is used in the first stage. The CHAMELEON partitioning algorithm generates

partitions of excellent quality with a minimal number of edge cuts. In order to join the subclusters and identify the actual clusters, the second phase employs an agglomerative or a bottom-up hierarchical clustering algorithm. The overall complexity for n items and m clusters is $O(nm + n \log n + m^2 \log m)$.

9.4.2.4 CACTUS

CACTUS (Clustering Algorithm for Categorical Data Using Summaries) [24] is a subspace clustering algorithm specifically designed for categorical data. CACTUS first summarizes the categorical data into a set of binary vectors called attribute summaries. These attribute summaries represent the presence or absence of each category within each attribute. Once the attribute summaries are created, CACTUS applies a modified K-means clustering algorithm to group the data points into clusters. In this modified K-means algorithm, the distance between two data points is calculated based on the Jaccard distance between their attribute summaries. After the clustering process is complete, CACTUS creates cluster projections to represent the characteristic of each cluster. The cluster projections are created by examining the attribute summaries of each cluster and identifying the most frequent categories within each attribute. These most frequent categories are then used to represent the characteristic of each cluster for that particular attribute.

For example, if a cluster has a high frequency of the "red" category in the "color" attribute, the cluster projection for that attribute would be "red". The resulting set of cluster projections can be used to gain insights into the characteristics of each cluster and to compare the clusters to each other.

CACTUS (CAtegorical data ClusTering Using Summaries) is a form of subspace clustering [24]. In an effort to cluster the projections of these tuples into just one set of characteristics, CACTUS splits the database vertically. By focusing on one attribute, CACTUS first determines the cluster projections on all pairs of attributes. Following this, it creates an distinctive set to represent the cluster projections on this characteristic for n clusters involving all the attributes. The database clusters are obtained by combining all of the cluster projections made on separate properties.

9.4.2.5 AMOEBA

AMOEBA [21] facilitates *Delaunay triangle*-based hierarchical spatial clustering. The source of analysis for locating clusters is the Delaunay triangle. When calculating proximity-related information, such as nearest neighbors, the Delaunay triangle is effective. There are two steps to the algorithm. The Delaunay diagram is first built. The algorithm receives a linked planar embedded graph that serves as the diagram. The points in the clusters are given recursively and are made up of the points in a related component. Passive edges and noise are eliminated after every edge has been compared to the criterion. Delaunay edges greater than or equal to some criterion function, $F(p)$, are considered *passive edges*. *Active edges* are Delaunay edges that are less than $F(p)$ at a specific level, and *noise* is a point that does not have an active edge meeting it. At each level of the hierarchy, proximity subgraphs are formed by active edges and their points. When the noise and passive edges are removed, the recursive algorithm calls AMOEBA repeatedly until no new linked components are produced.

9.4.3 Density-based clustering

It is difficult for the above two clustering approaches to detect the natural number of clusters without knowledge of the underlying data distribution. Also, they either fail to detect noise or outliers or use indirect approaches to do so. In addition, the above approaches are unable to detect nonspherical or nonconvex clusters. Density-based clustering has been developed with the following advantages.

- It can detect the number of natural clusters without prior knowledge or input.
- It can handle noise effectively.
- It can detect any irregular, nonspherical-shaped clusters.

The rationale behind the density-based methods of clustering is that each cluster has a typically higher point density than the surrounding areas. Additionally, there is less density in areas of noise than there is in any cluster. Density-based methods use two basic parameters or thresholds.

1. ϵ: A user-defined parameter that is regarded as the neighborhood's largest radius.
2. **MinPts**: The minimum number of objects within an object's ϵ neighborhood.

Let us introduce some of the basic concepts. To understand better, we use the illustration given in Fig. 9.5. We consider $MinPts$ as 4, and ϵ is the radius of the dotted circles.

Definition 9.4.1 (ϵ-Neighborhood (N_ϵ)). The neighborhood within a radius ϵ of a given object, say O_i is called the ϵ–neighborhood of the object. It can be defined as the set of all objects that are within a distance from O_i of less than or equal to ϵ.

$$N_\epsilon(O_i) = \{O_j | \forall O_j, dist(O_i, O_j) \le \epsilon\}. \tag{9.1}$$

For example (Fig. 9.5), there are four neighbors of object O_j, including O_i and O_k.

Definition 9.4.2 (Core Object). An object O_i is called a *core object* if it has $MinPts$ neighbors within radius ϵ:

$$Core(O_i) = \begin{cases} True & \text{if } |N_\epsilon(O_i)| \ge MinPts \\ False & \text{otherwise.} \end{cases} \tag{9.2}$$

In our example, objects O_i, O_j, and O_k are core objects, whereas O_p is not.

Definition 9.4.3 (Directly Density Reachable). An object is directly density reachable from another object if the object is within the neighborhood of a core object. More specifically, object O_j is directly density reachable from O_i, if $Core(O_i) = True$ and $O_j \in N_\epsilon(O_i)$.

O_p, in our example, is directly density reachable from O_k; however, the reverse is not true as O_p is not a core object.

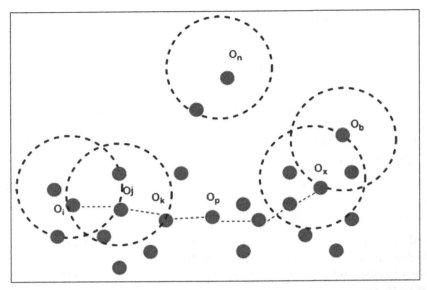

FIGURE 9.5 Illustration of density-based clustering. A dotted large circle is of radius ϵ centered around an object shown as a small solid circle. A connecting line between a pair of objects indicates that they are directly density reachable from each other.

Definition 9.4.4 (Density Reachable). Object O_k is density reachable from O_j, if O_k is reachable through a chain of directly density-reachable intermediate objects. For instance, if there is a chain of objects, $O_1, O_2, \cdots, O_n$ such that every O_{i+1} is directly density reachable from O_i and $O_1 = O_k$ and $O_n = O_j$, then O_k is density reachable from O_j.

In our example, O_i and O_p are density reachable via O_j and O_k.

Definition 9.4.5 (Density Connected). Object O_x is density connected to O_i if both O_x and O_i are density reachable from an intermediate object O_p.

Definition 9.4.6 (Density Cluster). A cluster is a set of density-connected objects that is maximal with regard to density reachability.

All objects in Fig. 9.5, except O_n and its immediate neighbor, form a cluster.

Definition 9.4.7 (Cluster Border). A border object (also known as cluster periphery) is a noncore object, but it belongs to the neighborhood of a core point.

O_b is a noncore object that is directly density reachable from core O_x; hence, a border object.

Definition 9.4.8 (Noise). Any object that does not belong to any cluster is noise. If an object neither satisfies a core nor a border object criterion, it is considered *noise* or *outlier*.

Both O_n and its immediate neighbor are examples of outliers that are neither core nor density connected to any core object.

9.4.3.1 DBSCAN

DBSCAN (Density-Based Spatial Clustering of Applications with Noise) [20] is a popular density-based clustering technique that starts with an arbitrary object O_i and obtains all objects density-reachable from O_i with respect to radius ϵ and $MinPts$, in order to form a cluster. This process produces a cluster if O_i is a core. When O_i is a border object, DBSCAN moves on to the next nonvisited object since no points are density reachable from O_i. The pseudocode is given in Algorithm 9.3.

The most expensive computation by DBSCAN or any density-based algorithm is the neighborhood query. If a spatial indexing data structure such as an R*-tree is used, the complexity of a neighborhood query is reduced from $O(n^2)$ to $O(\log_m n)$, where n is the size of the dataset and m is the number of entries on a node of the R*-tree. Therefore the DBSCAN algorithm's complexity is $O(n^2)$ or $O(\log_m n)$. Large datasets are manageable by the method if an efficient data structure is used. DBSCAN is intended to solve the noise problem and successfully disregard outliers. However, despite being able to handle hollow or arbitrary shapes, there is no evidence that it can handle issues with fluctuating density and shapes that are entirely nested. Since it uses global parameters like $MinPts$, DBSCAN may combine two clusters that are density connected into one cluster despite being of variable density.

9.4.3.2 OPTICS

OPTICS (Ordering Points To Identify Cluster Structure) [6] was introduced to address the issue of detecting clusters of variable densities. OPTICS orders the objects in the database according to their densities. The fundamental idea behind this approach is to assign each object a unique order in relation to the database's density-based structure. It is an extended DBSCAN algorithm with infinite distance parameters ϵ' smaller than ϵ ($0 \leq \epsilon' \leq \epsilon$). In addition to these common definitions used by other density-based approaches, it includes the following parameters to be associated with each object.

Definition 9.4.9 (Core distance). The core distance of an object O_i is the smallest ϵ' value that makes O_i a core object, i.e., the farthest distance from O_i to any object O_j within $N_\epsilon(O_i)$, when it satisfies the $MinPts$ constraints. If O_i is not a core object, the core distance of O_i is undefined:

$$\epsilon'(O_i) = \begin{cases} \text{Undefined} & \text{if } |N_\epsilon(O_i)| < MinPts \\ \max\{\texttt{dist}(O_i, O_j)|\forall O_j \in |N_\epsilon(O_i)| = MinPts\} & \text{otherwise.} \end{cases} \tag{9.3}$$

Definition 9.4.10 (Reachability distance). The reachability distance ($\texttt{dist}_r$) of an object O_j w.r.t. another object O_i is the greater value of the two distance measures, the core distance of O_i and the distance between O_i and O_j. If O_i is not a core object, the reachability distance between O_i and O_j is undefined:

$$\texttt{dist}_r(O_j, O_i) = \begin{cases} \max(\epsilon'(O_i), \texttt{dist}(O_i, O_j)) & \text{if } |N_\epsilon(O_i)| \geq MinPts \\ \text{Undefined} & \text{otherwise.} \end{cases} \tag{9.4}$$

Algorithm 9.3: DBSCAN algorithm.

Data: D: Database; ϵ; MinPts: Minimum neighbors for core constraint
Result: $C = \{C_1, C_2, \cdots, C_n\}$: Cluster Set

1   **Function** DBSCAN(**D**, ϵ, **MinPts**):
2      $C_i = 0$ // Initialize cluster id
3   **for** *each Unvisited point $P \in$ D* **do**
4         $O_i \leftarrow$ Visited
5         $O_{i\,Neigh} =$ RegionQuery $(O_i, \epsilon, $ **D**$)$ // Find ϵ neighbors of O_i
6   **if** $|O_{i\,Neigh}| <$ **MinPts then**
7            $O_i \leftarrow$ Noise
8   **end**
9   **else**
10   C_i++ // Increase cluster id
11   ExpandCluster $(O_i, O_{i\,Neigh}, C_i, \epsilon, $ **MinPts**$)$
12   **end**
13   **end**
14 **End Function**
15 **Function** ExpandCluster($O_i, O_{i\,Neigh}, C_i, \epsilon, $ **MinPts**):
16      $C_i = C_i \cup O_i$ // Add O_i to cluster C_i
17   **for** *each object $O_j \in O_{i\,Neigh}$* **do**
18   **if** $O_j = Unvisited$ **then**
19            $O_j \leftarrow$ Visited // Mark O_j as Visited
20            $O_{j\,Neigh} =$ RegionQuery $(O_j, \epsilon, $ **D**$)$
21   **end**
22   **if** $|P'_{Neigh}| \geq$ **MinPts then**
23            $O_{i\,Neigh} = O_{i\,Neigh} \cup O_{j\,Neigh}$
24   **end**
25   **if** O_j *not belongs to any cluster* **then**
26            $C_i = C_i \cup O_j$ // Add O_j to cluster C_i
27   **end**
28   **end**
29 **End Function**
30 **Function** RegionQuery($O_i, \epsilon, $ **D**):
31   **for** *each objects $O_j \in$ D* **do**
32   **if** dist $(O_i, O_j) \leq \epsilon$ **then**
33            $O_{i\,Neigh} = O_{i\,Neigh} \cup O_j$
34   **end**
35   **end**
36   **return** $O_{i\,Neigh}$
37 **End Function**

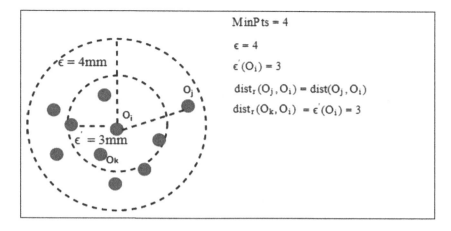

FIGURE 9.6 Illustration of core and reachability distance used in OPTICS.

An example of each of both measures is given in Fig. 9.6 for better understanding.

The two numbers, the core distance and a reachability distance, that this algorithm assigns to each point object represent its properties. In addition to storing the core distance and an appropriate reachability distance for each object, the algorithm generates an ordering of the objects in a database based on reachability distance. By scanning the cluster ordering and assigning cluster memberships based on the reachability distance and the core distance of the objects, we can easily extract any density-based cluster. Each item receives a cluster membership based on the scanning and extracting process, which can be thought of as an extended DBSCAN algorithm.

Although the OPTICS technique does not directly cluster a dataset, it serves as a sort of preprocessing step for other clustering algorithms like DBSCAN. OPTICS, as opposed to the DBSCAN approach, offers a solution to the problem of global density and variable density by providing each point object with an enhanced cluster ordering that contains information equivalent to density-based clustering that can work with a wide variety of parameter settings. The final clustering of the datasets takes more time because it is a preprocessing step that takes $O(n \log n)$ time if a spatial index is utilized. To extract clusters, it takes at least $O(n \log n) + O(n)$ time.

9.4.3.3 Extension of DBSCAN

DBSCAN has been the basis of numerous attempts to produce improved density-based clustering.

- **EnDBSCAN:** Many real datasets, including Magnetic-Resonance-image databases, gene-expression databases, and other databases, feature a pattern of embedded or nested cluster structures. DBSCAN cannot identify an underlying dense structure with variable densities since it only uses global density parameters. In contrast, OPTICS can identify irregularly shaped variable-density clusters, but it is unable to identify nested or embedded clusters. Additionally, OPTICS requires that objects be ordered before-hand based on their reachability distance, adding to the overall cost. EnDBSCAN [47], an improved version of DBSCAN and OPTICS, was proposed as a solution to these is-sues. Any intrinsic or embedded clusters can be effectively found with EnDBSCAN. It extends the concept of *core distance* of OPTICS and introduces the concept of *core neighborhood* that removes the need for global density parameter setting, required by DBSCAN. It also handles problems with varying density clusters as well as embedded clusters.

- **GDBSCAN:** The density-based approach is generalized by yet another clustering algo-rithm, Generalized DBSCAN (GDBSCAN) [50]. GDBSCAN can group point instances that have both categorical and numerical features. First, if the definition of the neigh-borhood is based on a binary predicate $NPred$ that is symmetric and reflexive, any idea of a neighborhood may be used in place of an ϵ-neighborhood. Secondly, differ-ent measurements can be employed to determine an equivalent of the *cardinality* of a neighborhood instead of just counting the things there.

- **DENCLUE:** DENCLUE (DENsity-based CLUstEring) [28] is a density-based clustering algorithm that works by identifying dense regions in the data space. The first step in the DENCLUE algorithm is to estimate the density of the data points in the feature space. This is typically done using a kernel-density estimator, which assigns a density value to each point based on the distance between that point and its nearest neighbors. Once the density estimation is complete, the algorithm initializes a cluster center at each point with a density value above a certain threshold. The algorithm then iteratively ex-pands each cluster by identifying nearby points with a high enough density value and adding them to the cluster. This process continues until there are no more points that can be added to the cluster. DENCLUE also includes a step for outlier detection. Any points that are not included in any cluster are considered outliers. Once the clustering is complete, postprocessing steps can be applied to improve the quality of the clusters. This might include merging nearby clusters with a high degree of overlap or removing outlier points that do not fit well with any of the clusters. Overall, DENCLUE is a power-ful density-based clustering algorithm that can effectively identify complex patterns in high-dimensional data. However, like any clustering algorithm, it is important to care-fully choose parameters and evaluate the quality of the resulting clusters to ensure that they are meaningful and useful.

9.5 High-dimensional data clustering

Real databases usually have high dimensions. Gene-expression datasets, protein data-bases, and even normal commercial relational databases are common examples. With high-dimensional data, it has been discovered that conventional low-dimensional clus-tering approaches are unable to produce adequate results. The *curse-of-dimensionality* issue connected with large-dimensional data is one of the main causes of such failure. *Richard Bellman* coined the term curse-of-dimensionality [8]. Recent theoretical findings [10] demonstrate that over a wide range of data distributions and distance functions, the distance between any pair of objects in high-dimensional space (e.g., 10–15 or higher) is almost constant. In such a high-dimensional space, the usual concepts of proximity and neighborhood no longer work; and hence "natural clusters" as in low dimensions are unlikely to exist in high-dimensional space. The following are some of the ways that have been used to cluster high-dimensional data.

9.5.1 Dimensionality reduction and feature selection

Applying a dimensionality-reduction approach to a dataset as a part of preprocessing is one option to deal with the dimensionality issue. Principal-component analysis (PCA) [1] and other techniques reduce the original data space's dimensions by creating dimensions that are linear combinations of the original dimensions. PCA is a statistical data-analysis methodology that can be used to (1) minimize the number of dimensions in the data without much loss of information, (2) find new, meaningful underlying composite dimensions or variables, (3) compress the data, and (4) visualize the data. The Karhunen–Loeve transformation [29] or singular-value decomposition (SVD) [19] are other names for essentially the same computation as PCA. These methods have a number of drawbacks when it comes to clustering, even though they may be effective at lowering dimensionality and useful for information compression and classification issues. Correlations between dimensions in practical applications may frequently be limited to certain clusters. Considering estimates computed for the entire database produces a significant information loss when dealing with localized clusters.

For supervised classification problems, a subset of dimensions is typically chosen using feature-selection techniques. Recently, a number of innovative feature-selection techniques have been created [17,18] to choose an *optimal* subset of features for unsupervised clustering or to create a pool of *dimension* subsets for cluster searching. Such techniques require optimization using techniques such as expectation maximization (EM) [15] or K-means. However, relying on a single clustering algorithm to assess potential dimension subsets can be counterproductive. The EM method depends on appropriate initialization and also favors circular or elliptical clusters. However, K-means tends to find equal-sized circular clusters and may fail to find arbitrary-shaped patterns.

Not all of the methods covered above are clustering methods. They are essentially preprocessing methods. Any relevant clustering methods, already discussed, can be used after dimension reduction or feature selection.

9.5.2 Projected clustering

Projected clustering techniques were created to prevent information loss from happening when feature-selection or dimensionality-reduction approaches are used. Projected clusters provide more insightful information regarding the underlying clustering structure. *A generalized projected cluster [4] is a set $\mathcal{E}$ of vectors together with a set $\mathcal{C}$ of data points such that the points in $\mathcal{C}$ are closely clustered in the subspace defined by the vectors $\mathcal{E}$.*

It is possible that the subspace defined by the vectors in $\mathcal{E}$ has fewer dimensions than the full-dimensional space. Two major projected clustering techniques are presented below.

- **PROCULUS:** This functions similarly to a K-means algorithm that has been expanded to include projected clustering [3]. First, a greedy method is used to choose a potential set of medoids, the centers of the clusters. It computes a correlated subspace for each cluster with the collection of medoids by looking at its location. The collection of data objects in a nearby area of a full-dimensional space is referred to as locality. To determine which single dimensions have closer average distances to the respective medoids, the projections of these data objects are computed on various single dimensions. These are picked as the correlated subspace's dimensions. Data objects are assigned to their closest medoid with the distance measured with respect to the appropriate subspace after the estimate of subspaces. These are the dimensions of the associated subspaces that are chosen. Following subspace estimation, data objects are assigned to their nearest medoid based on their distance from the appropriate subspace. The sum of the intracluster distances is used to assess the clustering's quality. In order to improve the clustering quality, medoids are substituted using a hill-climbing strategy. One issue is that the locality is formed using a complete set of dimensions, which is problematic. This might not include the actual neighbors in the associated subspace and might also include unrelated points. Finding neighbors in the high-dimensional space is actually not very useful.
- **ORCLUS:** This employs Singular Value Decomposition (SVD), a well-known approach for dimension reduction with minimal information loss. SVD, which is very similar to PCA, transforms the data to a new coordinate system (specified by a set of eigenvectors) that minimizes the data's correlations. ORCLUS [4], on the other hand, chooses the eigenvectors with the smallest spread (eigenvalue) to do the projection, in order to find the highest amount of similarity among the data points in the clusters. ORCLUS is a hierarchical merging approach that begins with a set of initial seeds. Using the SVD technique, the dimensionalities of the subspaces associated with each cluster are steadily lowered during the merging process. When the number of clusters and dimensionalities of subspaces meet the user input parameters, the merging is terminated.

9.5.3 Subspace clustering

Despite being more adaptable than dimensionality reduction, the projected clustering approach has the drawback of losing information about items that are clustered differently in multiple subspaces. To address this issue, the subspace clustering approach was created. *Subspace clustering* enables the simultaneous grouping of features and observations by generating both row and column clusters, respectively. It is also called *Biclustering* (discussed in detail in the next section). The basic assumption is that we are able to identify meaningful clusters that are specified by only a small number of dimensions. It is a way to overcome the problem of curse-of-dimensionality. We present some important subspace-clustering techniques in this section.

- **CLIQUE:** CLIQUE (Clustering In QUEst) [5], is a clustering strategy for high-dimensional datasets that performs automated subspace clustering of high-dimensional data based on density analysis and grid representation. The clustering technique restricts the search for clusters to subspaces of a high-dimensional data space rather than introducing new dimensions that combine the information in the existing dimensions. The actual clustering is carried out through a density-based strategy. A bottom-up method is used to divide each dimension of the space into intervals of equal length in order to approximate the density of the data points. As each partition has a constant volume, the density of each partition can be calculated from the number of points inside each partition. Automatic subspace identification is performed using these density estimates. Data points are divided based on the density function, and related high-density partitions within the subspace are then grouped to identify clusters in the subspace units. The clusters can be specified by a Disjunctive Normal Form (DNF) expression and are restricted to be axis-parallel hyperrectangles for simplicity. The union of several overlapping rectangles can be used to represent a cluster in a compact manner. Two input parameters are used to partition the subspace and identify the dense units. The input parameter ξ is the number of intervals of equal length into which each subspace is divided, and the input parameter τ is the density threshold.
- **MAFIA:** *Merging Adaptive Finite Intervals And is more than a clique* (MAFIA) [38] is a version of CLIQUE that operates faster and identifies clusters of a higher caliber. The key modification is the removal of the pruning method that restricts the number of subspaces evaluated and the adoption of an adaptive interval size that partitions a dimension based on the data distribution in that dimension. During the initial pass of the data, a histogram with the minimum number of bins for each dimension is created. If two bins are adjacent and have similar histogram values, they are combined. As the bin boundaries are not as stiff as they are in CLIQUE, the resulting clusters' shape can be greatly refined. MAFIA is 44 times quicker than the original CLIQUE and does a good job of handling both very large sizes and high dimensionality. The improved algorithm's computational complexity is $O(ck')$, where k' is the number of unique dimensions that the cluster subspaces in the dataset represent and c is a constant.
- **CBF:** Cell-Based Filtering (CBF) [11] aims to solve the scaling problems that many bottom-up algorithms have. Many bottom-up algorithms have the issue that when the number of dimensions rises, drastically more bins are formed. By constantly analyzing the minimum and maximum values in a given dimension, the cell-construction algorithm produces partitions while producing fewer bins (cells). Scalability in terms of the number of instances in the dataset is also addressed by CBF. In particular, when the dataset is too large to fit in the main memory, other techniques frequently perform poorly. The effective filtering-based index structure used by CBF to store the bins leads to better retrieval performance. Two variables have a significant impact on the CBF algorithm. The bin frequency of a dimension is determined by the section threshold. As the threshold value is raised, the retrieval time decreases because fewer records are accessed. Cell threshold, which controls the minimum density of data points in a bin,

is the other parameter. Bins with densities over this cutoff are chosen as prospective cluster members. CBF can locate clusters of various sizes and forms, just like the other bottom-up techniques. Additionally, it scales linearly as a dataset's instance count increases. It has been demonstrated that CBF is faster at retrieval and cluster creation but has poorer precision than CLIQUE.

9.6 Biclustering

Biclustering is a type of subspace clustering that is used frequently in analyzing gene-expression data [49]. It is also known as coclustering, block clustering, or two-mode clustering. J. A. Hartigan introduced the concept of biclustering in 1972. The first time this approach was used for biological gene-expression data was in 2000 by Y. Cheng and G. M. Church, who also presented a biclustering algorithm based on variance.

Definition 9.6.1 (Bicluster). Biclusters are a set of submatrices of the matrix $A = (N, M)$ with dimensions $I_1 \times J_1, \cdots, I_k \times J_k$ such that $I_i \subseteq N$, $J_i \subseteq M$, $\forall i \in \{1, \cdots, k\}$, where each submatrix (bicluster) meets a given homogeneity criterion.

There are four different types of biclusters that can be formed.

- **Constant Biclusters:** Biclusters or submatrices with the same values in rows and columns are called *constant biclusters*. They are also called *perfect biclusters* if all the values, $a_{ij} \in (I, J)$ are equal, i.e., $a_{ij} = \delta$.
  We can see an example in Fig. 9.7 A. Although perfect biclusters may exist in real life, they are usually masked by a noise factor, μ. Hence, the modified relationship for perfect biclusters can be written as $a_{ij} = \delta \pm \mu_{ij}$. To evaluate the quality of constant biclusters, one can use the variance of the submatrix [27]. A perfect bicluster has a variance equal to zero.
- **Biclusters with constant values in rows or columns:** A submatrix (I, J) is a perfect bicluster with constant rows, where any value in the bicluster can be found using one of the following expressions:

$$a_{ij} = \delta + \alpha_i \ \text{(Additive)}$$
$$a_{ij} = \delta \times \alpha_i \ \text{(Multiplicative)},$$

  where α_i the additive or multiplicative factor for row $i \in I$ depends on whether it is an additive or multiplicative perfect bicluster. For example, submatrices B and C in Fig. 9.7 are examples of constant rows and column biclusters, respectively. In the case of submatrix B, δ is 1, whereas $\alpha_{ij} = \{0, 1, 2\}$ (additive) or $\alpha_{ij} = \{1, 2, 3\}$ (multiplicative) for different rows.
- **Biclusters with coherent values:** A perfect bicluster (I, J) has coherent values, i.e., values within rows or columns are related to certain row- and column-specific factors. A perfect bicluster can be defined as a subset of rows and a subset of columns, whose

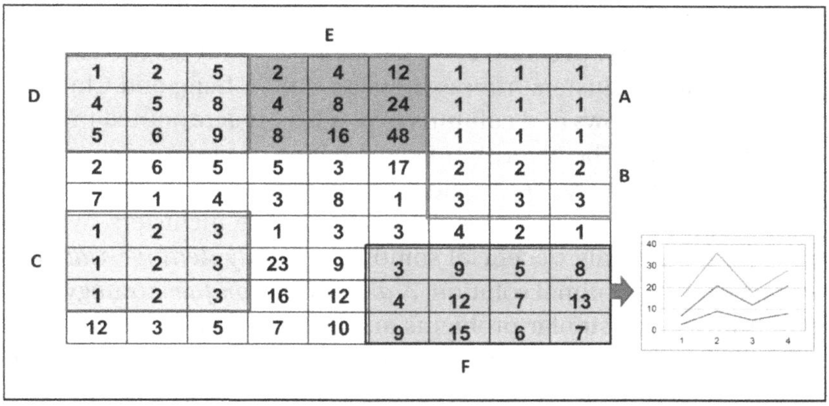

FIGURE 9.7 Example of different forms of biclusters. A. Constant bicluster, B. Constant Rows, C. Constant Columns, D. Coherent Additive, E. Coherent Multiplicative, F. Coherent Evolution.

values, a_{ij} can be derived using the following relations:

$$a_{ij} = \delta + \alpha_i + \beta_j \quad \text{(Additive)}$$
$$a_{ij} = \delta' \times \alpha'_i \times \beta'_j \quad \text{(Multiplicative)}.$$

Here, α_i is the coherence factor for row $i \in I$, and β_j is the coherence factor for column $j \in J$. The multiplicative model is equivalent to the additive model if in the multiplicative model, we substitute, $\delta = \log(\delta')$, $\alpha_i = \log(\alpha'_i)$, and $\beta = \log(\beta')$. The coherence of values can be observed in submatrices in Fig. 9.7 D (additive) and E (multiplicative) in the rows and columns of the bicluster, respectively. *Mean Squared Residue* (MSR) [12] can be used to measure the coherence of the rows and columns in the biclusters.

- **Biclusters with coherent evolutions:** Coherent evolution corresponds to a submatrix that contains coherent patterns irrespective of any particular value. A coherent evolution bicluster consists of a group of rows where the values in these rows cause a linear order to appear over a subset of the columns [9]. It is also called an order-preserving submatrix (OPSM). The values in coherent evolution biclusters and their relationships can be better explained with the help of row-wise value patterns, as shown in Fig. 9.7F. It is often difficult to draw certain uniformity across rows or columns just by looking into the matrix values, as in the case of other biclusters. However, if we look into the plotting of the values, it shows a pattern of similarity in terms of value change. Bicluster Detection of such biclusters is important in order to find coexpressed gene regulation from expression profiles [48].

9.6.1 Biclustering techniques

Biclustering techniques have gained in popularity after successful applications in gene-expression data. In order to capture the coherence of a subset of genes under a subset of situations, Cheng and Church [12] applied this clustering approach to gene-expression data. The approach was first described by Hartigan [27]. Several techniques have been proposed to find quality biclusters from expression data. Cheng and Church's algorithm greedily adds (or removes) rows and columns to reach a predetermined number of biclusters with a specified score. The algorithm uses the idea of mean squared residue (MSR) to quantify the degree of coherence. The lower the score, the stronger the coherence exhibited by the biclusters, and the better the quality of the biclusters. With the hope of eventually obtaining a globally beneficial solution, a *greedy iterative search*-based strategy [12] discovers a locally optimal solution. A *divide and conquer* strategy [27] breaks the problem down into smaller similar problems and solves them one at a time. In order to resolve the original issue, all of the solutions are finally combined. The best biclusters are found via exhaustive enumeration of all possible biclusters present in the data, in exponential time, in *exhaustive biclustering* [53]. To produce and iteratively obtain an ideal set of biclusters, a number of *metaheuristic-based* strategies, including evolutionary and multiobjective evolutionary frameworks, have also been investigated [7]. MSR is used by all of them as the merit function.

Identifying optimum maximal biclusters is successfully accomplished using an MSR-based method. Finding biclusters of subsets of genes with similar behavior but different values is of interest from a biological perspective. This metric simply takes into account expression values, not the patterns or tendencies of the gene-expression profile; hence, it may miss interesting and biologically significant patterns like shifting and scaling patterns. Finding these kinds of patterns is crucial since genes frequently exhibit identical behavior despite having expression levels that vary by different orders of magnitude or ranges. It has been observed that MSR is not a good measure to discover patterns in data when the variance among gene values is high, especially when the genes present scaling and shifting patterns. It has further been observed that coregulated genes also share negative patterns or inverted behaviors, which existing pattern-based approaches are unable to detect. *CoBi* [48] (Coregulated Biclustering) captures biclusters among both positively and negatively regulated genes as coregulated genes. It considers both up- and downregulation trends and similarity in degrees of fluctuation under consecutive conditions for expression profiles of two genes as a measure of similarity between the genes. It uses a new BiClust tree for generating biclusters in polynomial time with a single pass over the dataset.

9.7 Cluster-validity measures

During the last few decades, various methods have been developed for clustering. To evaluate and validate the results of a clustering algorithm, three approaches are available; *external evaluation*, *internal evaluation*, and *relative evaluation*. Both *external evaluation* and *internal evaluation* are statistical approaches that measure the degree to which a clustering confirms an a priori specified arrangement of data points and it requires a prior understanding of the characteristics of the cluster structure. In contrast, *relative evalu-*

ation strategy ranks and compares clusterings of the same dataset that are found using various algorithms and parameter settings.

9.7.1 External evaluation

External evaluation measures the degree of correspondence between the true clustering (i.e., ground truth) of the underlying dataset and the clustering that is obtained after applying a certain algorithm to the same dataset. The majority of real-world datasets lack ground-truth clusters; however, it is necessary for the external evaluation to be conducted on the dataset. Hence, external evaluation is mostly used in two scenarios: when i) real benchmark datasets with known cluster structure; ii) artificial datasets with known ground truth, are available.

Jaccard: Jaccard similarity [30] or Jaccard similarity coefficient calculates the degree of similarity between two clusters of data to identify which members are shared and distinct. It can vary from zero to one. A higher number corresponds to a higher ratio of common neighbors.

Given a graph G = (V, E), a set of neighbors for a node u is $N(u) = \{v \in V | \{u, v\} \in E\}$, Jaccard's index can be calculated as follows:

$$Jaccard(u, v) = \frac{|\Gamma(u) \cap \Gamma(v)|}{|\Gamma(u) \cup \Gamma(v)|},$$

where, $\Gamma(u)$ represents the set of extended neighbors for node u and can be defined as $\Gamma(u) = \{u\} \cup N(u)$.

Rand Index (RI): The Rand Index proposed by Rand [45], is a popular cluster validation index that measures the similarity between two data clusters. Mathematically, it can be defined as follows:

$$RI = \frac{TP + TN}{TP + FP + FN + TN},$$

where TP is true positive, TN is true negative, FP is false positive, and FN is false negative. The RI value lies in the range [0,1]. The upper bound is 1, which corresponds to a perfect match between the partitions, and the lower bound is 0, which indicates the opposite.

Fowlkes–Mallows Index: The Fowlkes–Mallows (FM) Index [23] determines how similar two groups are to one another. A higher FM value represents greater similarity between clusters. Mathematically, the Fowlkes–Mallows index can be defined as:

$$FM = \sqrt{\frac{TP}{TP + FP} \cdot \frac{TP}{TP + FN}}.$$

Purity: This index [13] compares each cluster produced by the clustering technique to the ground-truth one with the highest overlap and tries to determine the percentage of "correctly" clustered samples. For a given cluster $C = \{c_1, c_2.. \cdot c_n\}$ and its respective ground truth $C^g = \{c_1^g, c_2^g .. \cdot c_n^g\}$, purity can be defined as follows:

$$Purity(C_i, C^g) = \frac{1}{n} \sum_i \max_j |C_i \cap C_j^g|,$$

where n is the number of samples.

Normalized Mutual Information: Normalized Mutual Information (NMI) [14,34] is a measure to compare the similarity among clustering solutions. The Mutual Information (MI) score is normalized to produce the Normalized Mutual Information (NMI), which scales the outcomes from 0 (no mutual information) to 1 (perfect correlation). We can define the NMI score as follows:

$$NMI(p, r) = \frac{2MI(p, r)}{H(p) + H(r)},$$

where MI(p,r) signifies the mutual information between the real clustering solution r and the predicted clustering solution p and mathematically, can be defined as follows:

$$MI(p, r) = \sum_i \sum_j p(C_{pi} \cap C_{rj}) log \frac{p(C_{pi} \cap C_{rj})}{p(C_{pi})p(C_{rj})}.$$

Similarly, the cluster entropy is represented by H(p) and can be defined as follows:

$$H(p) = - \sum_j p(C_{pi}) log(p(C_{pi})),$$

where $p(C_{pi})$ represents the probability of a node being in the predicted cluster p. $p(C_{rj})$ represents the probability of a node within the actual cluster r.

9.7.2 Internal evaluation

An *Internal evaluation* method validates the relevance of the relationships between the identified cluster structure and the structural information (by using a proximity metric like similarity or distance matrix) inherent in the data. For an internal assessment of various cluster-mining techniques, see the research work in Leskovec et al. [35].

Dunn Index (DI): The Dunn index [16] is the ratio between the shortest separation between observations not in the same cluster to the greatest intracluster separation. In simple words, it measures the ratio between compactness and separation. It is computed as:

$$D = \frac{min(separation)}{max(diameter)} = \frac{min_{c_k, c_l \in \mathcal{C}, c_k \neq c_l}(min_{i \in c_k, j \in c_l} dist(i, j))}{max_{c_m \in \mathcal{C}} diam(c_m)},$$

where $\mathcal{C} = \{c_1 .. \cdot .. c_k\}$ is a set of k disjoint clusters. $dist(i, j)$ is the distance (e.g., Euclidean or Manhattan) between observations i and j. $diam(c_m)$ is the maximum distance between observations in cluster c_m. The Dunn index has a value between zero and ∞, and should be maximized for better clustering.

Silhouette Index: The Silhouette index [46] measures how each sample in one cluster is similar to samples in its own cluster compared to samples in its neighboring clusters. Let $x = \{1, 2, 3... \cdots ..n\}$ be the samples present within a cluster c, then for each observation i, the Silhouette index can be defined as follows:

$$S(i) = \frac{b_i - a_i}{\max(a_i, b_i)},$$

where a_i is the average distance between observation i and every other observation in the same cluster. b_i is the average separation between observation i and the observations in the neighboring cluster.

Connectivity: The Connectivity Index [26] is yet another typical internal validation metric. For a particular clustering partition $\mathcal{C} = \{c_1.. \cdots ..c_k\}$, the connectivity is defined as:

$$Conn(\mathcal{C}) = \sum_{i=1}^{n} \sum_{j=1}^{L} x_{i,nn_{i(j)}},$$

where, $nn_{i(j)}$ is the jth nearest neighbor of observation i. L determines the number of neighbors that contribute to the connectivity measure. The connectivity value lies between $[0, \infty]$ and should be minimized.

Modularity: Newman and Girvan [41] defined a quality function called modularity (Q) that measures the variance between the fraction of intracluster edges and the expected number of such edges when random connections between the nodes are made. Edges can be drawn between data points based on the properties of the points or other considerations. The modularity of a network of two communities can be defined as:

$$Q = \frac{1}{4m} \sum_{i,j} (A_{i,j} - \frac{k_i k_j}{2m})(s_i s_j), \tag{9.5}$$

where, $s_i = 1$, if node i is in cluster 1; -1 if node i is in cluster 2. $\frac{k_i k_j}{2m}$ is the expected number of edges between i and j if edges are placed randomly. k_i represent the degree of i. m is the total number of edges in the set of clusters. $A_{i,j}$ represents the adjacency matrix. Clusters that have dense associations among the nodes exhibit high modularity (Q) values. That means, a cluster with a higher modularity value signifies a better structure and conversely, a weak cluster structure causes a low Q value. Modularity has commonly been used in evaluating communities in social and other networks, where clusters correspond to communities.

9.7.3 Relative evaluation

Relative evaluation does not require any prior knowledge about the cluster structure. The goal is to determine which clustering solution is better suited for a specific task or dataset. These methods do not rely on external information or ground-truth labels but instead focus on the comparison of clustering outcomes. The key objective of this type of validity index is to select the best cluster detector from a set of clustering algorithms by using a predetermined objective function. For instance, specific to a particular dataset one may apply multiple clustering algorithms (e.g., K-Means, Hierarchical Clustering, DBSCAN) to the data and then compare their results. This can help identify which algorithm provides the most meaningful and interpretable clusters for a particular dataset.

9.8 Summary

Clustering is the most popularly used unsupervised learning technique. It has wide applications in real-life data analysis. We discussed a number of prominent clustering techniques, proximity measures, and how to validate the quality of clusters. New clustering algorithms are still being proposed to handle varying types of data. The techniques meant for moderate-sized data become ineffective and useless for large datasets. The dynamic nature of data also makes cluster analysis more challenging. There is a great need for new versatile clustering techniques that can work with small or large datasets with attributes of different data types, and datasets that may be static or generated dynamically with different generation speeds, including real-time or almost real-time generation.

References

[1] Hervé Abdi, Lynne J. Williams, Principal component analysis, Wiley Interdisciplinary Reviews: Computational Statistics 2 (4) (2010) 433–459.
[2] Basel Abu-Jamous, Rui Fa, David J. Roberts, Asoke K. Nandi, Yeast gene CMR1/YDL156W is consistently co-expressed with genes participating in DNA-metabolic processes in a variety of stringent clustering experiments, Journal of the Royal Society Interface 10 (81) (2013) 20120990.
[3] Charu C. Aggarwal, Joel L. Wolf, Philip S. Yu, Cecilia Procopiuc, Jong Soo Park, Fast algorithms for projected clustering, ACM SIGMOD Record 28 (2) (1999) 61–72.
[4] Charu C. Aggarwal, Philip S. Yu, Finding generalized projected clusters in high dimensional spaces, in: Proceedings of the 2000 ACM SIGMOD International Conference on Management of Data, 2000, pp. 70–81.
[5] Rakesh Agrawal, Johannes Gehrke, Dimitrios Gunopulos, Prabhakar Raghavan, Automatic subspace clustering of high dimensional data for data mining applications, in: Proceedings of the 1998 ACM SIGMOD International Conference on Management of Data, 1998, pp. 94–105.
[6] Mihael Ankerst, Markus M. Breunig, Hans-Peter Kriegel, Jörg Sander, Optics: ordering points to identify the clustering structure, ACM SIGMOD Record 28 (2) (1999) 49–60.
[7] Haider Banka, Sushmita Mitra, Evolutionary biclustering of gene expressions, Ubiquity 2006 (October 2006) 1–12, https://doi.org/10.1145/1183081.1183082.
[8] R. Bellman, Adaptive Control Process: A Guided Tour, Princeton University Press, ISBN 9780691652214, 1961.
[9] Amir Ben-Dor, Benny Chor, Richard Karp, Zohar Yakhini, Discovering local structure in gene expression data: the order-preserving submatrix problem, in: Proceedings of the Sixth Annual International Conference on Computational Biology, 2002, pp. 49–57.
[10] Kevin Beyer, Jonathan Goldstein, Raghu Ramakrishnan, Uri Shaft, When is "nearest neighbor" meaningful?, in: International Conference on Database Theory, Springer, 1999, pp. 217–235.
[11] Jae-Woo Chang, Du-Seok Jin, A new cell-based clustering method for large, high-dimensional data in data mining applications, in: Proceedings of the 2002 ACM Symposium on Applied Computing, 2002, pp. 503–507.
[12] Yizong Cheng, George M. Church, Biclustering of expression data, in: ISMB, vol. 8, 2000, pp. 93–103.

[13] Christopher D. Manning, Prabhakar Raghavan, Hinrich Schutze, Introduction to Information Retrieval, Cambridge University Press, 2008, https://doi.org/10.1017/CBO9780511809071.

[14] Leon Danon, Albert Diaz-Guilera, Jordi Duch, Alex Arenas, Comparing community structure identification, Journal of Statistical Mechanics: Theory and Experiment 2005 (09) (2005) P09008.

[15] Arthur P. Dempster, Nan M. Laird, Donald B. Rubin, Maximum likelihood from incomplete data via the em algorithm, Journal of the Royal Statistical Society, Series B, Methodological 39 (1) (1977) 1–22.

[16] Joseph C. Dunn, Well-separated clusters and optimal fuzzy partitions, Journal of Cybernetics 4 (1) (1974) 95–104.

[17] Jennifer G. Dy, Carla E. Brodley, Feature subset selection and order identification for unsupervised learning, in: Intl. Conf. Machine Learning (ICML), Morgan Kaufmann Publishers Inc., 2000, pp. 247–254.

[18] Jennifer G. Dy, Carla E. Brodley, Visualization and interactive feature selection for unsupervised data, in: Proceedings of the Sixth ACM SIGKDD International Conference on Knowledge Discovery and Data Mining, 2000, pp. 360–364.

[19] Carl Eckart, Gale Young, The approximation of one matrix by another of lower rank, Psychometrika 1 (3) (1936) 211–218.

[20] Martin Ester, Hans-Peter Kriegel, Jörg Sander, Xiaowei Xu, et al., A density-based algorithm for discovering clusters in large spatial databases with noise, in: KDD, vol. 96, 1996, pp. 226–231.

[21] Vladimir Estivill-Castro, Ickjai Lee, AMOEBA: hierarchical clustering based on spatial proximity using Delaunay diagram, in: Proceedings of the 9th International Symposium on Spatial Data Handling, Beijing, China, 2000, pp. 1–16.

[22] Edward W. Forgy, Cluster analysis of multivariate data: efficiency versus interpretability of classifications, Biometrics 21 (1965) 768–769.

[23] Edward B. Fowlkes, Colin L. Mallows, A method for comparing two hierarchical clusterings, Journal of the American Statistical Association 78 (383) (1983) 553–569.

[24] Venkatesh Ganti, Johannes Gehrke, Raghu Ramakrishnan, Cactus—clustering categorical data using summaries, in: Proceedings of the Fifth ACM SIGKDD International Conference on Knowledge Discovery and Data Mining, 1999, pp. 73–83.

[25] Marta C. Gonzalez, Cesar A. Hidalgo, Albert-Laszlo Barabasi, Understanding individual human mobility patterns, Nature 453 (7196) (2008) 779–782.

[26] Julia Handl, Joshua Knowles, Douglas B. Kell, Computational cluster validation in post-genomic data analysis, Bioinformatics 21 (15) (2005) 3201–3212.

[27] John A. Hartigan, Direct clustering of a data matrix, Journal of the American Statistical Association 67 (337) (1972) 123–129.

[28] Alexander Hinneburg, Daniel A. Keim, et al., An Efficient Approach to Clustering in Large Multimedia Databases with Noise, vol. 98, Bibliothek der Universität Konstanz, 1998.

[29] Yingbo Hua, Wanquan Liu, Generalized Karhunen–Loeve transform, IEEE Signal Processing Letters 5 (6) (1998) 141–142.

[30] Paul Jaccard, The distribution of the flora in the Alpine zone. 1, New Phytologist 11 (2) (1912) 37–50.

[31] Leonard Kaufman, Peter J. Rousseeuw, Partitioning around medoids (program PAM), in: Finding Groups in Data: an Introduction to Cluster Analysis, vol. 344, 1990, pp. 68–125.

[32] Leonard Kaufman, Peter J. Rousseeuw, Finding Groups in Data: An Introduction to Cluster Analysis, John Wiley & Sons, 2009.

[33] Nevan J. Krogan, Gerard Cagney, Haiyuan Yu, Gouqing Zhong, Xinghua Guo, Alexandr Ignatchenko, Joyce Li, Shuye Pu, Nira Datta, Aaron P. Tikuisis, et al., Global landscape of protein complexes in the yeast *Saccharomyces cerevisiae*, Nature 440 (7084) (2006) 637–643.

[34] Andrea Lancichinetti, Santo Fortunato, János Kertész, Detecting the overlapping and hierarchical community structure in complex networks, New Journal of Physics 11 (3) (2009) 033015.

[35] Jure Leskovec, Kevin J. Lang, Michael Mahoney, Empirical comparison of algorithms for network community detection, in: Proceedings of the 19th International Conference on World Wide Web, 2010, pp. 631–640.

[36] J. MacQueen, Classification and analysis of multivariate observations, in: 5th Berkeley Symp. Math. Statist. Probability, 1967, pp. 281–297.

[37] Hazel Nicolette Manners, Swarup Roy, Jugal K. Kalita, Intrinsic-overlapping co-expression module detection with application to Alzheimer's disease, Computational Biology and Chemistry 77 (2018) 373–389.

[38] Harsha Nagesh, Sanjay Goil, Alok Choudhary, MAFIA: Efficient and scalable subspace clustering for very large data sets, Technical report, Northwestern University, 1999.

[39] Keshab Nath, Swarup Roy, Sukumar Nandi, InOvIn: a fuzzy-rough approach for detecting overlapping communities with intrinsic structures in evolving networks, Applied Soft Computing 89 (2020) 106096.

[40] Tamás Nepusz, Haiyuan Yu, Alberto Paccanaro, Detecting overlapping protein complexes in protein–protein interaction networks, Nature Methods 9 (5) (2012) 471–472.

[41] Mark E.J. Newman, Modularity and community structure in networks, Proceedings of the National Academy of Sciences 103 (23) (2006) 8577–8582.

[42] Raymond T. Ng, Jiawei Han, CLARANS: a method for clustering objects for spatial data mining, IEEE Transactions on Knowledge and Data Engineering 14 (5) (2002) 1003–1016.

[43] Timo Ojala, Matti Pietikäinen, David Harwood, A comparative study of texture measures with classification based on featured distributions, Pattern Recognition 29 (1) (1996) 51–59.

[44] Shuye Pu, Jessica Wong, Brian Turner, Emerson Cho, Shoshana J. Wodak, Up-to-date catalogues of yeast protein complexes, Nucleic Acids Research 37 (3) (2009) 825–831.

[45] William M. Rand, Objective criteria for the evaluation of clustering methods, Journal of the American Statistical Association 66 (336) (1971) 846–850.

[46] Peter J. Rousseeuw, Silhouettes: a graphical aid to the interpretation and validation of cluster analysis, Journal of Computational and Applied Mathematics 20 (1987) 53–65.

[47] Swarup Roy, Dhruba K. Bhattacharyya, An approach to find embedded clusters using density based techniques, in: International Conference on Distributed Computing and Internet Technology, Springer, 2005, pp. 523–535.

[48] Swarup Roy, Dhruba K. Bhattacharyya, Jugal K. Kalita, CoBi: pattern based co-regulated biclustering of gene expression data, Pattern Recognition Letters 34 (14) (2013) 1669–1678.

[49] Swarup Roy, Dhruba K. Bhattacharyya, Jugal K. Kalita, Analysis of gene expression patterns using biclustering, in: Microarray Data Analysis, Springer, 2015, pp. 91–103.

[50] Jörg Sander, Martin Ester, Hans-Peter Kriegel, Xiaowei Xu, Density-based clustering in spatial databases: the algorithm GDBSCAN and its applications, Data Mining and Knowledge Discovery 2 (2) (1998) 169–194.

[51] Erich Schubert, Peter J. Rousseeuw, Fast and eager k-medoids clustering: $O(k)$ runtime improvement of the PAM, CLARA, and CLARANS algorithms, Information Systems 101 (2021) 101804.

[52] Robert R. Sokal, A statistical method for evaluating systematic relationships, The University of Kansas Science Bulletin 38 (1958) 1409–1438.

[53] Amos Tanay, Roded Sharan, Ron Shamir, Discovering statistically significant biclusters in gene expression data, Bioinformatics 18 (suppl_1) (2002) S136–S144.

[54] Xindong Wu, Vipin Kumar, J. Ross Quinlan, Joydeep Ghosh, Qiang Yang, Hiroshi Motoda, Geoffrey J. McLachlan, Angus Ng, Bing Liu, Philip S. Yu, et al., Top 10 algorithms in data mining, Knowledge and Information Systems 14 (1) (2008) 1–37.

[55] Tian Zhang, Raghu Ramakrishnan, Miron Livny, BIRCH: an efficient data clustering method for very large databases, ACM SIGMOD Record 25 (2) (1996) 103–114.

10

Ensemble learning

10.1 Introduction

The outcome of a single learning model or framework may not always be conclusive due to the relative merits and demerits of each framework. Single-model learning can be limited in its ability to achieve high accuracy and robustness in predictions for several reasons:

1. **Overfitting**: A single model may learn to fit the training data too closely and fail to generalize well to new, unseen data.
2. **Underfitting**: A model may not have enough complexity to capture the underlying patterns in the data, resulting in poor performance.
3. **Biases and errors**: A model may have biases and errors that are inherent in its design or training data, which can lead to inaccurate predictions.

Ensemble Learning is a powerful alternative learning technique that combines the outcomes of multiple models to improve the accuracy and robustness of the final outcome. In recent years, ensemble learning has gained increasing popularity due to its ability to achieve state-of-the-art performance on a wide range of tasks, including classification, regression, and clustering.

Ensemble learning takes into account the opinions of a number of learners to obtain the final decision [17]. This ensures that individual drawbacks of learners are overcome and overall performance is improved. Based on whether existing knowledge is used or not, ensemble-learning algorithms are categorized as supervised and unsupervised. Another category of ensemble learning is metaensemble learning, where an ensemble of ensembles is built to further avoid biases of individual ensembles.

In this chapter, we explore various types of ensemble-learning methods, such as supervised, unsupervised, semisupervised, and metaensemble learning, and discuss their strengths and weaknesses.

10.1.1 What is ensemble learning?

An ensemble-learning technique for a given task consists of a diverse set of learning algorithms, whose findings are combined in some manner to improve the performance of an individual learner. Ensemble learning builds on traditional machine learning. Conventionally, only one model is used to learn the hidden characteristics in a given dataset; the end performance of such a model may often be average. The primary notion in ensemble learning is that a group of decision makers is more reliable than a single decision maker. In the literature, such grouping of decision makers is known by different names such as

Fundamentals of Data Science. https://doi.org/10.1016/B978-0-32-391778-0.00017-X

Blending [6], Ensemble of Classifiers [4], Committee of Experts [14], Perturb and Combine [3], among many others.

The general idea behind such techniques is that the decision taken by a committee of models is likely to be better than the decision taken by a single model. This is because each model in the committee may have learned some distinct inherent properties in the data that other models may have missed. Hence, each model may contribute some unique knowledge, impacting the overall performance positively. Therefore, by way of explanation, ensemble learning can be thought of as an opportunity to improve the decision-making process by choosing a diverse group of superior to average learners rather than an individual learner who may not give the best possible performance for a given task. This usually indicates that the generalizing power of a group of models is usually better than the generalizing power of an individual model. We introduce basic terms that are helpful in understanding the concepts well. Some of these have been presented in an earlier chapter.

(a) **Generalization**: This describes the capability of a machine-learning model to work well with unseen data after it has gained experience by learning from labeled examples. Being able to learn a generalized model that performs well with examples that have never been seen before requires the model to learn the essential characteristics of the data, eschewing any incidental or accidental properties that may exist in the training dataset [10].

(b) **Bias**: Bias in machine learning comes from the prior assumptions that the learner makes about the target function to learn. A low-bias model makes very few assumptions about the function to learn; for example, the model may be open to learn any nonlinear function of the dependent variables. On the other hand, a high-bias model makes strict or narrow assumptions about the function to learn; for example, the learner may have a prior notion that the function to learn is always linear. A high value of bias means that the model may not perform well as it is not successful in learning important intrinsic patterns in the data.

(c) **Variance**: This is a measure of how varied the predictions are made by the model on the same set of observations. Another way to look at it is by observing how different the learned models are when the training dataset changes only slightly. Decision trees built from slightly different datasets (e.g., only one or two examples are different) may be very different; decision trees are high-variance models. On the other hand, in some approaches, slightly or moderately different datasets may produce models that are very or exactly similar. For example, changing one or two data points (unless they are outliers) may not change a linear regressor much at all, signifying it is a low variance model. High variance leads to overfitting the training population.

(d) **Overfitting**: This describes the situation when a model learns the training data too well but subsequently underperforms on previously unseen data. In other words, the model learns the patterns very well when taught by examples, but fails to perform well when taking a test, when slightly different problems are posed.

10.1.2 Building an ensemble

An ensemble can be constructed in several ways at different levels. The end goal of an ensemble is to enhance the performance of the decision-making process so that the overall outcome is improved. Below, we discuss some popular approaches to construct an ensemble.

(A) Constructing different training sets: From a given pool of sample training data, several random training subsample sets are formed. A learning algorithm is then trained on each of these training sets to obtain different classifiers. Techniques such as resampling and reweighting are used to obtain sets of random training subsamples.

  (a) Drawing samples randomly from a dataset in a repetitive manner.

  (b) Reweighting of training examples is performed in creating an ensemble of decision trees called boosting trees.

(B) Random Forests sample features of data examples randomly, in addition to randomly sampling data examples with replacement.

(C) An ensemble can be built from a number of Convolutional Neural Networks (CNNs) such that the CNN models have different numbers of layers, and/or different numbers of filters and/or different hyperparameter values.

10.1.3 Categories of ensemble learning

Depending on how the individual models learn, ensemble-learning techniques can be of five types, as shown in Fig. 10.1. Note that all individual learners use the same learning approach.

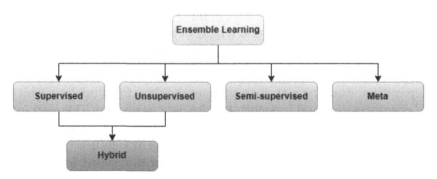

FIGURE 10.1 Categorization of various ensemble-learning techniques.

Below, we introduce these.

- **Supervised ensemble learning**: A set of base learners, say B_i^s, $i = 1\ to\ n$, are first trained on a given dataset D_s, and then tested on a previously not seen set of samples T_s. Supervised ensemble learning is the process of aggregating the respective predictions made by each B_i^s through a combination function f_s to obtain the final output.

- **Unsupervised ensemble learning**: A set of base learners, say B_i^u, $i = 1\ to\ n$, is applied on a given dataset D_s to group the instances into individual sets of clusters C_k without the use of any prior knowledge. Unsupervised ensemble learning is the process of reaching a unanimous decision to aggregate the individual groupings through a consensus function f_u to obtain a final set of clusters.
- **Semisupervised ensemble learning**: A set of base learners, say B_i^s, $i = 1\ to\ n$, are first trained using limited knowledge on a partially labeled dataset D_s. These base learners pseudotag the previously unlabeled data so as to incorporate them in the training set.
- **Metaensemble learning**: This is a learning technique where an ensemble of ensembles is constructed and the predictions of individual ensembles are aggregated to give the best possible final output.
- **Hybrid ensemble learning**: This is a learning technique where an ensemble of both supervised and unsupervised ensembles is constructed to generate the best possible final output by aggregating the decisions given by both ensembles.

10.2 Ensemble-learning framework

A typical ensemble-learning framework performs a multilevel task. The top levels include a selection of base-learning models, followed by consensus decision making based on the outcomes of the base models. The base learners may be any classifier or clustering models depending on the nature of the task, like supervised [12] or unsupervised ensembling [9]. It is important to note that before using any learning technique, the original data may undergo various preprocessing methods such as normalization, missing-value estimation, discretization, feature selection, or other necessary data transformations. The performance of a learning technique also largely depends on the preprocessing methods used. A typical ensemble-learning framework for classification or clustering can be broken down into several steps:

- **Data Preparation**: The first step is to prepare the data for modeling, which involves cleaning, preprocessing, and transforming the data as needed. This step may include tasks such as feature selection, dimensionality reduction, and normalization.
- **Model Selection**: The next step is to select a set of base models to be used in the ensemble. These models can be diverse in terms of the algorithms used, hyperparameters, or subsets of the data used for training. Common algorithms used in ensemble learning include decision trees, random forests, support-vector machines, and neural networks.
- **Ensemble Generation**: Once the base models are selected, the next step is to generate an ensemble by training each of the base models on either the entire dataset or a different subset of the data, or using different algorithms or hyperparameters. This can be done using techniques such as bagging, boosting, or stacking.
- **Model Combination**: The final step is to combine the predictions of the base models into a single, final prediction. This can be done using techniques such as averaging,

voting, or weighted voting. For clustering, a consensus clustering approach can be used to combine the results of individual clustering models into a final clustering.

- **Evaluation**: The ensemble is evaluated on a validation set to determine its performance. Common performance metrics for classification include accuracy, precision, recall, and F1-score, while for clustering, metrics like the silhouette score, Dunn index, or Davies–Bouldin index can be used.
- **Deployment**: Once the ensemble is evaluated and its performance is satisfactory, it can be deployed to make predictions on new, unseen data.

Ensemble learning can be accomplished at different levels, such as attribute (feature) level, or decision level based on the outputs given by the individual machine-learning technique. For example, at the feature level, the feature ranks given by the individual feature-selection algorithms can be combined using combination rules to obtain an aggregated list of the best possible relevant features. The same learning models can be trained on random full-featured subsets of a dataset or versions of the same dataset with sampled subsets of features. The learning models are then tested on previously unseen samples. The final prediction of the models can be combined using different combination rules. The main goal is to increase the generalization capability of an ensemble model such that the final predictions are highly accurate. Fig. 10.2 illustrates a generic framework for ensemble learning.

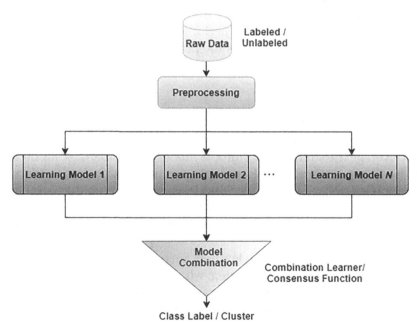

FIGURE 10.2 Typical ensemble-learning framework for supervised or unsupervised decision making. The learning models are either selected classifiers or clustering methods depending on the task. Combination learners are used during classification, while consensus is used during clustering.

10.2.1 Base learners

Several individual learners participate in forming an ensemble. Such learners are usually from distinct families and are called base learners. In a supervised framework, these base learners take as input a set of labeled samples (training data) and learn significant characteristics from the data to classify the samples with appropriate labels. The output of the base learners is combined to obtain an aggregated prediction, the ensemble's output. It is worth noting that the output of the individual classifier models can be either class labels or continuous outputs, or even class probabilities. The combination function is chosen depending on the output given by the base learners. For example, majority voting and weighted majority voting work when combining class labels, whereas the sum or mean rule work when combining continuous outputs.

In ensemble clustering, base-clustering algorithms are used to partition the data into clusters. These base-clustering algorithms can be different clustering algorithms or the same algorithm with different parameters or random seeds. Each base-clustering algorithm provides a different clustering solution.

The output of the base-clustering algorithms is then combined to form an ensemble clustering. Various methods can be used to combine the base-clustering algorithms, such as consensus clustering or cluster-level ensemble methods.

Similar to base learners in classification, the diversity and complementary nature of the base-clustering algorithms are important for the effectiveness of ensemble clustering. Ensemble clustering can improve clustering performance and robustness by combining the strengths of different clustering algorithms and providing a more stable clustering solution that is less sensitive to the choice of initial parameters or random seeds.

10.2.2 Combination learners

Combination learning refers to the process of combining multiple models or learners to improve the overall performance of a machine-learning system. The goal of combination learning is to leverage the strengths of multiple models to compensate for their individual weaknesses, leading to a more accurate and robust prediction or classification.

A combination learner is a type of machine-learning algorithm that combines the outputs of multiple base learners, which are usually simpler or weaker models, to form a more powerful or complex model. The combination learner can be used in both supervised and unsupervised learning problems.

In supervised learning, combination learners are typically used for ensemble methods, such as random forests, boosting, or bagging, which combine the predictions of multiple base classifiers to improve classification accuracy or reduce overfitting. The base classifiers can be different types of algorithms, such as decision trees, support-vector machines, or neural networks.

In unsupervised learning, combination learners are often used for clustering or anomaly-detection problems, where the goal is to identify hidden patterns or anomalies in the data. Ensemble-clustering methods, such as consensus clustering or cluster-level ensemble methods, can be used to combine the results of multiple base-clustering algorithms to improve the clustering performance and robustness.

A combination function tries to aggregate the outputs of learning models participating in the ensemble process. The outputs to be combined can be in class label form or continuous outputs. Depending on the type of output of the learning model, the combination function must be chosen.

10.2.2.1 Class-label combination

To combine the class-label outputs of learning models, two simple strategies that can be used are *Majority voting* and *Weighted majority voting*, as discussed below.

1. **Majority voting**: According to this technique, the ensemble chooses the class label with the most votes. That is, all the learning models have equal opportunities to choose their respective class labels (these are counted as votes), and the label with the most votes is chosen as the winner by the ensemble. For example, let us say we have ten learning models, and six learning models say that a particular instance, say X_1 belongs to class A and four models say X_1 belongs to class B. Then, the final output will be class A. This technique can have three different cases:

   (a) Case 1: All learning models agree unanimously without any conflicts in predicting a single class label. This is a *consensus voting* decision.

   (b) Case 2: At least one more than half the number of learning models agree to predict a class label. This is a *majority voting* decision.

   (c) Case 3: The class label with the highest number of votes is chosen as the winner. This is also called a *first-past-the-post* or *winner-take-all* voting decision.

   Although very popular, the majority or first-past-the-post voting techniques has limitations because it may so happen that certain learning models are more suitable for a given task compared to others. In that case, weights must be assigned to the learning models.

2. **Weighted majority voting**: According to this technique, weights are assigned to the learning models when predicting a class label for a given task. This is because, out of all the participating learning models in the ensemble, some models may be more suitable compared to others. In this case, the suitable members of the ensemble are assigned higher weights than others. Let us assume that, a learning model l_t makes a decision $d_{t,j}$ and chooses class c_j for an instance X_i. Hence, $d_{t,j} = 1$, if c_j is chosen, otherwise 0. Subsequently, all the learning models are assigned weights based on their performance such that learning model l_t has weight w_t. For a class c_j the total weight can be given by the sum total of the product of individual weights of the learning models and respective decisions given by them. The final output is the class with the highest weighted vote. Therefore the final output is class J according to the assumption if Eq. (10.1) holds true:

$$J = argmax_{j=1}^{C} \sum_{t=1}^{T} w_t d_{t,j},$$ (10.1)

where J is the class assigned to instance X_i, T is the member of classifiers in the ensemble, and C is the number of classes.

10.2.2.2 Combining continuous outputs

Instead of deciding the class label as output, some learning models may output the probability with which an instance belongs to a class. This probability can be thought of as the degree of support shown by the learning model towards a class.

Assume there are C classifiers in the system. For the instance x_i, $s_{c,j}(x_i)$ is the support received by the jth class from the cth classifier. w_j is the weight of the jth classifier and $\mu_j(x_i)$ is the total support for the jth class for instance x_i. One can combine the decisions in the following ways.

1. *Sum rule*: According to this rule, the individual supports from all the learning models are added to obtain the final support for a particular class as shown in Eq. (10.2). The final output of the ensemble is the class with the highest support:

$$\mu_j(x_i) = \sum_{c=1}^{C} s_{c,j}(x_i). \tag{10.2}$$

2. *Mean rule*: According to this rule, after adding the individual supports from all the learning models, the total sum is normalized by the total number of learning models $(\frac{1}{C})$ as shown in Eq. (10.3):

$$\mu_j(x_i) = \frac{1}{C} \sum_{c=1}^{C} s_{c,j}(x_i). \tag{10.3}$$

3. *Weighted sum rule*: Each learning model is assigned a weight and the total support is the total sum of the product of the learning model's weights and their supports as shown in Eq. (10.4):

$$\mu_j(x_i) = \sum_{c=1}^{C} w_t s_{c,j}(x_i). \tag{10.4}$$

4. *Product rule*: According to this rule, for a particular class the supports provided by the learning models are multiplied to obtain the final output:

$$\mu_j(x_i) = \prod_{c=1}^{C} s_{c,j}(x_i). \tag{10.5}$$

5. *Maximum rule*: According to this rule, for a particular class, the maximum support given by the participating learning models is selected as shown in Eq. (10.6):

$$\mu_j(x_i) = \max_{c=1}^{C} s_{c,j}(x_i). \tag{10.6}$$

6. *Minimum rule*: As the name suggests, for a particular class this rule selects the minimum support given by the participating learning models as shown in Eq. (10.7):

$$\mu_j(x_i) = \min_{c=1}^{C} s_{c,j}(x_i).$$ (10.7)

7. *Generalized mean rule*: A generalized mean rule is given in Eq. (10.8):

$$\mu_{j,n}(x_i) = \left[\frac{1}{C} \sum_{c=1}^{C} s_{c,j}(x_i)^n\right]^{\frac{1}{n}},$$ (10.8)

where n is a positive integer.

10.2.2.3 Consensus clustering

Consensus clustering is an ensemble-clustering technique that combines multiple clustering solutions to produce a more robust and accurate clustering result. It does this by calculating the level of agreement or similarity between the individual clustering solutions and combining them to maximize the agreement between the solutions. The consensus function is used to measure the agreement between two clustering solutions. Given two clustering solutions, A and B, the consensus function $c(A, B)$ measures the degree of similarity between A and B. A high value of $c(A, B)$ indicates a high degree of similarity between A and B, while a low value indicates dissimilarity. Several consensus functions can be used, including the Jaccard coefficient, adjusted Rand index, and normalized mutual information.

Once the consensus function is defined, the consensus matrix is constructed by calculating the pairwise consensus values between all pairs of clustering solutions. The consensus matrix C is an $n \times n$ matrix, where n is the number of clustering solutions. The diagonal elements of C are set to one, as the consensus between a clustering solution and itself is always perfect. The offdiagonal elements of C are the consensus values between pairs of clustering solutions.

Suppose we have three clustering solutions A, B, and C for a dataset containing four objects $\{o_1, o_2, o_3, o_4\}$. Assume that the clustering solutions are:

$$A : \{\{o_1, o_2\}, \{o_3\}, \{o_4\}\}, B : \{\{o_1, o_3\}, \{o_2, o_4\}\}, C : \{\{o_1, o_4\}, \{o_2\}, \{o_3\}\}.$$

The consensus matrix is then calculated by applying the consensus function to all pairs of clustering solutions. For example, the consensus value between A and B is 0.33, as only one cluster is the same between A and B ($\{o_3\}$). We can calculate the complete consensus matrix, C, as follows.

$$
\begin{array}{c c c c}
 & A & B & C \\
A & \begin{pmatrix} 1.00 \\ 0.33 \\ 0.00 \end{pmatrix} & \begin{matrix} 0.33 \\ 1.00 \\ 0.33 \end{matrix} & \begin{pmatrix} 0.00 \\ 0.33 \\ 1.00 \end{pmatrix}
\end{array}
$$

$$
\begin{array}{cc}
 & \begin{matrix} A & B & C \end{matrix} \\
\begin{matrix} A \\ B \\ C \end{matrix} & \begin{pmatrix} 1.00 & 0.33 & 0.00 \\ 0.33 & 1.00 & 0.33 \\ 0.00 & 0.33 & 1.00 \end{pmatrix}
\end{array}.
$$

Once the consensus matrix is obtained, various clustering algorithms can be applied to obtain a consensus clustering solution. For example, hierarchical clustering can produce a dendrogram that groups together clustering solutions with high consensus values.

Consensus clustering can be helpful in many applications with uncertainty or noise in the data. Combining multiple clustering solutions can produce a more robust and accurate clustering result that is less sensitive to the choice of clustering algorithm or parameter settings.

10.3 Supervised ensemble learning

Supervised ensemble learning uses existing knowledge to classify new instances into respective classes (categories). Let us assume a set of input instances $x_1, x_2, x_3, \cdots, x_n$ such that each $x_i \in X$. Each x_i is associated with y and is characterized by a feature set F such that $f_1, f_2, f_3, \cdots, f_n \in F$. X and y are related through a function f in such a way that $y = f(X)$. The x_is are now fed to the learning algorithm with their respective ys and are called training instances. This process yields a prediction model or an inducer. The primary aim of the learning algorithm is to learn a hypothesis h such that h is the best possible approximation of the function f [5]. The model attempts to learn the regions in the feature space for the instances. Once the training phase is complete, the model is tested with previously unseen instances called the Test set.

Individual classifier models, however, sometimes may not be successful in truly learning the feature space as they may miss some local regions. Hence, they will constantly misclassify instances from such regions [11]. That is why ensemble learning takes the decision of a set of base models to reach a conclusion. Each base model has its own strengths and weaknesses. The outputs of base classifier B_i^ss are combined using a combination function f_s to generate the best possible outcome.

10.3.1 Requirements for base-classifier selection

There are two primary requirements on how to select the base classifiers (learners).

1. *Diverse nature:* Classifiers based on a similar learning technique will tend to make similar or the same kind of mistakes. In other words, if one of the classifiers overlooks an inherent pattern in the data, there is a high probability that a similar classifier will also overlook that pattern. Such patterns may correspond to a distinguishing or interesting characteristic, in which case the classifiers will have poor performance. This is why, when creating an ensemble, the classifiers are usually chosen from a diverse set of families [18]. This also makes sure that the outputs of individual models are less correlated with each other, and hence, overall performance is improved.

2. *Accuracy in the classifier performance:* The chosen base learners should be high performers in terms of accuracy individually or at least perform better than a random learner. If there are C classes in a dataset, a random classifier will have an average accu-

racy of $\frac{1}{C}$. On the other hand, a combination of weak learners, each performing worse than a random learner, may not give the desired quality outcomes.

10.3.2 Ensemble methods

Over the years, several ensemble-construction approaches have been developed. Some prominent ones are described next.

(a) Bagging [3]: From the original dataset, samples are chosen at random with replacement to obtain different versions of the same dataset. This is known as a bootstrap technique and the samples are called bootstrap samples. A learning algorithm is then trained with these different sets of samples in parallel to obtain classifier models that assign class labels to the samples. The combination scheme used is majority voting, that is, if the majority of the base learners predict a sample to be of a particular class, then the corresponding label is assigned to it. Fig. 10.3 illustrates an example of Bagging.

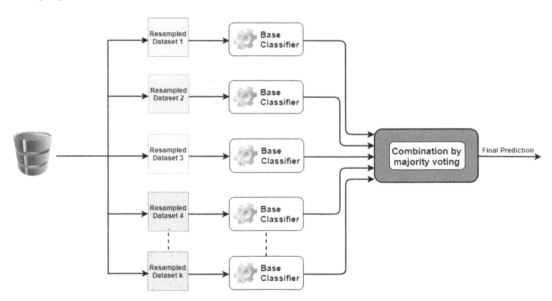

FIGURE 10.3 Working principle of Bagging approach.

(b) Boosting [8]: Boosting is an ensemble approach that works sequentially. It focuses on the misclassified instances in every iteration. In the first iteration, all the instances are assigned equal weights. In subsequent iterations, the misclassified samples of the previous iteration are given more weight, or are said to be boosted. Such a mechanism works well with weak learners as several weak learners can be combined to solve some difficult task. For boosting to work, weak learners must perform better than random learners. *Adaboost* [7] is a popular boosting model with which we could use other learning models such as Decision trees or Naive Bayes as a base learner. Fig. 10.4 illustrates an example of Boosting.

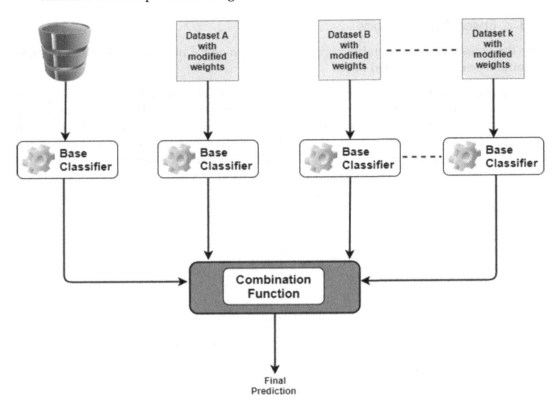

FIGURE 10.4 Typical Boosting steps.

(c) **Stacked Generalization** [16]: This is an ensemble approach where learning takes place in two levels. The layer-1 base classifiers are trained with a level-0 bootstrapped dataset. Their outputs are used as input by the next-level metalearner. As the name suggests, one layer of the dataset and classifier is stacked over another layer of the dataset and metaclassifier. It is called a metaclassifier because it learns from the behavior of a set of classifiers above it. Fig. 10.5 illustrates an example of Stacked Generalization.

In addition to the three ensemble methods discussed above, another method that has gained popularity over the years is the Random Subspace method. In this method, features are randomly sampled so as to construct a set of feature subspaces. On these subspaces, base classifiers are trained and evaluated so that their results can be combined into a final result. It is important to note that, based on the use of prior knowledge, a random subspace method may be supervised or unsupervised.

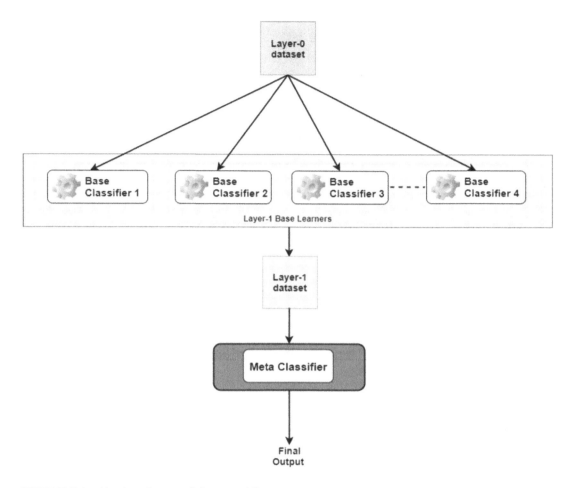

FIGURE 10.5 Stacking-based approach for ensembling.

10.4 Unsupervised ensemble learning

Unsupervised ensemble learning works by generating a set of clustering groups using clustering algorithms and combining these groups to produce a consensus output. These ensemble methods have better performance in terms of accuracy, robustness, and stability when compared to single clustering algorithms because they can make full use of the information provided by the various clustering algorithms. Fig. 10.6 shows an unsupervised ensemble-learning framework.

Different clustering models are combined to create a consensus cluster model that is superior to the individual clustering models. The resultant consensus is less sensitive to

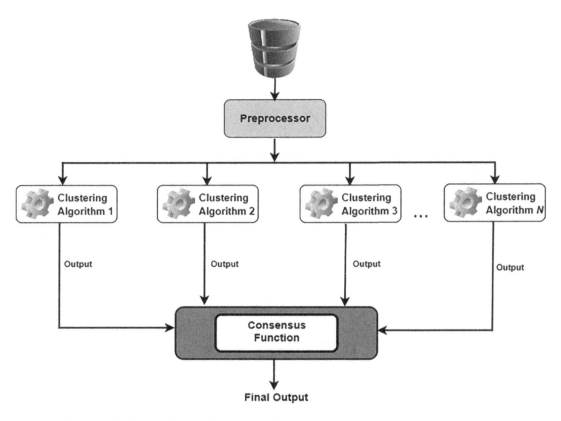

FIGURE 10.6 Unsupervised Ensemble-Learning Framework.

outliers and noise and is more stable and robust. There are two primary stages for creating clustering ensembles, generation and integration [15]. In the first stage, generation, the given dataset is used to train different clustering models and their respective clustered groups are obtained. In the second stage, these clustered groups are integrated into a single consensus clustering to obtain the final result.

For generating the clustering ensemble, different mechanisms may be employed, including resampled instances or varying feature subspaces, different base-clustering models, different hyperparameters to the clustering models, or data projected onto different subspaces [20]. The hardest is the integration stage, i.e., integrating the base-clustering models to obtain a consensus. For integration, the most common technique is the voting approach, where for a given instance, a vote is taken from each base-clustering model, and the instance is finally assigned to the cluster that receives the highest votes.

10.5 Semisupervised ensemble learning

In recent years, semisupervised ensemble learning has garnered tremendous attention. In semisupervised ensemble learning, the training set is expanded to incorporate previously unlabeled instances. New patterns and informative data can be introduced to the existing distribution. The main issue in many application domains is the unavailability of labeled data. Often only a small quantity of labeled data is available, it being the main motivation for proposing most semisupervised methods in various domains. It is worth noting that some works in the literature deal with only positive and unlabeled instances although the results are not always good [1]. Nevertheless, whenever the amount of labeled data is insufficient, extensive experimental results demonstrate that semisupervised ensemble learning performs better than the supervised ensemble-learning techniques. Fig. 10.7 describes a semisupervised ensemble-learning framework. Below, we discuss a few popular methods.

A graph-based semisupervised ensemble classification [19] method can effectively handle the underlying structures of high-dimensional data. High-dimensional data are hard to characterize directly by a graph method directly. The method constructs a neighborhood graph in each feature subspace and then, for integration purposes, a semisupervised linear classifier is trained on the learned graph. The authors also proposed a multiobjective subspace selection process to generate the optimal combination of feature subspaces and used unlabeled datasets to generate an additional training set based on sample confidence to improve the performance of the classifier ensemble.

A semisupervised ensemble learning-based feature ranking method has been proposed in [2]. For feature selection, a permutation-based out-of-bag feature importance measure and an ensemble of bagging and random subspaces methods are used. The Apriori algorithm, which is a frequent itemset mining technique, has been used to work with a small number of positive samples and a large number of unlabeled samples [1]. Negative feature set construction and labeling the unlabeled data are two significant steps in the system, called FIEL (Frequent itemset and ensemble Learning), shown in Fig. 10.8. In addition, an ensemble of Naive Bayes, LibSVM, and Random Forest was constructed with the majority voting scheme.

A novel semisupervised algorithm for classification based on a rough set and ensemble learning has been proposed in [13]. For 2-class classification, only positive and unlabeled data are considered (no negatively labeled samples). A rough set-approximation technique was used to approximate the values of negative samples and extract the initial collection of such samples. Next, an ensemble was created with Naive Bayes, SVM, and Rocchio as base classifiers to work on the margins between positive and negative samples in an iterative manner. The approach helped improve the approximation of the negative samples, which in turn helped converge the class boundary.

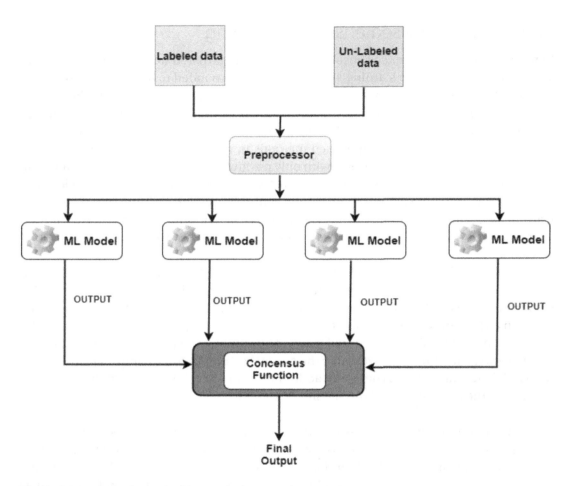

FIGURE 10.7 Semisupervised Ensemble-Learning Framework.

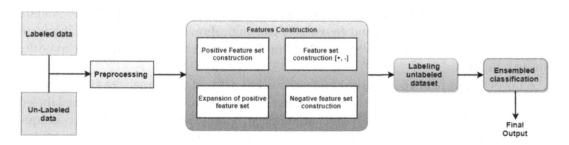

FIGURE 10.8 Frequent itemset and ensemble-Learning framework.

10.6 Issues and challenges

Despite significant advancement in ensemble learners, a few open issues are yet to be addressed. We present a few such issues below:

1. How do the learners work when there is a mixture of class labels and continuous outputs that need to be investigated? In such a case, what would be a useful combination function needs to be explored?
2. The problem of how to reach a fair decision in a high-dimensional dataset needs further discussion. Some learners have an inbuilt capability for dimensionality reduction, which the other learners participating in the ensemble may not have. In this case, how to reach a fair decision is an open question.
3. In the case of weighted majority combination, how to decide the weights is not clear.
4. How to handle mixed data also remains an issue. Data may be categorical as well as numerical. In the case of categorical features, there has to be exact matching, and for numerical, there has to be thresholding.

10.7 Summary

Ensemble learning is a relatively well-studied area in machine learning. Ensembles are flexible in structure and composition. An ensemble can be combined with other approaches in a seamless manner with some effort. It can be used not only with traditional machine-learning methods like supervised and unsupervised learning but also with more advanced concepts like deep learning and transfer learning. An ensemble for supervised learning can be constructed at different levels, for example at the sample level, or feature level or even the output level. Most of the research in the literature is focused on building ensembles at the feature level and at the output level. For unsupervised ensemble learning, two main stages are considered; first, the generation of clustering models and then the integration stage. In the literature, the construction of both stages has been given equal importance. Semisupervised learning, however, is a little different from the other two as labeled data are scarcely available for many tasks. It can be thought of as a more data-centric approach compared to the other two.

References

[1] Ishtiaq Ahmed, Rahman Ali, Donghai Guan, Young-Koo Lee, Sungyoung Lee, TaeChoong Chung, Semi-supervised learning using frequent itemset and ensemble learning for SMS classification, Expert Systems with Applications 42 (3) (2015) 1065–1073.
[2] Fazia Bellal, Haytham Elghazel, Alex Aussem, A semi-supervised feature ranking method with ensemble learning, Pattern Recognition Letters 33 (10) (2012) 1426–1433.
[3] Leo Breiman, Bias, variance, and arcing classifiers, Technical report, Tech. Rep. 460, Statistics Department, University of California, Berkeley, 1996.
[4] Thomas G. Dietterich, Machine-learning research, AI Magazine 18 (4) (1997) 97.
[5] Thomas G. Dietterich, et al., Ensemble learning, in: The Handbook of Brain Theory and Neural Networks, vol. 2, 2002, pp. 110–125.

[6] J. Elder, Daryl Pregibon, A statistical perspective on KDD, in: Advances in Knowledge Discovery and Data Mining, 1996, pp. 83–116.

[7] Yoav Freund, Robert Schapire, Naoki Abe, A short introduction to boosting, Journal-Japanese Society For Artificial Intelligence 14 (771–780) (1999) 1612.

[8] Yoav Freund, Robert E. Schapire, et al., Experiments with a new boosting algorithm, in: ICML, vol. 96, Citeseer, 1996, pp. 148–156.

[9] Monica Jha, Swarup Roy, Jugal K. Kalita, Prioritizing disease biomarkers using functional module based network analysis: a multilayer consensus driven scheme, Computers in Biology and Medicine 126 (2020) 104023.

[10] Mehryar Mohri, Afshin Rostamizadeh, Ameet Talwalkar, Foundations of Machine Learning, MIT Press, 2018.

[11] Omer Sagi, Lior Rokach, Ensemble learning: a survey, Wiley Interdisciplinary Reviews: Data Mining and Knowledge Discovery 8 (4) (2018) e1249.

[12] Abdullah Sheneamer, Swarup Roy, Jugal Kalita, A detection framework for semantic code clones and obfuscated code, Expert Systems with Applications 97 (2018) 405–420.

[13] Lei Shi, Xinming Ma, Lei Xi, Qiguo Duan, Jingying Zhao, Rough set and ensemble learning based semi-supervised algorithm for text classification, Expert Systems with Applications 38 (5) (2011) 6300–6306.

[14] Dan Steinberg, P. Colla, Cart—Classification and Regression Trees, Interface Documentation, Salford Systems, San Diego, CA, 1997.

[15] Sandro Vega-Pons, José Ruiz-Shulcloper, A survey of clustering ensemble algorithms, International Journal of Pattern Recognition and Artificial Intelligence 25 (03) (2011) 337–372.

[16] David H. Wolpert, Stacked generalization, Neural Networks 5 (2) (1992) 241–259.

[17] Zhi-Hua Zhou, Zhi-Hua Zhou, Ensemble Learning, Springer, 2021.

[18] Anders Krogh, Jesper Vedelsby, Neural network ensembles, cross validation, and active learning, in: NIPS'94, Denver, Colorado, 1994, pp. 231–238.

[19] Guoxian Yu, Guoji Zhang, Carlotta Domeniconi, Zhiwen Yu, Jane You, Semi-supervised classification based on random subspace dimensionality reduction, Pattern Recognition 45 (3) (2012) 1119–1135, Elsevier.

[20] Reza Ghaemi, Md Nasir Sulaiman, Hamidah Ibrahim, Norwati Mustapha, A survey: clustering ensembles techniques, International Journal of Computer and Information Engineering 3 (2) (2009) 365–374.

11

Association-rule mining

11.1 Introduction

Consider a chain of thriving department stores that are always trying to increase revenue and profit. One idea that is likely to help accelerate overall sales growth is to be able to read the minds of customers, and predict their buying behaviors and moods or buying tendencies. What high-value items are customers purchasing in large amounts? Which items are frequently bought during Sundays, holidays, or during festive seasons? What are other items that a customer purchases together with a specific item? The last query is important from a marketing point of view. For example, business analytics of a grocery store-based analysis of daily transactions for the past year may show that in 90% cases whenever a customer buys powdered milk, the customer also purchases tea packs. With this new finding, the management may immediately take a decision to keep all milk powder and tea packs close to each other to enhance sales. Inspired by this simple idea, Agrawal et al. introduced the concept of association-rule mining in 1993 [1]. Association rules infer potential relationships among purchasable items in a large transaction database in the form of rules that are presented in the form of a cause and effect. A large number of algorithms have been proposed to address the popular and computationally expensive task of association-rule mining. Efforts have been made to improve the computational throughput of the traditional algorithms to handle large transactional databases. There are also attempts to make association-rule mining applicable to different databases. It has been observed that traditional association mining based on a support-confidence framework may sometimes discover statistically nonexistent significant relationships between item pairs. To overcome this weakness, correlation mining has been recently used to provide an alternative framework for finding statistically interesting relationships. This chapter introduces the basic concepts related to association-rule mining. Popular algorithms for frequent itemset mining are illustrated with easy-to-understand examples. Various extensions of traditional algorithms or improvements to them for handling different data types or improving computational efficiency are also discussed in this chapter.

11.2 Association analysis: basic concepts

Association mining is an important method in unsupervised data mining. It plays a vital role in finding interesting patterns in databases, such as association rules, correlations, sequences, episodes, and clusters. The mining of association rules is a popular problem in association mining. The original motivation for searching for association rules came from the need to analyze customer buying behavior from supermarket-transaction data.

Fundamentals of Data Science. https://doi.org/10.1016/B978-0-32-391778-0.00018-1

Starting from market-basket data analysis, association mining is now widely used in applications in domains like machine learning, soft computing, network security, information retrieval, and computational biology. Association-rule mining usually is a two-step process consisting of frequent mining and rule generation. Rule generation is a relatively straightforward task, and researchers mainly concentrate on frequent mining activity.

11.2.1 Market-basket analysis

The concept of association mining was introduced to facilitate market-basket analysis. A market-basket analysis technique models the buying behavior of customers. It is based upon the likelihood of a customer who has purchased certain items, also purchasing another set of items together. For example, if she purchases bread and jam, she is more likely to buy tea bags at the same time. The set of items a customer has collected in her basket for purchase is termed an *itemset*, and market-basket analysis retrieves relationships among itemsets in the basket. An association or relationship between the itemsets is represented as dependency rules of the following form:

$$\{Bread, Jam\} \rightarrow \{Tea\ Bags\}.$$

The interpretation of the above association rule is that if a customer purchases bread and jam, it implies she will possibly purchase teabags also. Market-basket analysis may offer an idea to read the customers' buying patterns. Typically market-basket data are represented from daily transaction databases. For example, the simplistic structure of the transaction data can be seen in Table 11.1.

Table 11.1   Sample Transaction records of a store.

Transaction id	Items Purchased
T1	Milk, Bread, Sugar, Jam, Teabags, Biscuits
T2	Milk, Bread, Jam, Teabags
T3	Milk, Sugar, Jam, Biscuits
T4	Sugar, Coffee, Teabags

The market-basket conversion of the above transaction database is a binary representation of the records with respect to different items per transaction. A transaction is represented as a bit vector with the presence or absence of an item in the transaction. As shown in Table 11.2, item Coffee is denoted as 0 and Milk is 1 to indicate the absence or presence of the items in the T1 record in a market-basket representation of the transaction database.

Market-basket analysis may help management decide to promote sales of any items or goods in the department stores or their chains in different locations. They can even determine the location of the display of an item or set of items inside the store for better promotion of the items. For example, tea bags can be displayed close to the bread or jam display. They may even offer to buy one get one free deal to popularize particular items. If

Table 11.2 Market-basket conversion of transaction records.

	Milk	Bread	Sugar	Teabags	Biscuits	Jam	Coffee
T1	1	1	1	1	1	1	0
T2	1	1	0	1	0	1	0
T3	1	0	1	0	1	1	0
T4	0	0	1	1	0	0	1

purchasers of potatoes are more likely to buy onions, then onions can be placed near the potatoes. Stores may offer promotional sales in the form of buying two kilos of potatoes and getting one kilo of onions free. This may help attract customers to the store itself. For large-scale decision making, a chain of department stores may go for a differential market-basket analysis. Instead of analyzing the customers buying patterns in a particular store, differential market-basket analysis can be extended to different stores for large varieties of customers from different locations and cultures, between different periods or days of the week, different festive seasons in the year, etc. A few useful insights into the analysis may improve company sales. For example, a particular association rule may reveal that it is applicable in the majority of the chain stores, however, there may be deviations in some particular stores in different geographical locations. It may give a clue to the management for better investigation of the deviation in sales patterns for such stores for a set of products. Perhaps, it may give a better idea to organize and display the items in a novel and more lucrative way.

Although Market-Basket Analysis started with an analysis of shopping carts and supermarket shopper's behavior, the applicability of market-basket analysis has been extended to other domains of analysis such as **(i)** Potential credit-card applicants, **(ii)** Mobile-phone calling patterns, **(iii)** ATM fraudulent detection, **(iv)** Network-anomaly detection, and many more.

11.2.2 Sources of market-basket databases

A market basket can easily be created from transactions or other structured data sources. Market baskets can also be synthesized with the help of data generators. One popular data generator is the IBM data generator (www.almaden.ibm.com/cs/quest/). This generator can generate datasets with varying data distributions. As a convention, the datasets are named based on the parameters used in generating data. For example, in the case of "T10I4D100K", "T" means the average number of items per transaction; "I" means the average length of the maximal pattern; and "D" means the total number of transactions in the dataset. Therefore "T10I4D100K" refers to a dataset with the following characteristics: the average number of items per transaction is 10; the average length of the maximal pattern is 4; the total number of transactions is 100 000; and the total number of items is 1000. Another synthetic data generator, ARMiner (http://www.cs.umb.edu/~laur/ARMiner/) follows the basic spirit of the well-known IBM synthetic data generator. A few data generators and sources are listed in Table 11.3.

Table 11.3 Sources of Market-Basket Data for Experimentation.

Tool	Type	Source
IBM Quest Synthetic Data Generator	Data Generator	www.almaden.ibm.com
ARMiner	Data Generator	http://www.cs.umb.edu/~laur/ARMiner
ARTool	Data Generator	http://www.cs.umb.edu/~laur/ARtool
SyntheticDataGenerator	Data Generator	https://synthdatagen.codeplex.com/
FIMI	Real and Synthetic Data source	http://fimi.ua.ac.be/data/

11.2.3 Interestingness measures

Formally, a market-basket database is represented as a matrix $\mathbf{D} \times \mathbf{I}$, where $\mathbf{I} = \{\mathbf{i}_1, \mathbf{i}_2, \cdots, \mathbf{i}_n\}$ is a set of n items and database $\mathbf{D} = \{\mathbf{t}_1, \mathbf{t}_2, \cdots \mathbf{t}_m\}$ is a collection of m transaction records. Each transaction $\mathbf{t}_i = \{\mathbf{i}_1, \mathbf{i}_2, \cdots, \mathbf{i}_n\}, \forall i_j \in I$, is normally a binary vector identified by a unique transaction id.

Definition 11.2.1 (Itemset). A set $\mathbf{P} = \{\mathbf{i}_1, \mathbf{i}_2, \cdots, \mathbf{i}_k\} \subseteq I$ is called an itemset. It is also called a k-itemset or k-size itemset as $\mathbf{P}$ contains k items, i.e., $|\mathbf{P}| = k$.

Definition 11.2.2 (Association Rule). Given a market-basket database $\mathbf{D} \times \mathbf{I}$, an association rule $\mathbf{P} \rightarrow \mathbf{Q}$ is a dependency relationship between the itemsets $\mathbf{P}$ and $\mathbf{Q}$ such that $\mathbf{P} \cap \mathbf{Q} = \phi$ and $\mathbf{P}, \mathbf{Q} \subset \mathbf{I}$. $\mathbf{P}$ is commonly called antecedent and $\mathbf{Q}$ is consequent.

Given a large transaction database, the association-rule mining problem can derive a large number of such rules. All of them may not be important or interesting from a user's perspective. To find highly relevant rules by preventing the generation of unnecessary rules, user-specific interestingness measures can be used. In order to select interesting rules from the set of all possible rules, two best-known constraints are generally used, namely minimum support and minimum confidence.

Definition 11.2.3 (Support). Support of an itemset $\mathbf{P} = \{i_j, i_k\}$ is defined as the percentage of records that contain itemset $\mathbf{P}$ considering the total number $|D|$ of records in the database D.

$$Support(\mathbf{P}) = \frac{SupportCount(i_j \cup i_k)}{|D|}.$$

The measure of support is reflexive in nature, i.e., joint support of itemsets $\mathbf{P}$ and $\mathbf{Q}$, $Support(\mathbf{P}, \mathbf{Q})$ is the same as $Support(\mathbf{Q}, \mathbf{P})$. The $SupportCount$ for the itemset is increased by one every time the itemset is encountered in a different transaction in database D. For example, in our market-basket data, the support count for Milk and Bread together is two, whereas the support count of item Milk alone is three. The support of itemset Milk and Bread is 0.5 or 50%. Support is simply the joint probability of the form $Prob(\mathbf{P} \cup \mathbf{Q})$ in a discrete domain. Support is further used to measure the interestingness of a given rule, called *Confidence*.

Definition 11.2.4 (Confidence). *Confidence* of an association rule **P** → **Q** is defined as the ratio of the number of transactions that contain **P** and **Q** to the total number of records that contain **P**.

$$Confidence(\mathbf{P} \rightarrow \mathbf{Q}) = \frac{Support(\mathbf{P}, \mathbf{Q})}{Support(\mathbf{P})}.$$

Confidence(**P** → **Q**) can also be interpreted as conditional probability, **P**(**Q**|**P**). Unlike Support, Confidence is nonreflexive, i.e., *Confidence*(**P** → **Q**) is not necessarily the same as *Confidence*(**Q** → **P**). It is a measure of the strength of association rules. To find the confidence for the rule {*Bread, Jam*} → {*Tea Bags*} in our example market-basket data, we have to calculate the *Support*(*Bread, Jam, Teabags*) following which we compute *Support*(*Bread, Jam*). This gives the confidence of the above rule as 100%. On the other hand, the confidence of the reverse rule, i.e., {*Tea Bags*} → {*Bread, Jam*}, comes out to be 66%. This happens as in transaction T4; although the Teabags are available; however, Bread and Jam do not appear together. As a result, it penalizes the overall confidence in the rule.

The goal of association-rule mining is to discover all frequent itemsets and dependency rules that satisfy the predefined threshold *minimum support* and *confidence* (more often denoted in short as **minsup** and **minconf**) for a given database.

Definition 11.2.5 (Frequent Itemset). An itemset **P** is a frequent itemset, if **Support(P)** > **minsup**.

A rule that satisfies both a minimum support threshold and a minimum confidence threshold is called a **Strong Rule**. Taking **minsup** and **minconf** as 80% and 100%, respectively, the rule {*Bread, Jam*} → {*Tea Bags*} appears to be a Strong rule; however, the reverse rule {*Tea Bags*} → {*Bread, Jam*} is not. The illustration of the overall concept is shown with the help of another similar example in Fig. 11.1.

In addition to the above interestingness measures, a few more measures have been proposed (but rarely used) to evaluate the validity of an association rule. Quality or interestingness measures are summarized in Table 11.4.

11.2.4 Association-rule mining process

The overall process of mining association rules from market-basket data is usually decomposed into two subproblems:

i. The first subproblem is to find frequent itemsets whose occurrence frequency exceeds a predefined threshold (Support) in the database. Such itemsets are also called large itemsets.
ii. The second subproblem is to generate association rules from large itemsets with the constraint of minimal confidence.

Since the second subproblem is quite straightforward, most research focuses on the first subproblem. The first subproblem can be further divided into two subproblems: Generat-

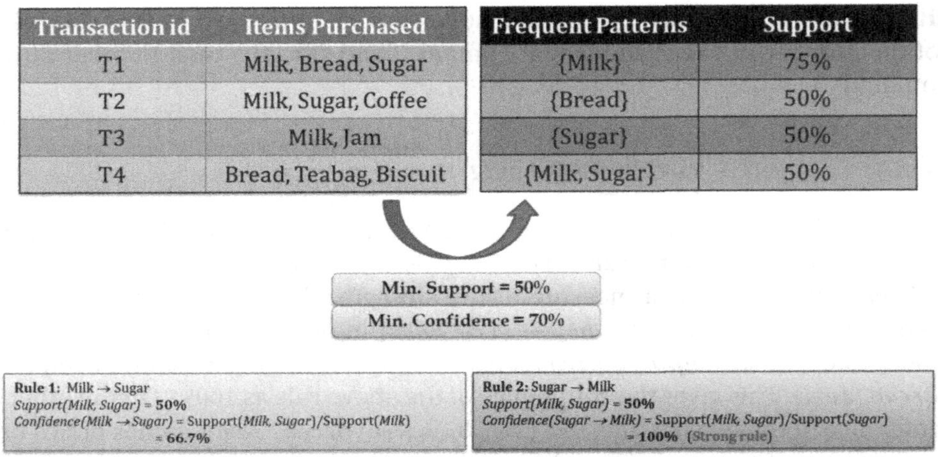

Transaction id	Items Purchased		Frequent Patterns	Support
T1	Milk, Bread, Sugar		{Milk}	75%
T2	Milk, Sugar, Coffee		{Bread}	50%
T3	Milk, Jam		{Sugar}	50%
T4	Bread, Teabag, Biscuit		{Milk, Sugar}	50%

Min. Support = 50%
Min. Confidence = 70%

Rule 1: Milk → Sugar
Support(Milk, Sugar) = **50%**
Confidence(Milk → Sugar) = Support(Milk, Sugar)/Support(Milk)
= **66.7%**

Rule 2: Sugar → Milk
Support(Milk, Sugar) = **50%**
Confidence(Sugar → Milk) = Support(Milk, Sugar)/Support(Sugar)
= **100%** (Strong rule)

FIGURE 11.1 Association-rule mining: an illustration.

Table 11.4 Different interestingness measures used to validate the quality of predicted association rules.

Measures	Equation	Range of values	Score	Significance
Lift (P → Q)	$\dfrac{Support(P,Q)}{Support(P) \times Support(Q)}$	$[0, +\infty)$	1	P, Q independent
			< 1	P, Q negatively dependent
			> 1	P, Q positively dependent
Conviction (P → Q)	$\dfrac{1 - Support(Q)}{1 - Conviction(P \to Q)}$	$(0, +\infty)$	1	P, Q independent
			< 1	P, Q negatively dependent
			> 1	P, Q positively dependent
Leverage (P → Q)	$Support(P,Q) -$ $Support(P) \times Support(Q)$	$[-0.25, 0.25]$	0	P, Q independent
			< 0	P, Q negatively dependent
			> 0	P, Q positively dependent
Gain (P → Q)	$Conviction(P \to Q) -$ $Support(Q)$	$[-0.5, 1]$	Monotonic	The higher the score the higher will be the Gain
Accuracy (P → Q)	$Support(P,Q) +$ $Support(\to P, \to Q)$	$[0, 1]$	Monotonic	The higher the score the higher will be the Accuracy

Note: $Support(\to P, \to Q)$ represents negative support of P and Q. It can be represented as (1- Support(P)-Support(Q)+Support (P,Q)).

ing candidate sets and extracting frequent itemsets. A diagrammatic representation of the association-mining technique is shown in Fig. 11.2.

11.3 Frequent itemset-mining algorithms

Finding all frequent itemsets in a database is a difficult and computationally expensive task since it involves searching for all possible combinations of itemsets. For a database with n items, there are a total of $(2^n - 1)$ item subsets to be explored for finding frequent item-

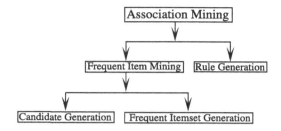

FIGURE 11.2 Major steps in association-rule mining.

sets. For a large database, exhaustively generating all such item subsets, called *candidate sets*, is impossible. Most frequent itemset-mining algorithms concentrate on improving the computational efficiency of the itemset-generation process by adopting a few heuristic alternatives to reduce and ignore a few item subsets to be generated during the process.

Popular association-mining algorithms consider the size of databases to be semilarge so that they can usually be accommodated in the main memory. They are static in nature and are sometimes referred to as *sequential association mining*.

The AIS (Agrawal, Imielinski, and Swami) algorithm [1] is the first algorithm proposed for mining-association rules. The main drawback of the AIS algorithm is that it generates too many candidate itemsets, out of which all are not actually frequent. Moreover, AIS requires too many passes over the whole database, which is considered the most expensive task in association mining.

11.3.1 Apriori algorithm

Among the popular algorithms for finding large itemsets, the Apriori algorithm [3] stands at the top because of its simplicity and effectiveness. Apriori exploits an itemset lattice to generate all possible subsets of items. An example lattice structure based on our market-basket dataset is shown in Fig. 11.3. Instead of generating all possible subsets in the lattice, the Apriori algorithm adopts two itemset-closure properties to avoid the generation of unnecessary subsets, which are not going to be frequent, and finally, based on a user-defined minimum support threshold.

A. Downward-Closure Property: If an itemset **X** is frequent with respect to user-defined threshold *minsup*, then all of its subsets also become frequent, i.e., $\forall \mathbf{Y} \subset \mathbf{X}$ will satisfy frequent itemset constraint if **X** is frequent.

B. Upward-Closure Property: If an itemset **X** is not frequent (infrequent) with respect to user-defined threshold *minsup*, then its superset cannot be frequent.

During candidate generation, Apriori avoids generating all those higher-order candidates whose subsets are already infrequent. In this way, it prevents a large amount of computational expenses required to generate candidate sets and count their frequencies of occurrence in the database. Apriori performs level-wise generation of frequent itemsets

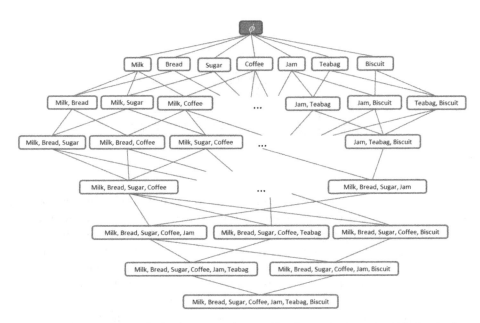

FIGURE 11.3 Itemset lattice of example database. Some intermediate item subsets are kept hidden due to space constraints.

following the lattice structure. It first generates frequent itemsets of size 1. It continues the process for k levels until it is not possible to further generate candidates of size $(k + 1)$.

To generate frequent itemsets from our transaction-database example given above, Apriori first generates all the one-element candidate subsets. Next, it computes the support or frequencies of occurrence for all the candidate sets. Considering $minsup$=50%, Apriori eliminates all infrequent itemsets $\{Jam, Teabag, and\ Biscuit\}$ marked in red color in Fig. 11.4. All itemsets that satisfy $minsup$ constraint are considered frequent itemsets of level 1. Next, it generates all the candidates of size 2. While generating candidates, it follows the upward-closure property and does not generate candidate sets any of whose subsets are infrequent. Before calculating the frequency count of a candidate set, Apriori checks whether any of the subsets of the candidate is infrequent. If any of the subsets are found to be infrequent, Apriori removes the candidate from the potential candidate list. In our example, Apriori never generates any candidates involving item Jam, Teabag, or Biscuit. It continues the process and stops after receiving frequent itemset $\{Milk, Sugar\}$. After that, it is not possible to generate any candidate sets at a higher level, and hence Apriori stops. The collection of all the frequent itemsets is the outcome of the Apriori algorithm.

11.3.2 Candidate generation

To generate a candidate set of size k, the algorithm considers only subsets of size $(k - 1)$. If $(k - 2)$ items are common in both the frequent itemsets, then a candidate of size k will be

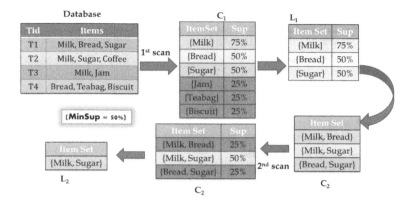

FIGURE 11.4 Frequent Itemset-generation process of Apriori.

generated by concatenating these common $(k-2)$ items with the $(k-1)$th items from both participating itemsets. In the lattice given in Fig. 11.3, candidate set {**Milk, Bread**, Sugar, Coffee} can be generated by concatenating itemsets {**Milk, Bread**, Sugar} and {**Milk, Bread**, Coffee}. This is because {**Milk, Bread**} is common in both the itemsets. A new candidate is generated by adding Sugar and Coffee from both the sets with common items {**Milk, Bread**}.

11.3.3 Cost of the Apriori algorithm

The number of whole database scans required by any algorithm is the major concern when computing the cost of any frequent itemset-mining algorithms. The original Apriori algorithm uses an efficient hash-based technique to compute the support counts of all the itemsets in level k, i.e., support counts for all the itemsets of size k. If for any database the maximum frequent itemset size is m, then Apriori needs m database scans. For our example given in Fig. 11.4, if the maximum itemset, say $\{Milk, Bread, Sugar, Coffee, Jam, Teabag, Biscuit\}$ which is of size 7, then Apriori requires a total of 7 rounds of database scans to compute all frequent itemsets. With the increase in the size of the database, the cost increases drastically. The performance of Apriori also depends on the choice of *minSup* values as with a low value of *minSup*, the number of frequent itemsets is higher, and it makes Apriori computationally expensive. The graph in Fig. 11.5 gives an idea regarding performance variations of Apriori with respect to varying *MinSup* thresholds for different market-basket databases.

Apriori has a number of drawbacks. As it scans the database during each iteration to produce the large itemsets, it is not easily scalable with the size of the database. Only a small percentage of the many candidate itemsets it generates is actually frequent itemsets. As a result, a very high ratio develops between the quantity of potential huge itemsets and the quantity of real frequent itemsets.

Algorithm 11.1: Apriori Algorithm.

Data: D =Database; σ = Minimum threshold
Result: F=Frequent Itemset

1   $C_1 := \{\{i\}|i \in I\}$;
2   $k := 1$;
3 **while** $C_k \neq \phi\}$ **do**
     `// Compute the supports of all candidate itemsets`
4   **for** $\forall\ transactions$(Tid, I) $\in$ D **do**
5   **for** $\forall\ candidate\ itemsets$ $X \in C_k$ **do**
6   **if** $(X \subseteq I)$ **then**
7         $X.support$ $++$;
8   **end**
9   **end**
10   **end**
     `// Extract all frequent itemsets`
11    $F_k := \{X|X.support > \sigma\}$;
     `// Generate new candidate itemsets`
12   **for** $\forall X, Y \in F_k$, $X[i] = Y[i]$, $for\ 1 \leq i \leq k - 1, and$ $X[k] < Y[k]$ **do**
13     $I = X \cup \{Y[k]\}$;
14   **if** $(\forall J \subset I, |J| = k : J \in F_k)$ **then**
15       $C_{k+1} := C_{k+1} \cup I$;
16   **end**
17   **end**
18   k++ ;
19 **end**

In the next section, we discuss an improved algorithm that eliminates the need for candidate generation, always requiring a constant number of and it is only two.

11.3.4 FP-growth

Frequent itemset mining using Apriori-like algorithms becomes expensive as the number of items increases. A transaction database with N items requires approximately 2^N candidates in total. The situation worsens when generating two-element itemsets.It requires $\binom{N}{2}$ candidates to be stored in the memory. In addition, with the increase in the number of levels, the number of potential frequent candidates decreases drastically. The closure properties used by Apriori help cut down a large number of nonpotential itemsets. The alternative is to avoid the candidate-generation phase while mining frequent itemsets.

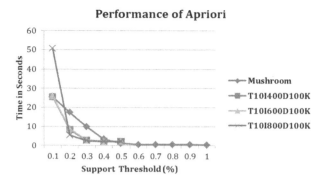

FIGURE 11.5 Running time for Apriori with varying supports.[1] Synthetic dataset generated using IBM synthetic data generator.[2] Real dataset Mushroom is collected from UCI data repository.

The Frequent Pattern Growth (FP-growth) [13] algorithm mines all frequent itemsets without candidate generation. Interestingly, it needs just two scans over the entire database to compute the frequent itemsets. Han et al. proposed a prefix tree-based method to create a compressed representation of the whole database into a Frequent Pattern Tree. Each path in the prefix tree represents a set of transactions that share the same prefix. (See Fig. 11.6.)

The method first performs a single scan of the database to identify the frequently occurring one-element items. Items in transactions are sorted in descending order of frequency, and infrequent items are deleted from the database. The least-frequent item is then chosen from all transactions and deleted from the transactions, leaving a smaller (projected) database. To identify common itemsets, this projected database is processed. The eliminated item is undoubtedly a prefix of all the frequent itemsets. The item is then deleted from the database, and the process is repeated with the following least-frequent item. It should be noted that the FP-tree contains all the data regarding the transactions and frequent itemsets. The tree must therefore be explored in order to obtain any information about the transactions and frequently used itemsets. One of the quickest methods for finding frequent itemsets is FP-growth, which is also reliable enough to locate all possible frequent itemsets. Despite the algorithm's many benefits, it also has two major drawbacks. (i) When the number of items is large, the FP-tree construction process takes a long time. (ii) Due to the large size of the tree, it eventually becomes almost identical to Apriori as minimal support is reduced and performance suffers. (See Fig. 11.7.)

Following FP-Growth, many attempts have been made to overcome the bottleneck of expensive database scans to generate candidate sets in a large-size transaction database. Several approaches concentrate on how to generate candidate itemsets in a single pass or avoid the step.

[1] Execution Environment: Java, Pentium-IV -2.1 GHz machine, 256 MB primary memory, Windows Xp environment.

[2] http://www.almaden.ibm.com/cs/quest/syndata.html.

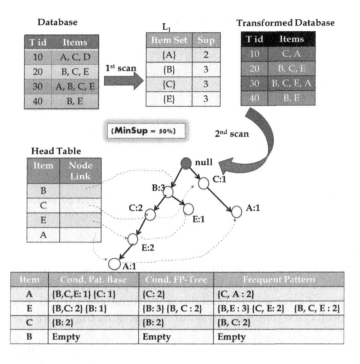

FIGURE 11.6 A two-step process of frequent-itemset generation using FP-Growth.

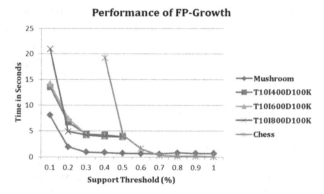

FIGURE 11.7 Time consumed by FP-Growth with respect to varying minimum supports for large transaction databases.[3] Synthetic dataset generated using IBM synthetic data generator.[4] Real dataset such as Mushroom and Chess are collected from UCI data repository.

[3] Execution Environment: Java, Pentium-IV -2.1 GHz machine, 256 MB primary memory, Windows Xp environment.

[4] http://www.almaden.ibm.com/cs/quest/syndata.html.

Inverted Matrix [9] is a disk-based approach evaluated with around 25 million transactions with 100,000 unique items. It uses a new database layout called Inverted Matrix and a relatively small independent tree called COFI-Tree is built summarizing cooccurrences. A simple and nonrecursive mining process is applied to reduce the memory requirements for candidate generation and counting. OPAM [25] finds frequent itemsets without generating any candidate sets. It uses a data structure called the correlogram matrix and a vertical database layout. OPAM adopts an integrated approach to solve the frequent-itemset finding problem in a single pass over the database.

11.4 Association mining in quantitative data

Market-based datasets are normally boolean in nature. The traditional association-mining techniques discussed above are limited to transactional market-basket analysis. However, most relational or other databases like medical, stock-market or biological databases contain attributes that are quantitative or categorical in nature. Traditional approaches are inadequate in finding effective rules from them. Hence, discovering meaningful association rules from such databases has become an important issue in learning patterns, understanding associations and thus predicting events. Therefore the conventional ARM techniques need to be overhauled. Hence, a new association-rule mining paradigm called Quantitative Association Rule Mining has come into the picture. The concept of Quantitative AM was introduced by Srikant et al. [1].

A Quantitative Association Rule (QAR) is a type of $X \Rightarrow Y$ implication that is created on the fly. QARs are similar to standard association rules, but they differ in that they can include both quantitative and qualitative/categorical features as antecedents or consequents. Unlike traditional ARs, QARs can also express negative connections, which show how the absence of certain attributes might affect the presence or absence of others in a database. These negative QARs can have negated items in the antecedent, the consequent, or both, and they are denoted by the negation symbol (¬). According to Alata et al. [4], negative QARs are useful for showing the negative relationships between attributes in a database.

Table 11.5 • Sample relational database.

Rec. id	Age	Married	NumCars
100	23	No	1
200	25	Yes	1
300	29	No	0
400	34	Yes	2
500	38	Yes	2

Given a sample relational database (Table 11.5), a possible quantitative association rule based on certain interestingness measures may specify that if a person age is between 30 to 39 years of age and he/she is already married, then he/she may have 2 cars.

$$(Age : [30\cdots39]) \ \& \ (Married : Yes) \Rightarrow (NumCars : 2).$$

This Quantitative Association Rule (QAR) involves three attributes, namely *Age*, *Num-Cars*, and *Married*, which are both quantitative and categorical. Specifically, *Age* and *Num-Cars* are quantitative, while *Married* is categorical, with possible categories being Yes and No. This QAR illustrates a positive association between these attributes. On the other hand, a negative QAR can also be constructed with negation present in the antecedent, and this is fairly easy to understand.

$$(Age : \neg[20\cdots29]) \ \& \ (Married : Yes) \Rightarrow (NumCars : 2).$$

11.4.1 Partitioning approach

In their groundbreaking research, Srikant and Agrawal [27] proposed a partitioning technique that can be used to convert quantitative attributes into boolean attributes. This approach involves breaking up a wide range of quantitative data into nonoverlapping intervals, and then assigning each (*attribute, interval*) combination to a boolean attribute, which is treated as a market-basket item. These partitions are then used to mine association rules. For instance, the sample database presented in Table 11.5 can be transformed into market-basket data, as demonstrated in Table 11.6.

Table 11.6 Market-basket representation of relational database.

Rec. id	Age: 20..29	Age: 30..39	Married: Yes	Married: No	NumCars: 0	NumCars: 1	NumCars: 2
100	1	0	0	1	0	1	0
200	1	0	1	0	0	1	0
300	1	0	0	1	1	0	0
400	0	1	1	0	0	0	1
500	0	1	1	0	0	0	1

From the above table, a rule of the form $(NumCars : 0) \Rightarrow (Married : No)$ having 100% confidence can be generated. However, if we had partitioned the attribute *NumCars* into intervals, $NumCars : 0$ and $NumCars : 1$ end up in the same partition $NumCars[0...1]$, then the detected rule is $NumCars : (0...1) \Rightarrow (Married : No)$, which only has 66.6% confidence. This signifies information loss. The loss of information caused by this technique becomes more pronounced as the size of the intervals increases. As a result, some Quantitative Association Rules (QARs) that depend on these attributes might be missed due to a lack of sufficient confidence. Additionally, the number of rules produced can be exceedingly large, which can lead to a time-consuming and computationally intensive rule-mining process.

11.4.2 Clustering approach

Clustering presents an effective approach for discovering association rules in quantitative databases by identifying meaningful quantitative regions. In particular, DRMiner [20] is a QARM method that employs the concept of *density* to handle quantitative attributes.

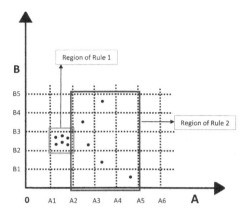

FIGURE 11.8 Mapping of transactions into 2D space where data points exhibit a dense-regions-in-sparse-regions property.

DRMiner is capable of detecting positive multidimensional association rules while avoiding trivial and redundant rules through the density measure. By mapping a database to a multidimensional space, the data points display the property of having dense regions in sparse regions. Therefore the problem of mining Quantitative Association Rules can be reformulated as the task of identifying regions with sufficient density and then mapping these dense regions to QA rules.

Assuming two quantitative attributes, A and B, each transaction in a database is mapped onto a two-dimensional space, as depicted in Fig. 11.8. The primary objective is to identify all association rules of the form $A \subseteq [x_1, x_2]) \Rightarrow B \subseteq [y_1, y_2]$, where $x_1, x_2 \in 0, A_1, A_2, A_3, A_4, A_5, A_6$ with $x_2 > x_1$ and $y_1, y_2 \in 0, B_1, B_2, B_3, B_4, B_5$ with $y_2 > y_1$. By setting the support threshold to 5 and the confidence threshold to 50%, the following rules can be generated:

1. $A[A_1, A_2] \Rightarrow B[B_2, B_3]$;
2. $A[A_2, A_5] \Rightarrow B[0, B_5]$;
3. $A[0, A_6] \Rightarrow B[0, B_5]$.

To ensure a meaningful rule set, it is important to avoid generating useless or redundant rules. When a range meets the minimum support requirement, any larger range containing it also meets the threshold, making it likely to generate many useless rules. Similarly, if the ranges in both dimensions are enlarged, the new ranges may also satisfy the confidence requirement, resulting in many redundant rules. Such rules are referred to as *trivial rules*. Rule (3) in the example covers all possible values of both attributes and is therefore useless, while Rule (2) covers almost all possible values, making it trivial. Redundant rules can be avoided by ensuring that each rule adds new information to the rule set.

Numerous clustering-based approaches have been proposed after DBMiner. DBMiner's [11] method scales well for mining quantitative association rules in high-dimensional databases using the concept of *density connected*. The N-dimensional search space is divided into multiple grids by partitioning each attribute equally, without any overlap between the grids. The high-density cells that are connected are merged to create clusters. Based on the dense grids concept, Junrui et al. [14] proposed another approach called MQAR (Mining Quantitative Association Rule), which solves the support-confidence conflict and eliminates noise and redundant rules. MQAR uses a DGFP-tree similar to an FP-tree [13] to mine association rules in high-dimensional databases.

The clustering-based approach primarily aims to generate intervals reasonably by utilizing density variations or dense regions. They attempt to resolve the support-confidence conflict and eliminate redundant and useless rules. Although the simpler clustering techniques are uncomplicated, the more complex ones such as [20], [11], and [14] use concepts such as hyperplanes and dense grids. While not all clustering techniques are scalable for high-dimensional data, most techniques can produce both single- and multidimensional QARs. The majority of these techniques employ minimum-support and minimum-confidence thresholds, but some may require additional user-specified thresholds. However, the generated rules by these techniques are generally positive.

11.4.3 Information-theoretic approach

The combination of quantitative attributes and their value intervals always gives rise to the generation of an explosively large number of itemsets, thereby degrading mining efficiency [17]. Moreover, the task of combining the consecutive intervals of a quantitative attribute to gain support and more meaningful intervals leads to another combinatorial explosion problem. Such combinatorial explosion can be handled using the information-theoretic approach.

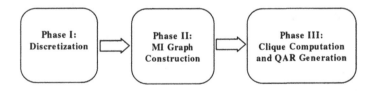

FIGURE 11.9 Three phases of MIC framework.

The MIC (Mutual Information and Clique) framework works in three phases (Fig. 11.9). First, the framework calculates the MI score between each pair of attributes and then it creates a Mutual Information Graph (MI Graph) representing attributes that have strong informative relationships w.r.t. a predefined minimum threshold. Next, each frequent itemset is represented by a Clique in the MI Graph. The cliques in the MI graph are used in the final phase to facilitate the computation of frequent itemsets. Hence, instead of joining the intervals for all attribute sets in the database, one only needs to focus on those attribute sets that form a clique in the MI graph. Thus both the number of attribute sets and the intervals to be joined are significantly reduced.

Table 11.7   Summary of QAR Mining Approaches.

Approach	Advantages	Disadvantages	Rule Dimension	Use Sup. & Conf.	Other Thresholds	Discretization	No. of Scans	(-)ve Rules
Partitioning [6,8,10,19, 27]	Easy to understand and implement	Information loss, many rules, redundant rules, Sup-Conf Conflict	Single	✓	✓	✓	Multiple	✗
Clustering [11,14,20]	Reasonable interval generation (strive to obtain best clusters as intervals)	Often lack high-dimensional scalability, Require many thresholds	Multiple	✓	✓	✓	Multiple	✗
Statistical [5,15]	No misleading rules, Generate subrules	Larger database scans, Uneven distributions appear in partitions	Multiple	✓	✗	✓	Multiple	✗
Fuzzy [30], [12,31]	Interpret intervals as fuzzy terms, Discover rules with crisp values	Require fuzzy thresholds, Execution time increases in the case of higher dimensions	Multiple	✓	✓	✓	Multiple	✗
Evolutionary [4,16], [22]	Mine +ve & -ve rules without discretization, Optimize Support and Confidence, Maximize rule comprehensibility	Comparatively difficult to implement, higher computational cost	Multiple	✗	✓	✗	Multiple	✓
Information-Theoretic [17]	Reduce combinatorial explosion of intervals that appear in most partitioning techniques	Not effective in smaller databases	Multiple	✓	✗	✓	Multiple	✗

In addition to the above approaches, several other techniques explore alternative measures for quantitative rule mining. In Table 11.7, various approaches are listed for quick reference.

Each technique for QAR mining has its advantages and disadvantages. While different methods handle quantitative data differently, the use of support and confidence to measure rule interestingness is prevalent in most approaches. Some modern QARM techniques incorporate newer measures for evaluating rules, as described in Martinez's analysis and selection studies [23,24]. However, several significant issues remain to be addressed in the future, such as the inability to mine both positive and negative QARs, the failure to efficiently generate multidimensional QARs, the dependence on selecting appropriate thresholds to generate rules, the generation of redundant, uninteresting, and misleading rules, poor scalability of mining techniques with respect to database dimensions and size, the need for extensive database scans to generate frequent itemsets and mine rules, and high computational costs or execution time.

11.5 Correlation mining

The traditional techniques used for association mining are based on the support-confidence framework. However, this framework can be misleading since it may identify a rule $A \implies B$ as strong or interesting, even when the occurrence of A does not necessarily imply the occurrence of B. In some cases, an item pair with a high support value may not necessarily have a statistically high correlation, while a highly correlated item pair may exhibit a low support value. Therefore the support and confidence measures are inadequate in filtering out uninteresting association rules. To overcome this limitation, correlation analysis can be performed to provide an alternative framework for identifying statistically significant relationships. This approach helps improve our understanding of the meaning of some association rules. In statistics, nominal measures of association such as Pearson's correlation coefficient and measures based on Chi-Square can be used to analyze relationships among nominal variables. The ϕ correlation coefficient is a computational form of Pearson's correlation coefficient for binary variables. A similar approach based on the computation of the ϕ correlation coefficient is introduced in [29] to find correlations between item pairs in a transaction database based on support count. For any two items X and Y in a transaction database, the support-based ϕ correlation coefficient can be calculated as:

$$\phi(X, Y) = \frac{Sup(X, Y) - Sup(X) * Sup(Y)}{\sqrt{Sup(X) * Sup(Y) * (1 - Sup(X)) * (1 - Sup(Y))}}, \tag{11.1}$$

where, $Sup(X)$, $Sup(Y)$ and $Sup(X, Y)$ are the individual supports and the joint support of item X, Y, respectively.

An all-pair-strongly correlated query is different from traditional association mining, as it aims to identify statistical relationships between pairs of items within a transaction database. The problem can be defined in the following way:

Definition 11.5.1 (Strongly correlated pair). Assume a market-basket database D with T transactions and N items. Each transaction T is a subset of I, where $I = \{X_1, X_2, \ldots, X_N\}$ is a set of N distinct items. Given a user-specified minimum correlation threshold θ, an all-strong-pairs correlation query (SC) finds a set of all item pairs (X_i, X_j) (for $i, j = 1 \ldots N$) with correlation, $Corr(X_i, X_j)$, above the threshold θ. Formally, it can be defined as:

$$SC(D, \theta) = \left\{ \{X_i, X_j\} | \{X_i, X_j\} \subseteq I, X_i \neq X_j \wedge Corr(X_i, X_j) \geq \theta \right\}. \qquad (11.2)$$

Determining the appropriate value of θ for a strongly correlated item pair query requires prior knowledge of the data distribution. Without such knowledge, users may struggle to set the correlation threshold to obtain the desired results. If the threshold is set too high, there may be too few results or even none at all, in which case the user would need to guess a smaller threshold and restart the mining process, which may or may not produce better results. Conversely, if the threshold is set too low, there may be too many results, which would take an excessive amount of time to compute and require additional effort to filter the answers.

The strongly correlated item pair query generates a list of pairs from the database where the $Corr$ value of a pair is greater than the user-specified θ. Similarly, the top-k correlated-pair query generates a sorted list of k pairs in the order of $Corr$ from the database. Fig. 11.10 provides an illustration of a correlated-pair query. In this example, the input is a market-basket database containing eight transactions and six items, with θ set to 0.05. Similarly, for the top-k problem, k is set to 8. As there are six items in the database, there are $\binom{6}{2} = 15$ item pairs for which the correlation coefficient ϕ is calculated. To compute $\phi(4, 5)$ using Eq. (11.1), we require the single-element supports $Sup(4) = 4/8$ and $Sup(5) = 3/8$, and joint support $Sup(4, 5) = 3/8$, to calculate the correlation coefficient, which is 0.77. Finally, all pairs that satisfy the θ constraint are extracted, and the list of strongly correlated pairs is generated as output.

Strongly-Correlated-Pairs Query				
Input: a) Market Basket	**Pair**	**Support**	**Corr**	**Output**
	{1,2}	0.37	-0.44	{4,5}
TID | Items	{1,3}	0.37	-0.66	{1,5}
1 | 1,2,4,5,6	{1,4}	0.37	0.25	{2,4}
2 | 2,4	{1,5}	0.37	0.6	{2,5}
3 | 2,3,6	{1,6}	0.25	0.06	{1,4}
4 | 1,2,4,5	{2,3}	0.37	-0.44	{1,6}
5 | 1,3,6	{2,4}	0.5	0.57	{3,6}
6 | 2,3	{2,5}	0.37	0.44	
7 | 1,3	{2,6}	0.25	-0.14	
8 | 1,2,3,4,5	{3,4}	0.12	-0.77	
	{3,5}	0.12	-0.46	
	{3,6}	0.25	0.06	
	{4,5}	0.37	0.77	
b) θ=0.05	{4,6}	0.12	-0.25	
	{5,6}	0.12	-0.06	

FIGURE 11.10 Illustration of Strongly Correlated Pairs Query Problem.

According to some studies [7], parametric techniques such as Pearson's correlation is affected by the distribution of data and may not be efficient when the data is noisy and binary. In situations where two metric variables like interval or ratio scale measures are to be correlated, a parametric method such as Pearson's correlation coefficient is suitable. However, this may not always be feasible. For nonparametric variables like nominal or ordinal measures, nonparametric correlation techniques like Chi-square, Point biserial correlation [28], Spearman's ρ [18], and Kendall's τ [21] may work better, especially in the presence of outliers. Among these techniques, Spearman's ρ is widely used as it is rank-order based. To calculate the coefficient, the raw scores are converted to ranks, and the differences between the ranks of the observations on the two variables are determined. When there are no tied ranks, Spearman's ρ is given by the following equation:

$$\rho = 1 - \frac{6 \sum D_i^2}{N(N^2 - 1)},$$ (11.3)

where, $D_i = x_i - y_i$, is the difference between the ranks of corresponding values X_i and Y_i, and N is the number of samples in each dataset (same for both sets). In cases where tied ranks exist, the rank assigned to each tied score is the average of all tied positions. This means that if a pair of scores are tied for the 2nd and 3rd ranks, both scores will be given a rank of 2.5, which is calculated by taking the sum of the tied ranks and dividing by the number of tied scores.

In situations where Spearman's ρ is being calculated for market-basket binary variables, a method has been proposed by Roy et al. [26]. In this method, since binary market-basket data only contains scores of 0 and 1, their natural ordering can be used to compute tied ranks. To determine the rank of tied cases, 0 may be assigned greater priority than 1, and simple frequency counts of 1 and 0 may be used.

Suppose we have a binary variable I with N values. The frequency of score 1 in variable I is represented as $f(I)$. To calculate the appropriate rank of tied cases, we sum the rank positions and divide them by the number of tied cases. Since binary variables have only two possible scores, the rank of tied cases for scores 0 and 1 can be calculated as:

$$Rank_0 = \sum_{i=1}^{N-f(I)} \frac{i}{N - f(I)}.$$ (11.4)

Similarly, the rank of score 1 can be calculated as:

$$Rank_1 = \sum_{i=f(I)}^{N} \frac{i}{f(I)}.$$ (11.5)

After calculating the tied ranks of 0 and 1 for the binary variables being analyzed, such as I_1 and I_2, their rank difference is used to determine D_i^2. For binary itemsets, the only possible score combinations are (0,0), (0,1), (1,0), and (1,1). By counting the frequency of these score patterns and using the ranks of 0 and 1 for both itemsets, the sum of square differences of ranks (D_i^2) can be computed easily:

$$\sum D_i^2 = P_{(00)}(Rank_0(I_1) - Rank_0(I_2))^2 + P_{(01)}(Rank_0(I_1) - Rank_1(I_2))^2$$
$$+ P_{(10)}(Rank_1(I_1) - Rank_0(I_2))^2 + P_{(11)}(Rank_1(I_1) - Rank_1(I_2))^2, \qquad (11.6)$$

where, $P_{(00)}$, $P_{(01)}$, $P_{(10)}$, and $P_{(11)}$ are the frequencies of (0,0), (0,1), (1,0), and (1,1) patterns, respectively. Here, $Rank_0(I_1)$ and $Rank_1(I_1)$ represent the ranks of 0s and 1s in item I_1, while $Rank_0(I_2)$ and $Rank_1(I_2)$ are the corresponding ranks for 0s and 1s in item I_2.

To calculate $P_{(00)}$, $P_{(01)}$, $P_{(10)}$, and $P_{(11)}$ for a long sequence of binary data in a large transaction dataset, it can be expensive to compute the frequencies for each item pair. In such cases, it would be more efficient to calculate these probabilities with minimal information. One approach to do this is when the frequency of score 1 in item I_1 and I_2 ($f(I_1)$ and $f(I_2)$, respectively) and the joint occurrences in both item pairs ($f(I_1, I_2)$) are given:

$$P_{(01)} = f(I_2) - f(I_1, I_2), \; P_{(10)} = f(I_1) - f(I_1, I_2),$$
$$P_{(11)} = f(I_1, I_2), \; P_{(00)} = N - (P_{(01)} + P_{(10)} + P_{(11)}).$$

For example, let us consider the following item pairs I_1 and I_2 with $N = 6$ transactions $(T_1, T_2, \cdots, T_6)$ with similar occurrence patterns (Table 11.8).

Table 11.8 Market-basket transactions for two items and six transactions.

	T1	T2	T3	T4	T5	T6
I_1	1	1	1	0	0	0
I_2	1	1	1	0	0	0

In the above dataset, it is obvious that the frequency of 1 in I_1 and I_2, and joint occurrences of 1 in both I_1 and I_2 are $f(I_1) = 3$, $f(I_2) = 3$, $f(I_1, I_2) = 3$, respectively. Using Eqs. (11.4) and (11.5), the value of $Rank_0(I_1)$ and $Rank_1(I_1)$ for I_1 become $Rank_0(I_1) = (1 + 2 + 3)/3 = 2$ and $Rank_1(I_1) = (4 + 5 + 6)/3 = 5$. Similarly, $Rank_0(I_2)$ and $Rank_1(I_2)$ of I_2 are 2 and 5, respectively.

The next step is to calculate $P_{(00)}$, $P_{(01)}$, $P_{(10)}$, and $P_{(11)}$ using the joint frequency count $f(I_1, I_2)$:

$$P_{(01)} = f(I_2) - f(I_1, I_2) = 3 - 3 = 0, \; P_{(10)} = f(I_1) - f(I_1, I_2) = 3 - 3 = 0,$$
$$P_{(11)} = f(I_1, I_2) = 3, \; P_{(00)} = N - (P_{(01)} + P_{(10)} + P_{(11)}) = 6 - (0 + 0 + 3) = 3.$$

The summation of square rank differences D_i^2 is:

$$\sum D_i^2 = P_{(00)}(Rank_0(I_1) - Rank_0(I_2))^2 + P_{(01)}(Rank_0(I_1) - Rank_1(I_2))^2$$
$$+ P_{(10)}(Rank_1(I_1) - Rank_0(I_2))^2 + P_{(11)}(Rank_1(I_1) - Rank_1(I_2))^2$$

$$= 3(2-2)^2 + 0(2-5)^2 + 0(5-2)^2 + 3(5-5)^2$$
$$= 0.$$

Thus Spearman's ρ can be calculated as: $\rho = 1 - (6 \times 0)/6(6^2 - 1) = 1 - 0 = 1$.

In a transaction database, computing $f(I_1)$, $f(I_2)$, and $f(I_1, I_2)$ for any item pair is simply finding their individual and joint supports (1- and 2-element itemsets).

11.6 Distributed and parallel association mining

With the advancements in communication technologies, complex and heterogeneous data are now stored on different computers connected via LANs or WANs. Traditional sequential mining techniques are not scalable and cannot handle geographically dispersed data. Therefore there is a need for high-performance parallel and distributed association-mining techniques. Researchers have proposed various techniques, which are typically extensions of existing sequential methods. Although parallel and distributed techniques are often used interchangeably, the main difference between them is the latency and bandwidth of the interconnection network. In distributed mining, the network is typically slower. However, the difference between the two is becoming less distinct. In the following sections, we briefly discuss some of the parallels and distributed AM techniques.

11.6.1 Count distribution

Parallelism is primarily achieved by distributing the task of counting frequent itemsets on physical partitions of a database to different computing units.

Count Distribution (CD) [2] is a parallel version of the Apriori algorithm. It expedites the costly task of candidate generation in traditional Apriori. It parallelizes the support count task of Apriori. Nonexchangeable database partitions $(D_1, D_2, \cdots D_n)$ are created from the original transaction database (D) and assigned to the local memory of each computing unit (see Fig. 11.11). The global candidate set (C_k) is also distributed among different units. The support for a candidate is computed on each local database partition (D_m). Each computing unit independently obtains partial support of the candidates from its local database partition. Next, the algorithm performs a sum reduction to obtain global counts by exchanging local counts with all other units. Once the global F_k has been determined, each unit generates the entire candidate set C_{k+1} in parallel using the Apriori candidate generation step. The same steps are repeated in k processing units until all frequent itemsets are found.

The communication in this algorithm is minimized by exchanging only the counts between processors. However, because the entire C_k is replicated on each unit, the algorithm does not effectively utilize the aggregate system memory. Since each processing unit counts the support of the same C_k, the algorithm can only count the same number of candidates in each pass, regardless of the increase in processing units.

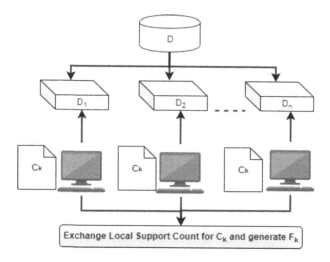

FIGURE 11.11 Count Distribution generates frequent itemsets based on local database partitions.

11.6.2 Data distribution

Unlike Count Distribution, Data Distribution (DD) [2] distributes the task of counting support with high utilization of the increased number of processing units. It helps count the support for a large number of candidates in a single pass. In DD, the partitions of the database, $D_1, D_2, \cdots, D_n$, are exchanged among the computing units (see Fig. 11.12). Unlike CD, each processing unit will have a different subset of candidate sets ($C_1, C_2, \cdots, C_k$). For every pass ($k > 1$), each processor counts the support for its local candidate set, C_k, based on all the database partitions by performing communication with each partition. After every round of support count computation, local frequent itemsets (F_k) are exchanged among the units to calculate the global frequent set. Every unit will have frequent itemsets for the kth pass and can generate the next level of candidate set, C_{k+1}. In the next pass, $1/n$ th of the total set of candidates are used by every processor. However, no further exchange of $1/n$ th of the candidate items needs to be performed among the processors as it can be easily done with initial agreements among the processors according to their sequence number or some other way to decide the appropriate $1/n$th part. Data Distribution needs more data exchange in comparison to Count Distribution, and hence is more expensive.

11.6.3 Parallel-rule generation

Parallel-rule generation is straightforward and does not require any access to the database. The frequent itemset can be generated using either the CD or DD approach, and hence each processing unit will have access to the global frequent itemset, F. The rule can be generated based on the downward closure property. Given a frequent itemset L, rule gen-

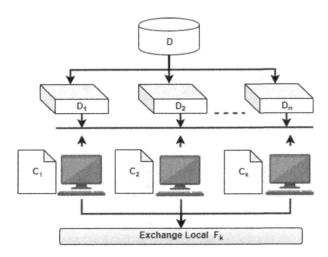

FIGURE 11.12 Data-Distribution exchange database partitions and local frequent itemsets.

eration examines each nonempty subset, $a \subset L$ to generate the rule $a \implies (L - a)$ based on support, $Sup(L)$ and confidence, $Sup(L)/Sup(a)$. Given a frequent itemset $ABCD$, if the rule $ABC \implies D$ does not have the minimum confidence, then a rule consisting of subset items of the above cannot be valid, e.g., the rule such as $AB \implies CD$ is invalid. During parallel-rule generation, each unit will consider a subset of global frequent items for parallel-rule generation using this simple idea. However, the approach requires the large-size frequent itemset to be available on all the processors during rule generation and pruning of invalid rules. In the case of DD and CD, it is not a problem as it is always available.

11.7 Summary

Association mining is a two-step process, frequent-itemset generation and rule generation based on frequent itemsets. Of these two steps, the second step is almost trivial. Most work focuses on frequent-itemset generation. Candidate generation is a preliminary step for frequent-itemset generation. Most algorithms are inefficient because they generate too many candidate sets, compared to frequent itemsets. The situation becomes more onerous during the second iteration, because to generate and count frequencies of all 2-element candidates, the complexity is quadratic. The reason for this is that we must consider all item pairs and no pruning is possible at this stage. Many approaches attempt to represent the original database in a concise representation such that the concise database can be placed in the main memory. Efficient implementation of these algorithms can improve the performance of the techniques many folds. A limited amount of work has focused on the problem of quantitative association mining. An effective and direct quantitative mining technique without converting into binary form is necessary. The need for applications that mine data streams, such as networks for traffic monitoring and web-click stream analysis, is growing rapidly. As a result, there is an increasing need to perform association mining on stream data. Interactive association mining is a need where depending on the varying requirements of users, data may be mined multiple times using different support levels to generate association rules with low effort.

Acknowledgment

The discussion on quantitative association mining is a part of the dissertation work by Mr. Dhrubajit Adhikary.

References

[1] Rakesh Agrawal, Tomasz Imieliński, Arun Swami, Mining association rules between sets of items in large databases, in: Proceedings of the 1993 ACM SIGMOD International Conference on Management of Data, 1993, pp. 207–216.

[2] Rakesh Agrawal, John C. Shafer, Parallel mining of association rules, IEEE Transactions on Knowledge and Data Engineering 8 (6) (1996) 962–969.

[3] Rakesh Agrawal, Ramakrishnan Srikant, et al., Fast algorithms for mining association rules, in: Proc. 20th Int. Conf. Very Large Data Bases, in: VLDB, vol. 1215, Citeseer, 1994, pp. 487–499.

[4] Alataş Bilal, Erhan Akin, An efficient genetic algorithm for automated mining of both positive and negative quantitative association rules, Soft Computing 10 (3) (2006) 230–237.

[5] Yonatan Aumann, Yehuda Lindell, A statistical theory for quantitative association rules, Journal of Intelligent Information Systems 20 (3) (2003) 255–283.

[6] Keith C.C. Chan, Wai-Ho Au, An effective algorithm for mining interesting quantitative association rules, in: Proceedings of the 1997 ACM Symposium on Applied Computing, ACM, 1997, pp. 88–90.

[7] G.W. Corder, D.I. Foreman, Nonparametric Statistics for Non-statisticians: A Step-by-Step Approach, Wiley, 2009.

[8] Li Dancheng, Zhang Ming, Zhou Shuangshuang, Zheng Chen, A new approach of self-adaptive discretization to enhance the apriori quantitative association-rule mining, in: Intelligent System Design and Engineering Application (ISDEA), 2012 Second International Conference on, IEEE, 2012, pp. 44–47.

[9] Mohammad El-Hajj, Osmar R. Zaiane, Inverted matrix: efficient discovery of frequent items in large datasets in the context of interactive mining, in: Proceedings of the Ninth ACM SIGKDD International Conference on Knowledge Discovery and Data Mining, 2003, pp. 109–118.

[10] Takeshi Fukuda, Yasuhido Morimoto, Shinichi Morishita, Takeshi Tokuyama, Mining optimized association rules for numeric attributes, in: Proceedings of the Fifteenth ACM SIGACT-SIGMOD-SIGART Symposium on Principles of Database Systems, ACM, 1996, pp. 182–191.

[11] Yunkai Guo, Junrui Yang, Yulei Huang, An effective algorithm for mining quantitative association rules based on high dimension cluster, in: Wireless Communications, Networking and Mobile Computing, 2008. WiCOM'08. 4th International Conference on, IEEE, 2008, pp. 1–4.

[12] Attila Gyenesei, A fuzzy approach for mining quantitative association rules, Acta Cybernetica 15 (2) (2001) 305–320.

[13] Jiawei Han, Jian Pei, Yiwen Yin, Mining frequent patterns without candidate generation, in: ACM SIGMOD Record, vol. 29, ACM, 2000, pp. 1–12.

[14] Yang Junrui, Zhang Feng, An effective algorithm for mining quantitative associations based on subspace clustering, in: Networking and Digital Society (ICNDS), 2010 2nd International Conference on, vol. 1, IEEE, 2010, pp. 175–178.

[15] Gong-Mi Kang, Yang-Sae Moon, Hun-Young Choi, Jinho Kim, Bipartition techniques for quantitative attributes in association rule mining, in: TENCON 2009-2009 IEEE Region 10 Conference, IEEE, 2009, pp. 1–6.

[16] Mehmet Kaya, Reda Alhajj, Novel approach to optimize quantitative association rules by employing multi-objective genetic algorithm, in: Innovations in Applied Artificial Intelligence, Springer, 2005, pp. 560–562.

[17] Yiping Ke, James Cheng, Wilfred Ng, MIC framework: an information-theoretic approach to quantitative association rule mining, in: Data Engineering, 2006. ICDE'06. Proceedings of the 22nd International Conference on, IEEE, 2006, pp. 112–112.

[18] E.L. Lehmann, H.J.M. D'Abrera, Nonparametrics: Statistical Methods Based on Ranks, Springer, New York, 2006.

[19] Jiuyong Li, Hong Shen, Rodney Topor, An adaptive method of numerical attribute merging for quantitative association rule mining, in: Internet Applications, Springer, 1999, pp. 41–50.

[20] Wang Lian, David W. Cheung, S.M. Yiu, An efficient algorithm for finding dense regions for mining quantitative association rules, Computers & Mathematics with Applications 50 (3) (2005) 471–490.

[21] J.T. Litchfield Jr., F. Wilcoxon, Rank correlation method, Analytical Chemistry 27 (2) (1955) 299–300.

[22] Diana Martin, Alejandro Rosete, Jesús Alcalá-Fdez, Francisco Herrera, A new multiobjective evolutionary algorithm for mining a reduced set of interesting positive and negative quantitative association rules, IEEE Transactions on Evolutionary Computation 18 (1) (2014) 54–69.

[23] Maria Martínez-Ballesteros, Francisco Martínez-Álvarez, A. Troncoso, José C. Riquelme, Selecting the best measures to discover quantitative association rules, Neurocomputing 126 (2014) 3–14.

[24] Maria Martínez-Ballesteros, J.C. Riquelme, Analysis of measures of quantitative association rules, in: Hybrid Artificial Intelligent Systems, Springer, 2011, pp. 319–326.

[25] Swarup Roy, D.K. Bhattacharyya, OPAM: an efficient one pass association mining technique without candidate generation, Journal of Convergence Information Technology 3 (3) (2008) 32–38.

[26] Swarup Roy, Dhruba Kr. Bhattacharyya, Mining strongly correlated item pairs in large transaction databases, International Journal of Data Mining, Modelling and Management 5 (1) (2013) 76–96.

[27] Ramakrishnan Srikant, Rakesh Agrawal, Mining quantitative association rules in large relational tables, in: ACM SIGMOD Record, vol. 25, ACM, 1996, pp. 1–12.

[28] S.L. Weinberg, K.P. Goldberg, Statistics for the Behavioral Sciences, Cambridge University Press, 1990.

[29] H. Xiong, S. Shekhar, P.N. Tan, V. Kumar, Exploiting a support-based upper bound of Pearson's correlation coefficient for efficiently identifying strongly correlated pairs, in: ACM SIGKDD'04. Proceedings of the 10th International Conference on, ACM, 2004, pp. 334–343.

[30] Weining Zhang, Mining fuzzy quantitative association rules, in: 2012 IEEE 24th International Conference on Tools with Artificial Intelligence, IEEE Computer Society, 1999, pp. 99–99.

[31] Hui Zheng, Jing He, Guangyan Huang, Yanchun Zhang, Optimized fuzzy association rule mining for quantitative data, in: Fuzzy Systems (FUZZ-IEEE), 2014 IEEE International Conference on, IEEE, 2014, pp. 396–403.

Big Data analysis

12.1 Introduction

The term "Big Data" can be used to refer to data that are too big in size, too fast in generation, too varied in nature, and too complex in context for existing tools to handle and process [24]. Storage, analysis, and visualization of such data are difficult because of size, complex structure, and heterogeneity [39]. With rapidly evolving technologies, data-generation sources have also been evolving. Today, data are generated from online transactions, emails, social media, sensors, mobile phones, and many other varied sources. Almost every device we use can connect through the Internet, and produce and record data such as details of device maintenance, location, and usage including data pertaining to personal health and bill payments. The worldwide explosion of digital data can be surmised from the fact that the total amount of data created till 2003 was 5 exabytes (10^{18} bytes), which expanded to 2.72 zettabytes (10^{21} bytes) in 2012, and further increased to 8 zettabytes of data by 2015 [39]. In 2022, 2.5 quintillion bytes of data were generated daily, approximately 90% of the total data created in the prior two years. Today's online platforms generate a huge amount of data every minute. For example, 41.6 million messages per minute on WhatsApp, 1.3 million video calls per minute across applications, 404 000 hours of Netflix streaming per minute, 494 000 Instagram and Facebook posts per minute, 2.4 million Google searches per minute and 3 million emails sent per second all contribute to the growing bulk of digital data.[1] Storage of this massive amount of data in databases is a major issue. Open-source database-management systems such as MySQL and Postgres lack scalability [24]. Cloud storage that transfers the problem to someone else may be an option, but storing huge amounts of valuable data at distant service providers is subject to uncertainties of network availability, power outage, or other disruptions [27]. All these data, generated incrementally in a geographically distributed manner, need to be processed fast, which is impossible for traditional data warehouses and their limited processing power.

Big Data analysis is concerned with the process of extracting valuable insights from large and complex datasets. The process involves several steps: collection, processing, storage, and analysis. In recent years, the ability to perform analysis of massive amounts of data has emerged as a crucial necessity for businesses and organizations. It allows them to gain new insights into customer behavior, market trends, and operational performance. One key challenge in Big Data analysis is dealing with the sheer volume of data that needs

[1] https://wpdevshed.com/how-much-data-is-created-every-day/.

to be processed. Traditional data-analysis methods are insufficient to handle such large datasets, which can contain millions or even billions of records. As a result, new tools and techniques have been developed to process and analyze large amounts of data more efficiently. One such approach is Hadoop, an open-source software framework that provides a distributed file system and tools for processing large datasets. Hadoop allows analysts to break down a large dataset into smaller, more manageable chunks, which can be processed in parallel across many computers. This approach can greatly reduce the time and resources required for data analysis. Another important technique in Big Data analysis is machine learning. This involves using algorithms to automatically identify patterns and insights in large datasets. Machine learning can be used for various tasks, including predictive modeling, anomaly detection, and natural language processing. As Big Data analysis continues to evolve, it is likely that we will see new tools and techniques emerge to help process and analyze data even more efficiently. With the increasing availability of large datasets and the growing demand for insights from businesses and organizations, Big Data analysis is set to play a crucial role in the future of data-driven decision making.

This chapter discusses Big Data and its analysis. It explores the challenges involved in processing and analyzing Big Data and the tools and techniques developed to address these challenges. Specifically, we delve into the Hadoop and Spark frameworks and how they enable analysts to process and analyze large datasets efficiently. By the end of this chapter, the reader should have a solid understanding of the key concepts and techniques involved in Big Data analysis and how they can be applied to real-world business problems.

12.2 Characteristics of Big Data

Big Data are characterized by five major attributes: variety, velocity, volume, veracity, and value, termed the 5Vs [3,15,36,39]. (See Fig. 12.1.)

1. **Variety** refers to the varied sources from where data are collected. The data are generally of three types: structured, semistructured, and unstructured. Structured data are tagged, can be easily sorted, and stored in a data warehouse. Unstructured data are random and difficult to analyze, and semistructured data do not have any specific fields but are tagged. The type of data must be known as the storage and analysis complexities vary.
2. **Velocity** refers to the increase in speed at which data are created, processed, stored, and analyzed. It also refers to the speed at which data are generated and moved around among databases. It is ideally necessary to process the bulk of data during the limited time during streaming, to maximize utilization and value. For example, Google processes videos of different sizes uploaded on YouTube, or emails sent in almost real-time.
3. **Volume** refers to the size of the data to be handled by an organization, which is increasing every day and has reached terabytes and petabytes for large organizations. This is among the first problems faced while handling Big Data. Organizations need to procure ample capacity to store and process huge volumes of data to gain maximum benefit.

4. **Veracity** refers to the reliability of the source from where the data are accumulated. The quality of the data must be of the utmost importance while simultaneously dealing with the large volume, velocity, and variety of data. The data must be correct and properly tagged. Different data sources are likely to have varied quality, and this must be taken into account while ensuring accuracy.

5. **Value** refers to utilizing the accumulated data such that the required information is extracted from it and used to the maximum value. Accessing and storing data become useless if they cannot be processed.

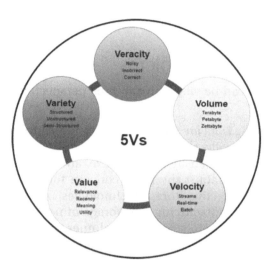

FIGURE 12.1 The five major challenges, denoted as 5V, that characterize Big Data. Every V is associated with key attributes shown.

12.3 Types of Big Data

As discussed in Chapter 2 and also mentioned earlier in this chapter, datasets that constitute Big Data may be structured, unstructured, and semistructured.

1. **Structured data:** Data items are properly tagged and placed in fixed fields within a record. All the data values within a certain field have the same set of properties, and thus it is easy to understand and process. Storing the data within fixed records enhances the integrity of the data; enables efficient and effective querying and analysis; and facilitates the accumulation of new data and transmission of data. Structured data can be considered relational data, to be handled using relational databases and SQL (structured query language). For example, customer data from e-commerce websites are structured data.

2. **Unstructured data:** Data items do not follow any specific structure or rules and do not conform to the syntax of a definite type of record. The arrangement of the data is un-

planned and cannot be handled without a prior preprocessing step. For example, text, image, and video data are unstructured as they might have many components that cannot be categorized into specific groups in a straightforward manner.

3. **Semistructured data:** Data components are not bound by any rigid schema for storage and handling. The datasets are not organized like relational data but have certain features that help categorize the data. Such data cannot be handled using SQL; thus, NoSQL or a similar data-serialization language may be used to exchange semistructured data across systems. Data-serialization languages like XML, JSON, and YML are used to process semistructured data. For example, the metadata associated with images and videos are semistructured data.

12.4 Big Data analysis problems

Issues related to Big Data can be viewed from three perspectives – data, processing, and management [10].

1. **Data:** Constituents of Big Data are characterized by high dimensionality and large sample sizes. Organizations collect these data from both internal and external sources. The volume, speed, quality, sensitivity, structure, and diversity of the data are major concerns. The volume of data is the main concern for storage and processing. Due to rapidly evolving cost-effective storage technologies, organizations can often handle data storage. In addition, outsourcing the storage of nonsensitive data to resourceful cloud-based providers can help store huge volumes of data. The inflow and outflow speeds of different types of data are different. The speed at which a text file is sent is much faster than the speed at which an audio file is sent. In addition to completeness and accuracy, quality must be maintained while working with Big Data. For example, missing values due to unavailability or undetectability may reduce the data quality. The sensitivity of the data also determines the data-handling requirements. Sensitive information, such as people's personal data, and organizations' confidential information requires encryption and secured storage. The structure of the data is required to decide the type of processing requirements. Structured data have a high degree of organization and are suitable for direct machine processing, whereas unstructured data require a preprocessing step before using conventional tools. The huge volume of Big Data can be very diverse, with different data types and varied sources. Data can be text, multimedia, or location data requiring different approaches to storage and processing. The diversity of the data adds enormously to the complexity.

2. **Processing:** Data processing is a major concern after collecting a large volume of data. The data need to be processed to extract maximal value. Availability of algorithms, scalability, actionable knowledge, and timelines are aspects of processing that need to be addressed. Suitable algorithms for processing the data are required to complete the required tasks with efficient use of space and time. Even with significant progress in technology, algorithms that can handle huge volumes of diverse data with ease are scarce. Scalable processing power is required to handle the ever-expanding amounts of

data to meet demands. The knowledge derived from the collected data after processing is actionable knowledge. Different organizations need to extract different information from the same datasets, thus requiring different processing techniques. Timelines refer to the time required for processing the data. Offline and online modes of processing require different processing times. Online processing can be real-time and starts as soon as data acquisition starts, whereas offline processing may be delayed. Real-time processing of Big Data requires higher computing power than offline mode.

3. **Management:** Challenges in managing the data abound during both acquisition and processing. Infrastructure, resource optimization, storage management, and database management are aspects of the management of Big Data. Efficient collection of the data, adequate storage capacities and computing power, and availability of high-speed Internet are all aspects of infrastructure challenges in handling Big Data. Optimizing resources, such as computing and communication equipment, manpower, time, and financial investments are also matters of concern. The large volume of data imposes huge demands on storage. As the volume of data increases each passing year, storage technologies need to progress at the same pace to accommodate the growing data. Storage management requires fast internal memories, with necessary storage capacity, and efficient backup options such as hard drives. Advanced database technologies are required for efficient processing and management of the data. Traditional databases that run on specialized single-rack high-performance hardware cannot meet the growing demands of Big Data. Data storage and management require both SQL and NoSQL database systems to handle the different data types.

12.5 Big Data analytics techniques

Big Data analytics refers to the process of extracting insights and knowledge from large and complex datasets. Approaches for Big Data analytics are constantly evolving, although there are a variety of existing techniques that can be used for analysis and interpretation. Some of the most common techniques used in Big Data analytics include the following:

- **Machine Learning:** Machine learning refers to the use of algorithms and statistical models to enable computer systems to learn from data without explicit programming. Machine learning is a powerful technique that can be used to extract insights from Big Data, such as identifying patterns and trends, making predictions, and detecting anomalies.
- **Data Mining:** Data mining is the process of analyzing large datasets to identify patterns and relationships. Data-mining techniques can be used to uncover hidden insights and identify trends. This can be useful in applications such as market-basket analysis, where patterns in customer purchasing behavior can be used to improve product recommendations and increase sales.
- **Predictive Modeling:** Predictive modeling involves using statistical techniques to make predictions about future events based on historical data. Predictive modeling can be

used in a variety of applications, such as fraud detection, risk management, and customer segmentation.

- **Visualization:** Data visualization refers to the creation of graphical representations of data. Visualization techniques can be used to simplify complex data and enable analysts to quickly identify patterns and trends. This can be useful in applications such as dashboard reporting, where real-time data is displayed in an easy-to-understand format.

12.6 Big Data analytics platforms

The analysis of Big Data is supported by several platforms used to handle the growing demand for processing. Peer-to-peer networks and Apache Hadoop[2] are two of the most well-known Big Data analytics platforms [41].

1. **Peer-to-peer networks:** A peer-to-peer network connects millions of machines. It is a decentralized and distributed network architecture where the nodes, called peers, provide and consume resources. The scheme used in the communication and exchange of data in a peer-to-peer network is called Message Passing Interface (MPI). The number of nodes in the network can be scaled to millions. Broadcasting messages in a peer-to-peer network is inexpensive, following the spanning-tree method to send messages, choosing any arbitrary node as the root. A major drawback is that the aggregation of data is expensive. MPI is a widely accepted scheme. A major feature of MPI is state preserving, whereby a process can live as long as the system runs and need not be read again and again. MPI is suitable for iterative processing and uses a hierarchical master–slave model. The slave machines can have dynamic resource allocation when there is a large volume of data to process. MPI has methods to send and receive messages, broadcast messages over all nodes, and methods to allow processes to synchronize. Although MPI is good for developing algorithms for Big Data analytics, it has no mechanism for fault tolerance. Users must implement their own fault-tolerance mechanisms [42].
2. **Apache Hadoop:** Hadoop is an open-source framework[3] for storing and processing huge volumes of data. Hadoop can scale up to thousands of nodes and is highly fault tolerant. It comprises two main components – Distributed File System (HDFS) and Hadoop YARN. HDFS is a distributed file system used to store data across clusters of commodity hardware and also provides high availability and high fault tolerance. Hadoop YARN is used for resource management and to schedule jobs across the cluster. The programming model used in Hadoop is allied with MapReduce. The process followed by MapReduce includes dividing the task into two parts, called *mappers* and *reducers*. The job of mappers is to read the data from the HDFS file system and process it to generate intermediate results for the reducers. Reducers aggregate the intermediate

[2] https://hadoop.apache.org/.
[3] https://hadoop.apache.org/.

results and generate the final result, which is written back to the HDFS. Several mappers and reducers are run across different nodes in the cluster to perform a job, helping perform parallel data processing. MapReduce *wrappers* are available to help develop source code to process the data. Wrappers such as Apache Pig and Hive provide an environment to programmers for developing code without handling complex MapReduce coding. DryadLINQ is a programming environment that provides flexibility for coding using C#. A major drawback of MapReduce is that it cannot handle iterative algorithms. Mappers must read the same data repeatedly, and each iteration must be written to the disk to pass them on to the next iteration. A new mapper and a new reducer are initialized for every new iteration. Thus the disk has to be accessed frequently, making it a bottleneck for performance [8]. A scalable and distributed database called HBase supports structured data for large database tables. It is a nonrelational database (NoSQL) and runs on HDFS. It allows transactional functions like updates, inserts, and deletions. It also allows fast real-time read–write access and has fault-tolerant storage capability. Hadoop also includes a scripting language called Apache PIG that enables users to write MapReduce transformations like summarizing, joining, and sorting. PIG also enables parallel processing to handle large datasets. Another important module of Hadoop is Cassandra, which supports an open-source distributed management system. It is decentralized, highly scalable, and has the capability of handling large volumes of data across multiple servers [3].

3. **Spark:** Spark is a data-analysis paradigm that has been developed as an alternative to Hadoop. Spark has the ability to perform in-memory computations and can overcome disk I/O limitations that were observed in Hadoop. Data can be cached in memory, thus eliminating repetitive disk access for iterative tasks. Spark can run on Hadoop YARN and can read data from HDFS. It supports Java, Scala, and Python and has been tested to be up to 100x faster than Hadoop MapReduce [41].

4. **High-Performance Computing (HPC) clusters:** An HPC cluster is a supercomputer with thousands of cores that offers a parallel computing platform. An HPC cluster provides powerful hardware that is optimized for speed and throughput. The high-end hardware is also fault tolerant, making hardware failures extremely rare. The communication scheme used for HPC is MPI, discussed earlier [44].

5. **Multicore CPU:** A multicore CPU has dozens of processing cores that share memory, and have only one disk. The CPUs have internal parallelism and there is a single motherboard that houses multiple CPUs, improving parallel computing. Parallelism is achieved by multithreading. Each task is broken down into threads and each thread is executed in parallel on different CPU cores [5].

6. **Graphics Processing Unit (GPU):** A GPU is a specialized hardware designed primarily for graphical operations such as image and video editing and accelerating graphics-related processing. A GPU has a large number of processing cores, much higher than a multicore CPU. It has a parallel architecture and a high throughput memory. In the past few years, GPU performance has increased significantly compared to CPUs. The CUDA framework has helped developers with GPU programming without the need to

access the hardware. Machine-learning algorithms implemented on GPU using CUDA run much faster as compared to multicore CPU. GPUs are extremely popular due to their high speed and parallel processing [32].

12.7 Big Data analytics architecture

The five major issues to be considered in developing any analytics architecture for Big Data are (a) scalability, (b) fault tolerance, (c) iterative computing, (d) complex data dependencies, and (e) heterogeneity. Several novel efforts have been made in developing appropriate architectures to support cost-effective Big Data analytics. Three popular architectures are described next.

12.7.1 MapReduce architecture

The MapReduce architecture was developed by Google to achieve data parallelism. It exploits multiple compute nodes to perform a given task on different data subsets in parallel. Following this data-parallel execution approach, several successful efforts have been made commercially as well as noncommercially. Apache Hadoop[4] is one such commercially successful open-source MapReduce implementation. The MapReduce approach includes two types of nodes: *master* and *compute* nodes. The master node is responsible for the tasks of configuration and control during the process of execution of a task. Fig. 12.2 depicts the schematic view of the MapReduce architecture.

Compute nodes are involved in executing a task iteratively following two major steps of operations: (i) map and (ii) reduce. Each of these two steps includes three distinct states: (a) accept input, (b) perform computation, and (c) generate output. Between two consecutive steps, synchronization is necessary. The local memory is cleared during synchronization and writes onto the global memory are performed. A basic difference between the master and the compute nodes is that the master is empowered to (i) read–write onto the global memory and (ii) communicate with the nodes without any time restriction, whereas the compute nodes can use the global memory for read–write during the synchronization only. These two distinct abilities of the two types of nodes are indicated in the figure with different types of directed lines (thin and thick). The tasks of data distribution, partial results generation, and storage are completed with the compute nodes during the map operations.

During reduce operations, the intermediate results are combined to generate the final results that are written onto global memory. In case there is a necessity for further processing of the partial results, the steps are iterated. This architecture gives an excellent performance especially when (i) data are voluminous and (ii) the scope for parallelism is high. However, if the computational dependencies among the data are high, the architecture fails to perform well. Further, the higher iterative computational requirements often

[4] https://hadoop.apache.org/.

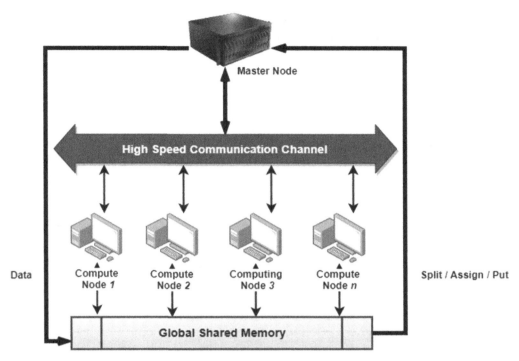

FIGURE 12.2 A MapReduce architecture with n computing nodes and a master node that is connected through a fast communication network and has access to a shared global memory.

lead to high I/O overhead, leading to poor performance. Efforts have been made to address these issues of the MapReduce architecture. A few such efforts are found in [9], [48], which use in-memory computation to avoid writing onto the distributed memory. However, a fault-tolerant architecture that can handle such iterative computational requirements for voluminous heterogeneous data with complex dependencies with cost-effective I/O operations is still a need of the hour.

12.7.2 Fault-tolerant graph architecture

To address the issue of handling iterative processing with complex data dependencies using fault-tolerant approaches, graph-based architectures are often suitable. Such a fault-tolerant architecture associated with a global shared memory performs the computation in a synchronous manner like MapReduce. It divides the computation among the compute nodes in a heterogeneous manner by assigning dissimilar (or particular) tasks to the nodes. A graph-based architecture of this type includes two components, (a) a directed graph with a set of compute nodes and (b) a distributed global shared memory. Fig. 12.3 depicts a schematic view of the architecture. The dependencies among the compute nodes are indicated by the directed dashed lines. A fast communication medium is used to improve communication with the nodes to achieve an improved overall performance.

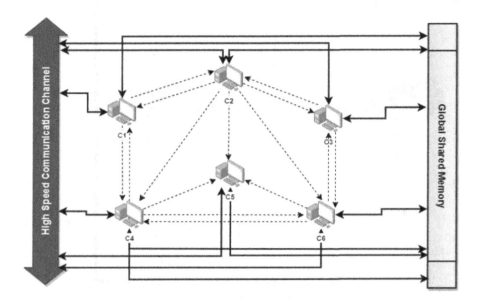

FIGURE 12.3 Fault-Tolerant Graph Architecture.

GraphLab [22] not only offers fault tolerance while supporting iterative computation using unreliable networks such as the Internet, but also can handle data with high dependency. Each node initially reads the instances from the shared repository and executes a compute task using its own and neighboring nodes' data. Results are gathered and merged, and written back onto the shared global repository for use in the subsequent cycle of execution. If a node fails during a cycle of operation, the corresponding compute task is repeated and as a consequence, one cycle is lost by the dependent node(s). Such loss of a cycle definitely causes degradation in efficiency, although, it guarantees fault tolerance. This architecture has a provision for replacing a node if it fails permanently. Additionally, it has other advantages such as (i) it can execute a problem with iterations and complex dependencies and (ii) it performs scalable distributed computation. However, two major disadvantages of GraphLab are: (a) high disk I/O overhead and (b) nonoptimized performance. Several variants of GraphLab [22] have also been introduced, including Pregel [25], Giraph,[5] GraphX [12], and Hema.[6]

---

[5] giraph.apache.org.
[6] hama.apache.org.

12.7.3 Streaming-graph architecture

The streaming-graph architecture was introduced to address the issue of high I/O overhead of previous architectures while handling streaming data. Although a good number of solutions have been introduced and consequently several packages such as Spark Streaming[7] have been introduced, most of these attempts handle stream data in batches. These solutions convert stream data internally to batches prior to downstream processing. However, stream data require in-memory processing for high bandwidth. This architecture (see Fig. 12.4) attempts to handle such stream data with the high processing speed, reliability, and high bandwidth.

The major characteristics of this architecture are as follows:

1. Unlike the previous architectures, it does not use a global shared memory.
2. It executes the operations in an asynchronous manner, allowing synchronous data flow only during merge operations.
3. It facilitates in-memory data processing at any scale rather than storing the data on a disk. It saves costs significantly as well as ensures high throughput.
4. This architecture also does not support fault tolerance, a major limitation. The failure of any node leads to restarting the whole process from the beginning, another serious drawback of this architecture.

12.8 Tools and systems for Big Data analytics

One major application area of Big Data analytics is Bioinformatics. Many efforts in this evolving area of research support the fast construction of coexpression and regulatory networks, salient module identification, detection of complexes from growing protein–protein interaction data, fast analysis of massive DNA, RNA, and protein-sequence datasets, and fast querying on incremental and heterogeneous disease networks. Extensive research in Bioinformatics has given birth to a number of tools and systems to support downstream analysis of gene-expression data. In addition to the development of tools and systems, there are efforts to provide adequate cloud-based bioinformatics platforms to support integrative analysis. Some such example platforms are: Galaxy [11] and Cloud-BLAST [28].

12.8.1 Bioinformatics tools

Microarrays are classical gene-expression data types commonly used for hidden knowledge extraction using statistical and machine-learning techniques. A large number of software tools have been introduced by computational biologists to perform analyses of microarray data. However, with the exponential growth of such data both in terms of volume and dimensionality, the time required to process disease queries on heterogeneous

---

[7] spark.apache.org/streaming.

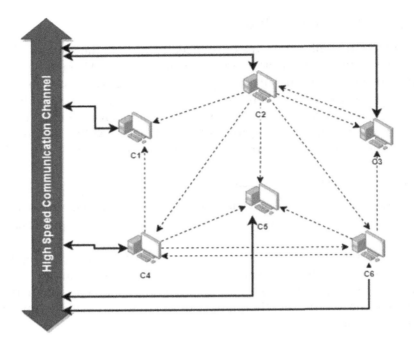

FIGURE 12.4 Streaming-graph architecture without any global memory.

data to find interesting patterns (relevant complexes, modules, or biomarker genes) is extremely high.

One commonly used tool for such analysis is *Beeline*,[8] which can handle Big Data by exploiting parallel computations and adaptive filtering (to minimize data size). Another tool named *caCORRECT* [43] performs quality assurance by removing artifactual noises from high-throughput microarray data. This tool helps improve the integrity and quality of microarray gene-expression and reproduced data for validation with a universal quality score. *OmniBiomarker* [35] is a popular web-based tool that helps find differentially expressed genes towards the identification of biomarkers from high-throughput gene-expression data using knowledge-driven algorithms. This tool is capable of identifying reproducible biomarkers with high stability.

The construction of a gene regulatory network using large gene-expression datasets (microarray or RNAseq) and subsequent analysis for the identification of crucial genes for critical diseases is a task of high computational requirement. A tool that enables network-based analysis using voluminous gene-expression datasets helpful to computational biologists and bioinformaticians. *FastGCN* [21] is one such tool that facilitates large-scale gene-network analysis using GPU-based distributed computing. Several other tools also

[8] https://www.illumina.com/techniques/microarrays/array-data-analysis-experimental-design/beeline.html.

support complex network analysis. *UGET* (UCLA Gene Expression Tool) [7] is useful in extracting interesting disease associations by performing coexpression analysis. A popular tool that facilitates large-scale weighted coexpression network analysis is *WGCNA* [17]. It works in a distributed computing environment. *NETRA* [52] is a comprehensive single-stop web platform that can perform inference, visualization, network analysis, and benchmarking of regulatory networks.

Finding interesting complexes or groups from large-scale PPI (protein–protein interactions) networks is a difficult computational task. In a traditional uniprocessor computing environment, the successful execution of such a data-intensive (involving millions of interactions) task often requires weeks. However, to perform such computational tasks quickly, user-friendly tools have been developed. Such tools extract interesting complexes, isolated as well as overlapped from large PPI networks. *ClusterONE* [29], *MCODE* [4], and *NeMo* [38], are some popular tools that can operate in a standalone mode as well as *Cytoscape*[9] Plugins. On the other hand, *PathBLAST* [16] is a useful tool to align PPI networks efficiently.

Tools have been developed to perform analytics on large-scale sequence data using the Hadoop MapReduce platform. *BioPig* [30] and *SeqPig* [40] are two such scalable tools for sequence analysis that can be ported directly to Hadoop infrastructures. *Crossbow* [18] is another effective tool that in turn uses an extremely fast and memory-efficient short-read aligner called *Bowtie* [19]. *SoapSNP* [20] is a scalable genotyper, which supports sequence analysis on large-scale whole genome-sequence datasets using cloud or Hadoop cluster platforms. There are several other scalable cloud-based tools such as *Stormbow* [50], *CloVR* [2], and *Rainbow* [51].

Pathway analysis is essential for validation of bioinformatics research findings. A faster and more accurate pathway analysis tool may be able to help identify pathway(s) for an identified gene during validation. Available pathway analysis tools include, *GO-Elite* [49] to provide description of a particular gene or metabolite, *PathVisio* [45] to support pathway analysis and visualization, *directPA* [47] to identify pathways by performing analysis over high-dimensional spaces, *Pathway Processor* [13] to analyze large expression datasets regarding metabolic pathways, *Pathway-PDT* [33] to perform analysis using raw genotypes in general nuclear families, and *Pathview* [23] to support integration of data based on pathway. A common disadvantage of these tools is that none can perform in a distributed computing environment to provide scalability.

12.8.2 Computer-vision tools

Computer-vision tools are developed to perform various computer-vision tasks like image processing, video analysis, object detection, etc. These tools are also included with various machine-learning and deep-learning models that can be applied on the computer-vision data. *OpenCV* [6] is a library for real-time computer-vision tasks. Similarly, *Pytorch* [34] and *GluonCV* [14] are two libraries that are built on top of two different frameworks that support computer-vision tasks. NVIDIA *CUDA-X*[10] is a GPU-accelerated library to perform computer-vision tasks with high performance. Similarly, *Insight Toolkit (ITK)*[11] is another crossplatform open-source library that includes a set of functions to perform image-processing tasks.

12.8.3 Natural language processing tools

Natural Language Processing tools analyze and process large amounts of natural language data. These tools are used for NLP tasks like POS tagging and NER (Named Entity Recognition). *Spacy*[12] is one such open source toolkit. *Fairseq* [31] is an open-source sequence modeling toolkit used for translation, summarization, language modeling, and other text-generation tasks. *Gensim* [37] is an open-source Python library for topic modeling, document indexing, and similarity retrieval with large corpora. *Flair* [1] is an open-source library built on the Pytorch framework, and includes state-of-the-art natural language processing (NLP) models. Another library is *HuggingFace Transformer* [46], which is backed by JAX, Pytorch, and Tensorflow. It contains various transformer models for NLP tasks. *CoreNLP* [26] is an open-source NLP tool in Java. A web-based tool *Voyant*[13] is used for textual analysis.

12.8.4 Network-security tools

Network-Security tools are used to analyze large amount of data collected from different sources. *NuPIC*[14] and *Loom*[15] are two network-security tools used for anomaly detection. *XPack*[16] is a machine-learning-based security tool that provides monitoring, alerting, reporting of security incidents. *Splunk*[17] handles and analyzes large amounts of data from real-time sources to detect and prevent threats. *QRadar*[18] and CISCO's *Stealthwatch*[19] are two data-driven malware analysis tools that continuously monitor network traffic and generate real-time alerts in case of any anomaly.

As seen in the above discussions, there are already a number of Big Data analytics tools to support specific problems in real-life domains. Although some of these tools are good enough in extracting relevant, interesting, and nontrivial knowledge from massive amounts of specialized data, we believe that there is a lack of generic tools that can help comprehensive analytical research over a wide range of problems.

[10] https://www.nvidia.com/en-in/technologies/cuda-x/.
[11] https://itk.org/.
[12] https://spacy.io/.
[13] https://voyant-tools.org/.
[14] https://numenta.com/.
[15] https://www.loomsystems.com/.
[16] https://www.elastic.co/.
[17] https://www.splunk.com/.
[18] https://www.ibm.com/qradar.
[19] https://www.cisco.com/.

12.9 Active challenges

Although researchers have been successful in developing platforms to handle Big Data in many real-life application domains, there are several research challenges that need to be addressed.

1. *I/O bottleneck*: Rapid growth of mobile-network connection speeds has made it difficult to handle data well in real time. The annual growth rate reported for the period (2014–2019) reveals that there was a significant increase from 1683 kbps to 4 Mbps during this period. Presently, 4G speeds are up to 1000 Mbps, much higher than that reported in 2014. 5G has been introduced as well. To handle such high-velocity data in real time, the major bottleneck happens to be the speed of I/O operations. Slower I/O operations limit the desired computational throughput.

2. *High-speed network accelerator*: With continued advances in networking technology, network accelerators have become extremely fast, contributing to the growth of generated data. Handling of such growing voluminous data is challenging, and it lags behind data growth.

3. *Heterogeneous semistructured data*: The nature of exploding data is mostly heterogeneous and semi- or unstructured. Traditional approaches such as data warehousing are often inadequate in handling such highly varied data, growing with high velocity. As such approaches attempt to organize the data in terms of a set of defined schemas, it is often impossible in the case of unstructured heterogeneous data. Thus development of methods that enable storage and update of unstructured heterogeneous data in real time remains a challenge.

4. *Increasing use of hand-held devices*: The increasing use of hand-held and the IoT devices and sensors has been contributing significantly in the growth of Big Data. Such data are voluminous not only in the number of instances but also in dimensionality. It is necessary to develop adequate methods and techniques to handle such growing data in real time.

12.10 Summary

In this chapter, we introduced the topic of Big Data along with the characteristics, types, and problems that large amounts of data introduce. We also discussed several analytics techniques and three major architectures for handling Big Data, and their pros and cons from the implementation perspective. Further, we presented a number of tools that have evolved in the recent past to support Big Data analytics. We summarize the chapter contents as follows:

- In recent times, there has been an exponential growth of data. Advancing technological developments are generating 2.5 quintillion bytes of data every day, which is almost 90% of the data generated in the entire prior two years.
- The five major characteristics of Big Data are: volume, velocity, veracity, variety, and value.

- Big Data can be structured, unstructured, or semistructured. To analyze such data, three major perspectives are necessary: data, processing, and management.
- Several Big Data analytics platforms have been introduced. The most commonly used platforms are: (i) peer-to-peer networks and (ii) Apache Hadoop.
- The three major architectures necessary to handle Big Data are: (i) MapReduce, (ii) Graph-based fault tolerance, (iii) streaming graph. These architectures have been developed to satisfy five major requirements, (i) scalability, (ii) fault tolerance, (iii) iterative computation, (iv) complex data dependencies, and (v) data heterogeneity.
- In the past decade, a large number of tools have been developed to facilitate Big Data analytics. These tools are designed to support analysis and/or visualization of massive amounts of data with the complex need for extraction of hidden knowledge in a user-friendly and cost-effective manner.

References

[1] Alan Akbik, et al., FLAIR: an easy-to-use framework for state-of-the-art NLP, in: Proc. North American Chapter of the Association for Computational Linguistics, 2019, pp. 54–59.
[2] Samuel V. Angiuoli, et al., CloVR: a virtual machine for automated and portable sequence analysis from the desktop using cloud computing, BMC Bioinformatics 12 (1) (2011) 1–15.
[3] J. Anuradha, et al., A brief introduction on Big Data 5Vs characteristics and Hadoop technology, Procedia Computer Science 48 (2015) 319–324.
[4] Gary D. Bader, Christopher W.V. Hogue, An automated method for finding molecular complexes in large protein interaction networks, BMC Bioinformatics 4 (1) (2003) 1–27.
[5] Ron Bekkerman, Mikhail Bilenko, John Langford, Scaling up Machine Learning: Parallel and Distributed Approaches, Cambridge University Press, 2011.
[6] Gary Bradski, The OpenCV library, Dr. Dobb's Journal of Software Tools for the Professional Programmer 25 (11) (2000) 120–123.
[7] Allen Day, et al., Disease gene characterization through large-scale co-expression analysis, PLoS ONE 4 (12) (2009) e8491.
[8] Jeffrey Dean, Sanjay Ghemawat, MapReduce: simplified data processing on large clusters, Communications of the ACM 51 (1) (2008) 107–113.
[9] Jaliya Ekanayake, et al., Twister: a runtime for iterative MapReduce, in: Proceedings of the 19th ACM International Symposium on High Performance Distributed Computing, 2010, pp. 810–818.
[10] Peter Géczy, Big data characteristics, The Macrotheme Review 3 (6) (2014) 94–104.
[11] Jeremy Goecks, Anton Nekrutenko, James Taylor, Galaxy: a comprehensive approach for supporting accessible, reproducible, and transparent computational research in the life sciences, Genome Biology 11 (8) (2010) 1–13.
[12] Joseph E. Gonzalez, et al., {GraphX}: graph processing in a distributed dataflow framework, in: 11th USENIX Symposium on Operating Systems Design and Implementation (OSDI 14), 2014, pp. 599–613.
[13] Paul Grosu, Jeffrey P. Townsend, Daniel L. Hartl, Duccio Cavalieri, Pathway processor: a tool for integrating whole-genome expression results into metabolic networks, Genome Research 12 (7) (2002) 1121–1126.
[14] Jian Guo, et al., GluonCV and GluonNLP: deep learning in computer vision and natural language processing, Journal of Machine Learning Research 21 (23) (2020) 1–7.
[15] Hirak Kashyap, Hasin Afzal Ahmed, Nazrul Hoque, Swarup Roy, Dhruba Kumar Bhattacharyya, Big data analytics in bioinformatics: architectures, techniques, tools and issues, Network Modeling Analysis in Health Informatics and Bioinformatics 5 (2016) 1–28.
[16] Brian P. Kelley, et al., PathBLAST: a tool for alignment of protein interaction networks, Nucleic Acids Research 32 (suppl_2) (2004) W83–W88.
[17] Peter Langfelder, Steve Horvath, WGCNA: an R package for weighted correlation network analysis, BMC Bioinformatics 9 (1) (2008) 1–13.

[18] Ben Langmead, Michael C. Schatz, Jimmy Lin, Mihai Pop, Steven L. Salzberg, Searching for SNPs with cloud computing, Genome Biology 10 (11) (2009) 1–10.

[19] Ben Langmead, Cole Trapnell, Mihai Pop, Steven L. Salzberg, Ultrafast and memory-efficient alignment of short DNA sequences to the human genome, Genome Biology 10 (3) (2009) 1–10.

[20] Ruiqiang Li, et al., SNP detection for massively parallel whole-genome resequencing, Genome Research 19 (6) (2009) 1124–1132.

[21] Meimei Liang, Futao Zhang, Gulei Jin, Jun Zhu, FastGCN: a GPU accelerated tool for fast gene co-expression networks, PLoS ONE 10 (1) (2015) e0116776.

[22] Yucheng Low, Joseph Gonzalez, Aapo Kyrola, Danny Bickson, Carlos Guestrin, Joseph M. Hellerstein, Distributed GraphLab: a framework for machine learning in the cloud, arXiv preprint, arXiv:1204. 6078, 2012.

[23] Weijun Luo, Cory Brouwer, Pathview: an R/Bioconductor package for pathway-based data integration and visualization, Bioinformatics 29 (14) (2013) 1830–1831.

[24] Sam Madden, From databases to big data, IEEE Internet Computing 16 (3) (2012) 4–6.

[25] Grzegorz Malewicz, Matthew H. Austern, Aart J.C. Bik, James C. Dehnert, Ilan Horn, Naty Leiser, Grzegorz Czajkowski, Pregel: a system for large-scale graph processing, in: Proceedings of the 2010 ACM SIGMOD International Conference on Management of Data, 2010, pp. 135–146.

[26] Christopher D. Manning, Mihai Surdeanu, John Bauer, Jenny Rose Finkel, Steven Bethard, David McClosky, The Stanford CoreNLP natural language processing toolkit, in: Proceedings of 52nd Annual Meeting of the Association for Computational Linguistics: System Demonstrations, 2014, pp. 55–60.

[27] Vivien Marx, The big challenges of big data, Nature 498 (7453) (2013) 255–260.

[28] Andréa Matsunaga, Maurício Tsugawa, José Fortes, CloudBLAST: combining MapReduce and virtualization on distributed resources for bioinformatics applications, in: 2008 IEEE Fourth International Conference on eScience, IEEE, 2008, pp. 222–229.

[29] Tamás Nepusz, Haiyuan Yu, Alberto Paccanaro, Detecting overlapping protein complexes in protein-protein interaction networks, Nature Methods 9 (5) (2012) 471–472.

[30] Henrik Nordberg, Karan Bhatia, Kai Wang, Zhong Wang, BioPig: a Hadoop-based analytic toolkit for large-scale sequence data, Bioinformatics 29 (23) (2013) 3014–3019.

[31] Myle Ott, et al., fairseq: a fast, extensible toolkit for sequence modeling, arXiv preprint, arXiv:1904. 01038, 2019.

[32] John D. Owens, Mike Houston, David Luebke, Simon Green, John E. Stone, James C. Phillips, GPU computing, Proceedings of the IEEE 96 (5) (2008) 879–899.

[33] Yo Son Park, Michael Schmidt, Eden R. Martin, Margaret A. Pericak-Vance, Ren-Hua Chung, Pathway-PDT: a flexible pathway analysis tool for nuclear families, BMC Bioinformatics 14 (1) (2013) 1–5.

[34] Adam Paszke, et al., Pytorch: an imperative style, high-performance deep learning library, Advances in Neural Information Processing Systems 32 (2019).

[35] John H. Phan, Andrew N. Young, May D. Wang, omniBiomarker: a web-based application for knowledge-driven biomarker identification, IEEE Transactions on Biomedical Engineering 60 (12) (2012) 3364–3367.

[36] V. Rajaraman, Big data analytics, Resonance 21 (8) (2016) 695–716.

[37] Radim Řehůřek, Petr Sojka, et al., Gensim—statistical semantics in Python. Retrieved from genism.org, 2011.

[38] Corban G. Rivera, Rachit Vakil, Joel S. Bader, NeMo: network module identification in Cytoscape, BMC Bioinformatics 11 (1) (2010) 1–9.

[39] Seref Sagiroglu, Duygu Sinanc, Big data: a review, in: 2013 International Conference on Collaboration Technologies and Systems (CTS), IEEE, 2013, pp. 42–47.

[40] André Schumacher, Luca Pireddu, Matti Niemenmaa, Aleksi Kallio, Eija Korpelainen, Gianluigi Zanetti, Keijo Heljanko, SeqPig: simple and scalable scripting for large sequencing data sets in Hadoop, Bioinformatics 30 (1) (2014) 119–120.

[41] Dilpreet Singh, Chandan K. Reddy, A survey on platforms for big data analytics, Journal of Big Data 2 (1) (2015) 1–20.

[42] Ralf Steinmetz, Klaus Wehrle, Peer-to-Peer Systems and Applications, vol. 3485, Springer, 2005.

[43] Todd H. Stokes, Richard A. Moffitt, John H. Phan, May D. Wang, chip artifact CORRECTion (caCORRECT): a bioinformatics system for quality assurance of genomics and proteomics array data, Annals of Biomedical Engineering 35 (6) (2007) 1068–1080.

[44] Rajkumar Buyya, High Performance Cluster Computing: Architectures and Systems, vol. 1, Pearson Education India, 1999.

[45] Martijn P. van Iersel, et al., Presenting and exploring biological pathways with PathVisio, BMC Bioinformatics 9 (1) (2008) 1–9.

[46] Thomas Wolf, et al., HuggingFace's transformers: state-of-the-art natural language processing, arXiv preprint, arXiv:1910.03771, 2019.

[47] Pengyi Yang, et al., Direction pathway analysis of large-scale proteomics data reveals novel features of the insulin action pathway, Bioinformatics 30 (6) (2014) 808–814.

[48] Matei Zaharia, et al., Resilient distributed datasets: a {Fault-Tolerant} abstraction for {In-Memory} cluster computing, in: 9th USENIX Symposium on Networked Systems Design and Implementation (NSDI 12), 2012, pp. 15–28.

[49] Alexander C. Zambon, et al., Go-elite: a flexible solution for pathway and ontology over-representation, Bioinformatics 28 (16) (2012) 2209–2210.

[50] Shanrong Zhao, Kurt Prenger, Lance Smith, Stormbow: a cloud-based tool for reads mapping and expression quantification in large-scale RNA-Seq studies, International Scholarly Research Notices (2013).

[51] Shanrong Zhao, Kurt Prenger, Lance Smith, Thomas Messina, Hongtao Fan, Edward Jaeger, Susan Stephens, Rainbow: a tool for large-scale whole-genome sequencing data analysis using cloud computing, BMC Genomics 14 (1) (2013) 1–11.

[52] Sumit Dutta, Binon Teji, Sourav Dutta, Swarup Roy, NetRA: An Integrated Web Platform for Large-Scale Gene Regulatory Network Reconstruction and Analysis, Preprints: 2023100820, 2023.

13

Data Science in practice

Data Science is being adopted to solve an exponentially growing number of real-life problems in diverse domains. The use of Data Science is expanding across disciplines where bulk data are available and there is a need to extract hidden or unknown facts. For smart decision making and intelligent support-system development, machine learning, and more recently deep learning have become indispensable due to the ability for accurate prediction. We discuss, in this chapter, a few areas where Data Science is playing a large role.

13.1 Need of Data Science in the real world

In today's technology-driven world, Data Science enables smart problem solving, sometimes it even becomes a life saver. Government agencies and business organizations can use Data Science to make people's lives comfortable. Properly applied Data Science may help resolve many problems within an organization, in turn making life easier and more productive for individuals. Data Science helps people make the right decisions in advance and build effective strategies. For example, to travel to an unknown destination, one can depend on the Google Map application that guides in real time which roadways to follow. You can check in advance the weather conditions at the destination for the next five hours or days using a weather-prediction system.

With the assistance of Data Science-enhanced software technologies, progress in healthcare has accelerated, leveraging simultaneous advancements in actual medical science. It is now possible to investigate the root causes of many terminal diseases quickly, and hypothesize remedies for the same. During COVID-19, the world handled the pandemic rapidly, leveraging Data Science [4]. Starting from characterizing the SARS-CoV2 virus [1–3], its origin, evolution, spreading pattern, and drug and vaccine development, Data Science played a major role in effective and rapid solution development. With the evolution of superfast communication and dissemination technologies, the world has become more dependent on network science and engineering. Ubiquitous technologies such as IoT, sensor, and wireless networks have positively impacted on Data Science. Alongside this, networks and networked computers also have become more vulnerable to security threats. Intruders have also become smart and often adopt the attack patterns to fool preinstalled detectors. Such zero-day attacks are a growing threat. Effective machine learning is important to address such challenges in a timely manner. Recently, Google has become successful in attracting more users by developing easy-to-use techniques. For example, Google facilitates voice-over command through natural language processing (NLP). Ama-

Fundamentals of Data Science. https://doi.org/10.1016/B978-0-32-391778-0.00020-X

zon's *Alexa* is the popular name for a smart and artificially intelligent software tool where a machine is interacting with users and assisting in human-like interactions. Speech synthesis and NLP work together to make flexible and natural interactions a reality where users can avoid rigid command structures to work with a machine by typing on the keyboard, or clicking mouse buttons, or by speaking.

Data Science and associated analytics can play a critical role in achieving sustainable development goals (SDGs) by providing insights into complex systems, identifying trends and patterns, and developing models to predict future outcomes. Data Science and analytics provide powerful tools that can support sustainable development by providing insights, enabling evidence-based decision making, identifying inequalities, predicting the impacts of climate change, and tracking financial flows. By using Data Science to inform policy and decision making, we can achieve SDGs and create a more sustainable future for all.

1. **Data Science can help monitor and measure progress towards SDGs**. By collecting, analyzing, and interpreting data from various sources, we can assess progress towards the SDGs and identify areas that require attention.
2. **Data Science can enable the tracking of financial flows to support sustainable development. By using data analytics, we can monitor the flow of resources and identify areas where investments are needed to achieve SDGs**. By using data-driven approaches, we can identify the most effective policies and interventions for achieving SDGs and evaluate their impact over time.
3. **Data Science can help identify and address inequality and exclusion**. By analyzing data on poverty, health, education, and other social indicators, we can identify vulnerable groups and develop targeted interventions to reduce inequalities.
4. **Data Science can aid in understanding complex environmental systems and predicting the impacts of climate change**. By analyzing environmental data, we can develop models to predict future outcomes and develop strategies to mitigate the effects of climate change.
5. **Data Science can enable the tracking of financial flows to support sustainable development**. By using data analytics, we can monitor the flow of resources and identify areas where investments are needed to achieve SDGs.

Next, we introduce how various Data Science tasks can be performed practically using Python, a popular scripting language.

13.2 Hands-on Data Science with Python

Demonstrating the various steps involved in Data Science using Python[1] is helpful for several reasons. First, Python is a popular and widely used programming language in the Data Science community, and it offers a rich set of libraries and tools that are specifically designed for data analysis and manipulation. Therefore showcasing Data Science tech-

[1] https://www.python.org/.

niques using Python can help learners and practitioners become more familiar with these powerful tools and improve their data-analysis skills. Python is easy to learn and understand, even for individuals without a strong background in programming. This accessibility makes it an ideal language for demonstrating Data Science steps to a broad audience, including students, researchers, and industry professionals. Furthermore, Python offers a wide range of visualization and data-presentation tools, which can be used to create meaningful and impactful representations of complex datasets. By demonstrating Data Science techniques using Python, learners and practitioners can develop skills in data visualization and communication, which are essential for effectively disseminating insights to stakeholders and decision makers.

Demonstrating Data Science techniques using Python can help learners and practitioners stay up-to-date with the latest advancements in the field. Python is a dynamic language that is constantly evolving, and new libraries and tools are regularly developed to improve data-analysis capabilities. By showcasing Data Science steps using Python, learners and practitioners can stay informed about the latest developments and trends in the field and retool their skills and knowledge accordingly.

In this section, we demonstrate a step-by-step guide through a small end-to-end Data Science pipeline for a sample project using Python. We assume that the reader has a basic idea of coding in a Python environment. First, it is necessary to check the Python version that is installed in our system.

```
[1]: !python -version
```

```
Python 3.9.12
```

13.2.1 Necessary Python libraries

We use a set of libraries (already installed) for our Data Science project. First, we import these libraries with their required functions for our downstream activity.

```
[2]: import warnings      # suppresses the library warnings
     warnings.filterwarnings("ignore")

     import numpy as np
     import pandas as pd
     import matplotlib.pyplot as plt
     import seaborn as sns

     from sklearn.preprocessing import LabelEncoder
     from sklearn.feature_selection import SelectKBest
     from sklearn.feature_selection import chi2
     from sklearn.model_selection import train_test_split
     from sklearn.model_selection import cross_val_score, StratifiedKFold
```

```
from sklearn.linear_model import LogisticRegression
from sklearn import svm
from sklearn.metrics import accuracy_score, confusion_matrix, auc, roc_curve
from sklearn.cluster import KMeans, AgglomerativeClustering
import scipy.cluster.hierarchy as shc

import tensorflow as tf
import keras
from scikeras.wrappers import KerasClassifier
```

13.2.2 Loading the dataset

We load the dataset using the *Pandas* dataframe. The dataset of interest can be downloaded from the UCI Machine Learning Repository,[2] which contains information on 303 patients with 55 features. Out of 303 instances, the dataset contains the status of 216 patients with coronary artery disease (CAD) and 88 patients with normal status. We further use the Pandas library functions to explore the dataset and understand it indepth.

[3]:
```
data = pd.read_excel(open('Z-Alizadeh sani dataset.xlsx', 'rb'),
    ↪sheet_name='Sheet 1 - Table 1')
```

13.2.3 A quick look at the dataset

[4]: `data.head() # displays the first 5 rows of the dataset`

[4]:

	Age	Weight	Length	Sex	BMI	DM	HTN	Current Smoker	EX-Smoker	\
0	53	90	175	Male	29.387755	0	1	1	0	
1	67	70	157	Fmale	28.398718	0	1	0	0	
2	54	54	164	Male	20.077335	0	0	1	0	
3	66	67	158	Fmale	26.838648	0	1	0	0	
4	50	87	153	Fmale	37.165193	0	1	0	0	

	FH	...	K	Na	WBC	Lymph	Neut	PLT	EF-TTE	Region RWMA	VHD	\
0	0	...	4.7	141	5700	39	52	261	50	0	N	
1	0	...	4.7	156	7700	38	55	165	40	4	N	
2	0	...	4.7	139	7400	38	60	230	40	2	mild	
3	0	...	4.4	142	13000	18	72	742	55	0	Severe	
4	0	...	4.0	140	9200	55	39	274	50	0	Severe	

[2] https://archive.ics.uci.edu/dataset/412/z+alizadeh+sani.

```
        Cath
0        Cad
1        Cad
2        Cad
3     Normal
4     Normal


[5 rows x 56 columns]
```

13.2.4 Checking dataset header

```
[5]: data.columns # displays the column label of the dataset
```

```
[5]: Index(['Age', 'Weight', 'Length', 'Sex', 'BMI', 'DM', 'HTN', 'Current Smoker',
           'EX-Smoker', 'FH', 'Obesity', 'CRF', 'CVA', 'Airway disease',
           'Thyroid Disease', 'CHF', 'DLP', 'BP', 'PR', 'Edema',
           'Weak Peripheral Pulse', 'Lung rales', 'Systolic Murmur',
           'Diastolic Murmur', 'Typical Chest Pain', 'Dyspnea', 'Function Class',
           'Atypical', 'Nonanginal', 'Exertional CP', 'LowTH Ang', 'Q Wave',
           'St Elevation', 'St Depression', 'Tinversion', 'LVH',
           'Poor R Progression', 'BBB', 'FBS', 'CR', 'TG', 'LDL', 'HDL', 'BUN',
           'ESR', 'HB', 'K', 'Na', 'WBC', 'Lymph', 'Neut', 'PLT', 'EF-TTE',
           'Region RWMA', 'VHD', 'Cath'],
          dtype='object')
```

13.2.5 Dimensions of the dataset

```
[6]: data.shape # shows the number of rows and columns in the dataset
```

```
[6]: (303, 56)
```

13.3 Dataset preprocessing

Next, we look into the dataset, to find if there exists any nonnumeric columns. We ignore such feature columns of the dataset.

13.3.1 Detecting nonnumeric columns

```
[7]: cols = data.columns
     num_cols = data._get_numeric_data().columns
     categorical_cols = list(set(cols) - set(num_cols))
     print("Categorical Columns : ", categorical_cols)
```

Categorical Columns : ['Poor R Progression', 'Dyspnea', 'Sex', 'LVH', 'BBB',
'Atypical', 'Weak Peripheral Pulse', 'CHF', 'Thyroid Disease', 'Airway disease',
'Obesity', 'Cath', 'CVA', 'LowTH Ang', 'Diastolic Murmur', 'Exertional CP',
'Systolic Murmur', 'Lung rales', 'VHD', 'Nonanginal', 'CRF', 'DLP']

13.3.2 Encoding nonnumeric columns

```
[8]: label_encoder = LabelEncoder()

     for i in categorical_cols:
         data[i] = label_encoder.fit_transform(data[i])
     print("Encoded data")
     data.head()
```

Encoded data

[8]:

	Age	Weight	Length	Sex	BMI	DM	HTN	Current Smoker	EX-Smoker	\
0	53	90	175	1	29.387755	0	1	1	0	
1	67	70	157	0	28.398718	0	1	0	0	
2	54	54	164	1	20.077335	0	0	1	0	
3	66	67	158	0	26.838648	0	1	0	0	
4	50	87	153	0	37.165193	0	1	0	0	

	FH	...	K	Na	WBC	Lymph	Neut	PLT	EF-TTE	Region RWMA	VHD	Cath
0	0	...	4.7	141	5700	39	52	261	50	0	1	0
1	0	...	4.7	156	7700	38	55	165	40	4	1	0
2	0	...	4.7	139	7400	38	60	230	40	2	3	0
3	0	...	4.4	142	13000	18	72	742	55	0	2	1
4	0	...	4.0	140	9200	55	39	274	50	0	2	1

[5 rows x 56 columns]

13.3.3 Detecting missing values

```
[9]: any(X.isnull().sum())
```

```
[9]: False
```

13.3.4 Checking the class distribution

```
[10]: print(data.groupby('Cath').size()) # class distribution
```

```
Cath
0     216
1     87
dtype: int64
```

13.3.5 Separating independent and dependent variables

```
[11]: X = data.iloc[:,0:54] # Input feature variables
      y = data.iloc[:,-1]   # Class label
```

13.4 Feature selection and normalization

For predictive modeling, we perform the feature selection on input variables in order to use the most important (selective) variables for our task. We select the top-40 features for our task.

```
[12]: # Selecting the top-k features
      BestFeatures = SelectKBest(score_func=chi2, k=40)
      fit = BestFeatures.fit(X,y)
```

```
[13]: df_scores = pd.DataFrame(fit.scores_)
      df_columns = pd.DataFrame(X.columns)
```

```
[14]: f_Scores = pd.concat([df_columns,df_scores],axis=1)    # feature scores
      f_Scores.columns = ['Specification','Score']
```

```
[15]: top_features = f_Scores.nlargest(40,'Score')
```

```
[16]:   feat = {}
        for i, (v1,v2) in enumerate(zip(top_features['Specification'],
        ↪top_features['Score'])):
            feat[v1] = v2

        print(feat)
```

{'WBC': 1167.316183337504, 'TG': 381.0171907573125, 'FBS': 290.38280449892636,
'ESR': 125.49382868553845, 'Age': 70.67536529000986, 'Region RWMA':
62.344116837857655, 'BP': 47.262173693860326, 'PLT': 45.376214084848584,
'Typical Chest Pain': 40.97898560882161, 'Atypical': 36.32810118238372, 'EF-
TTE': 27.90445792173911, 'Nonanginal': 21.575700431034484, 'Lymph':
14.93885035810506, 'DM': 13.622818220519367, 'Tinversion': 11.957221157939546,
'HTN': 10.267987863610093, 'PR': 9.048198654830475, 'Neut': 8.015657437627791,
'BUN': 6.6005254923295755, 'Q Wave': 6.444444444444445, 'Diastolic Murmur':
6.3338122605364, 'St Elevation': 5.638888888888889, 'St Depression':
4.8392490961092225, 'Function Class': 4.570597682087642, 'Poor R Progression':
3.625, 'Dyspnea': 2.649543232115285, 'Weight': 2.625385886318694, 'CRF':
2.416666666666666, 'Airway disease': 2.0691396725879483, 'Weak Peripheral
Pulse': 2.0138888888888893, 'LDL': 2.0028214668383395, 'HDL': 1.517846316747344,
'Current Smoker': 1.2966687952320137, 'BMI': 1.1383731806674868, 'Edema':
0.8507343550446993, 'LowTH Ang': 0.8055555555555556, 'LVH': 0.74176245210728,
'Thyroid Disease': 0.6841817186644772, 'Lung rales': 0.5960031347962381, 'Sex':
0.5708050330895151}

```
[17]:   X = X[top_features['Specification']] # putting top-40 features in X
```

13.4.1 Correlation among features

Additionally, we check the correlation of the selected features among themselves. The diagonal column shows very high correlation of a variable with itself.

```
[18]:   # Checking the pairwise correlation of the top features columns with y-label

        df = pd.concat([X, y], axis=1)

        correlation = df.drop(['Cath'], axis=1).corr()
        ax = sns.heatmap(correlation, annot=False, annot_kws={'size': 25})
        ax.figure.axes[-1].yaxis.label.set_size(18)
```

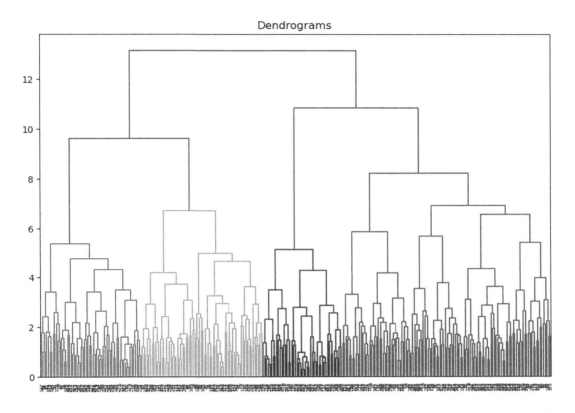

Dendrograms

13.4.2 Normalizing the columns

```
[19]: for column in X.columns:
          X[column] = (X[column] - X[column].min()) / (X[column].max() - X[column].
      ↪min())
```

13.4.3 Viewing normalized columns

```
[20]: X.head() # Top five rows after normalization
```

[20]:

	WBC	TG	FBS	ESR	Age	Region	RWMA	BP \
0	0.139860	0.210267	0.082840	0.067416	0.410714		0.0	0.2
1	0.279720	0.268509	0.053254	0.280899	0.660714		1.0	0.5
2	0.258741	0.065153	0.068047	0.101124	0.428571		0.5	0.1
3	0.650350	0.025666	0.047337	0.842697	0.642857		0.0	0.1
4	0.384615	0.131293	0.124260	0.292135	0.357143		0.0	0.2

	PLT	Typical Chest Pain	Atypical	...	LDL	HDL \

```
0  0.329149              0.0    0.0  ...  0.640187  0.148265
1  0.195258              1.0    0.0  ...  0.481308  0.211356
2  0.285914              1.0    0.0  ...  0.242991  0.305994
3  1.000000              0.0    0.0  ...  0.172897  0.116719
4  0.347280              0.0    0.0  ...  0.429907  0.358570
```

	Current Smoker	BMI	Edema	LowTH	Ang	LVH	Thyroid Disease \
0	1.0	0.494721	0.0	0.0	0.0		0.0
1	0.0	0.451314	1.0	0.0	0.0		0.0
2	1.0	0.086105	0.0	0.0	0.0		0.0
3	0.0	0.382846	0.0	0.0	0.0		0.0
4	0.0	0.836058	0.0	0.0	0.0		0.0

	Lung rales	Sex
0	0.0	1.0
1	0.0	0.0
2	0.0	1.0
3	0.0	0.0
4	0.0	0.0

```
[5 rows x 40 columns]
```

Following the required transformation of the dataset for the downstream activity, we analyze the dataset before applying a few supervised and unsupervised modeling techniques. We employ a few classification and clustering algorithms to analyze predictive outcomes and observe various patterns that reside within the input data. We perform the classification task with some popular algorithms.

13.5 Classification

Classification is a supervised activity. We split the transformed dataset into requisite subsets for training and evaluation.

13.5.1 Splitting the dataset

```
[21]: # Dataset splitting into 80% train and 20% test set

X_train, X_test, y_train, y_test = train_test_split(X, y, test_size=0.
 ↪2,random_state=1)
print(X_train.shape, y_train.shape)
print(X_test.shape, y_test.shape)
```

```
(242, 40) (242,)
(61, 40) (61,)
```

13.5.2 Logistic regression

[22]:
```python
logreg = LogisticRegression(solver='liblinear')

# Using 10-fold cross-validation
kfold = StratifiedKFold(n_splits=10)
logreg_cv_results = cross_val_score(logreg, X_train, y_train, cv=kfold,
 ↪scoring='accuracy')

print("Logistic Regression accuracy: ", logreg_cv_results.mean())
```

```
Logistic Regression accuracy:  0.8513333333333334
```

13.5.3 Support-vector machine

[23]:
```python
svm_rbf = svm.SVC(kernel='rbf')


# Using 10-fold cross-validation
kfold = StratifiedKFold(n_splits=10)
svm_cv_results = cross_val_score(svm_rbf, X_train, y_train, cv=kfold,
 ↪scoring='accuracy')

print("SVM accuracy: ", svm_cv_results.mean())
```

```
SVM accuracy:  0.8758333333333332
```

13.5.4 Artificial neural network (ANN)

```
[24]:
def model_creation():
    model = keras.Sequential([
    keras.layers.Flatten(input_shape=(40,)),
    keras.layers.Dense(32, activation=tf.nn.relu),
    keras.layers.Dense(16, activation=tf.nn.relu),
    keras.layers.Dense(1, activation=tf.nn.sigmoid),
    ])

    model.compile(optimizer='adam',
            loss='binary_crossentropy',
            metrics=['accuracy'])

    return model

model = KerasClassifier(model=model_creation, epochs=15, batch_size=5, verbose=0)

# Using 10-fold cross-validation
kfold = StratifiedKFold(n_splits=10)
ann_cv_results = cross_val_score(model, X_train, y_train, cv=kfold)
print("ANN accuracy: ", ann_cv_results.mean())
```

```
ANN accuracy:  0.8846666666666666
```

13.5.5 Predictions on test data with a high-performing ANN model

```
[25]:
# Fitting X_train and y_train for prediction on the test data

model.fit(X_train, y_train)
y_pred = model.predict(X_test)
```

13.5.6 Evaluating the model

```
[26]:
# Test accuracy

print("Test Accuracy for ANN : ", accuracy_score(y_test, y_pred))
```

```
Test Accuracy for ANN :   0.8524590163934426
```

13.5.7 Performance measurement

```
[27]: # Confusion Matrix

      sns.heatmap(confusion_matrix(y_test, y_pred), annot=True, fmt='d')
```

```
[27]: <AxesSubplot:>
```

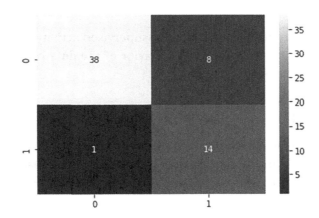

13.5.8 Curve plotting

```
[28]: # ROC Curve

      fpr, tpr, thresholds = roc_curve(y_test, y_pred)

      plt.plot(fpr, tpr, label="Test AUC ="+str(auc(fpr, tpr)))
      plt.legend()
      plt.xlabel("FPR")
      plt.ylabel("TPR")
      plt.title("AUC Score")
      plt.grid()
      plt.show()
```

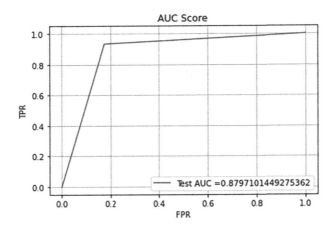

Next, we perform clustering, which is an unsupervised activity by employing a couple of algorithms to understand the grouped behavior of the data points that share similar properties or similar behavior.

13.6 Clustering

First, we use the K-means algorithm. As a preliminary activity for K-means, we use the elbow method to search for optimal clusters given the input dataset. More precisely, we compute inertia, which is the sum of squared distances of the samples nearest to the cluster center divided by the count of clusters (the fewer the better).

13.6.1 K-means—using the elbow method

```
[29]: inertia = []
      for i in range(1, 11):
          kmeans = KMeans(n_clusters = i, init = 'k-means++', max_iter = 300, n_init =
          ↪10, random_state = 16)
          kmeans.fit(X)
          inertia.append(kmeans.inertia_)
      plt.plot(range(1, 11), inertia)
      plt.title('The Elbow Method')
      plt.xlabel('Number of clusters')
      plt.ylabel('Inertia')
      plt.show()
```

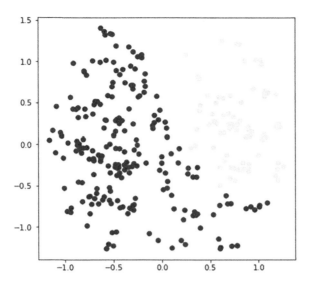

13.6.2 Fitting the data

```
[30]: # Using Cluster Count = 2

      kmeans = KMeans(n_clusters=2,random_state=16)
      kmeans.fit(X)
```

```
[30]: KMeans(n_clusters=2, random_state=16)
```

13.6.3 Validation with labels

```
[31]: # Finding how many samples were correctly sampled

      labels = kmeans.labels_
      correct_labels = sum(y == labels)

      print("Correctly sampled %d out of %d." % (correct_labels, y.size))
      print('Accuracy score: {0:0.2f}'. format(correct_labels/float(y.size)))
```

```
Correctly sampled 70 out of 303.
Accuracy score: 0.23
```

13.6.4 Agglomerative clustering—using dendrograms to find optimal clusters

```
[32]: plt.figure(figsize=(10, 7))
      plt.title("Dendrograms")
      dend = shc.dendrogram(shc.linkage(X, method='ward'))
```

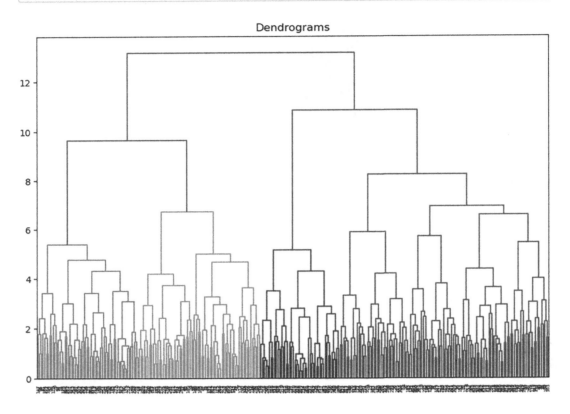

13.6.5 Finding optimal clusters

```
[33]:

      unique_colors = set(dend['color_list'])

      optimal_clusters = len(unique_colors) - 1
      print(optimal_clusters)
```

4

13.6.6 Fitting the data

```
[34]:  # Agglomerative Clustering

       ag = AgglomerativeClustering(n_clusters=optimal_clusters, affinity='euclidean',␣
       ↪linkage='ward')
       ag.fit_predict(X)
```

```
[34]:  array([2, 1, 3, 0, 0, 3, 3, 1, 0, 0, 3, 1, 1, 0, 0, 0, 3, 2, 1, 1, 2, 1,
              1, 0, 1, 3, 0, 2, 1, 1, 3, 0, 1, 1, 0, 0, 0, 0, 0, 1, 0, 2, 0, 0,
              0, 3, 0, 1, 0, 0, 0, 1, 1, 3, 0, 1, 0, 1, 2, 1, 2, 3, 2, 0, 0, 1,
              1, 3, 0, 1, 0, 3, 0, 0, 1, 1, 0, 0, 3, 0, 1, 3, 0, 0, 1, 3, 0, 1,
              1, 1, 3, 2, 0, 3, 3, 3, 3, 0, 3, 3, 0, 3, 2, 1, 3, 1, 0, 0, 0, 1,
              2, 0, 1, 0, 1, 1, 0, 1, 1, 3, 2, 0, 0, 2, 0, 3, 3, 3, 2, 0, 2, 1,
              3, 3, 1, 3, 0, 2, 0, 0, 1, 1, 0, 0, 3, 0, 1, 3, 2, 1, 3, 0, 0, 0,
              3, 3, 0, 0, 0, 3, 0, 1, 3, 0, 0, 0, 0, 1, 2, 0, 2, 1, 1, 1, 1, 3,
              2, 0, 3, 0, 0, 0, 0, 0, 3, 0, 0, 0, 2, 3, 2, 1, 2, 0, 2, 1, 0, 0,
              0, 2, 0, 0, 3, 0, 3, 0, 1, 0, 0, 1, 2, 0, 0, 3, 3, 1, 0, 0, 2, 2,
              0, 0, 0, 2, 0, 0, 1, 3, 1, 1, 1, 1, 0, 0, 2, 3, 0, 3, 2, 0, 1, 2,
              3, 1, 0, 2, 1, 0, 1, 3, 1, 2, 3, 3, 3, 0, 2, 2, 0, 3, 0, 2, 3, 0,
              0, 0, 0, 3, 2, 1, 2, 3, 0, 0, 1, 1, 2, 1, 1, 1, 2, 0, 3, 0, 0, 2,
              0, 2, 0, 0, 2, 2, 0, 2, 0, 0, 0, 2, 3, 0, 0, 0, 1])
```

13.6.7 Validation with labels

```
[35]:  labels = ag.labels_
       correct_labels = sum(y == labels)

       print("Correctly sampled %d out of %d." % (correct_labels, y.size))
       print('Accuracy score: {0:0.2f}'. format(correct_labels/float(y.size)))
```

```
Correctly sampled 76 out of 303.
Accuracy score: 0.25
```

13.7 Summary

The chapter focuses on the practice of Data Science. It also emphasizes hands-on learning. The chapter provides step-by-step tutorials that guide readers through the process of implementing Data Science techniques using Python. These tutorials cover a wide range of topics, including data cleaning and preprocessing, feature engineering, model selection and evaluation, and data visualization using Python, covering popular libraries such as NumPy, Pandas, Scikit-learn, and Matplotlib. The chapter provides practical examples of how to preprocess data, perform feature selection, and train and evaluate machine-learning models.

References

[1] Jayanta Kumar Das, Swarup Roy, Comparative analysis of human coronaviruses focusing on nucleotide variability and synonymous codon usage patterns, Genomics 113 (4) (2021) 2177–2188.
[2] Jayanta Kumar Das, Swarup Roy, A study on non-synonymous mutational patterns in structural proteins of SARS-CoV-2, Genome 64 (7) (2021) 665–678.
[3] Jayanta Kumar Das, Antara Sengupta, Pabitra Pal Choudhury, Swarup Roy, Characterizing genomic variants and mutations in SARS-CoV-2 proteins from Indian isolates, Gene Reports 25 (2021) 101044.
[4] Jayanta Kumar Das, Giuseppe Tradigo, Pierangelo Veltri, Pietro H. Guzzi, Swarup Roy, Data science in unveiling COVID-19 pathogenesis and diagnosis: evolutionary origin to drug repurposing, Briefings in Bioinformatics 22 (2) (2021) 855–872.

14

Conclusion

Data Science is an interdisciplinary field that involves the extraction of meaningful insights from complex and structured as well as unstructured data. This Data Science book has provided a comprehensive overview of the fundamental concepts and techniques in Data Science and Analytics. This book has presented an exploration of the entire Data Science process, including data generation, preprocessing, machine learning, and Big Data analysis. Topics covered include regression, artificial neural networks, feature selection, clustering, and association mining, which are fundamental topics in Data Science.

This book has emphasized that understanding the data-generation process and data sources is crucial to the success of any Data Science endeavor. Preprocessing is an essential step that involves cleaning, transforming, and normalizing the data to ensure its quality and suitability for downstream analysis. Machine-learning algorithms such as regression, decision trees, random forests, support-vector machines, and artificial neural networks help predict outcomes for previously unseen data. Feature selection is necessary to identify the most important variables that affect the outcome of a particular model so that predictions are accurate and efficient. Clustering is another important Data Science technique that helps group similar data points providing a means for early exploration of unlabeled data. Association mining is a method used to identify patterns in large datasets, making it possible to discover cooccurrence relationships among components of data generated by big corporations. The book also has introduced Big Data analysis, which involves processing and analyzing very large and complex datasets. Lastly, a chapter has been devoted to discussing practical applications of Data Science using Python, a popular programming language, to perform hands-on exercises. Various Python libraries, such as NumPy, Pandas, matplotlib, and Scikit-learn, commonly used in Data Science, have been introduced. By understanding the concepts and techniques discussed in this book, readers can confidently work on Data Science projects and make informed decisions based on data. The authors hope readers find this book informative and useful in their Data Science journey.

Data Science is constantly evolving, and most areas still require further research. In particular, since the volume of data continues to increase, there is a need for more efficient algorithms and techniques for data processing and analysis. Additionally, as new technologies, such as AI and machine learning emerge, data scientists will have opportunities to develop innovative solutions for even more complex problems. Moving forward, future research in Data Science should focus on exploring these emerging technologies and developing new data-analysis and -modeling techniques. There is also a pressing need to address ethical concerns related to data privacy and security and develop methods for dealing with biased data. Overall, the field of Data Science is full of exciting opportunities

Fundamentals of Data Science. https://doi.org/10.1016/B978-0-32-391778-0.00021-1

295

and challenges, and we hope that this book has provided readers with a solid foundation for further exploration and research.

It is essential to ensure that the next generation of professionals is adequately prepared to meet the challenges. Thus there is a need to update the curricula of educational institutions across the board to include the basics of AI/ML-based Data Analytics. By thoughtfully incorporating Data Science into the curricula, we believe every student can acquire the necessary knowledge and skills to work with data effectively, regardless of their pursued major. This will enable the student to leverage the power of Data Science in their respective fields and make informed decisions based on data-driven insights, enhancing their disciplinary as well as interdisciplinary intellectual growth and problem-solving skills. Furthermore, including Data Science in the curricula will help bridge the gap between academia and industry, as students will be better prepared to meet the demands of the job market in the current and future technology-mediated society, which is increasingly focused on data-driven decision making. As the demand for Data Science and Analytics professionals continues to grow, it is imperative to equip every student with the necessary skills and knowledge to succeed in this field. Data Science training can be incredibly valuable for students looking to become self-sustained, start their own businesses, and become entrepreneurs. By providing students with the skills and knowledge to collect, analyze, and interpret data, Data Science training can help students develop innovative business ideas, discover current market gaps, and create sustainable businesses that benefit them and at the same time positively impact society.

Index

A

Actionable
  information, 10
  knowledge, 3, 9, 262, 263
Activation functions, 138
Active edges, 196
Agglomerative, 194
  clustering, 194, 292
Analysis of Variance (ANOVA), 67
Apriori algorithm, 229, 239–241, 254
Artificial
  datasets, 209
  examples, 58
  neurons, 121, 188
Artificial Intelligence (AI), 47
Artificial Neural Network (ANN), 12, 87, 88, 93,
    121, 122, 124, 130, 133, 134, 139, 141,
    145, 159, 161, 162, 288, 295
  classifier, 137
Artificially generated datasets, 24
Association
  analysis, 233
  mining, 245
Association-rule mining process, 237
Asymmetric binary attribute, 16
Attribute
  binary, 16
  categorical, 16
  categorization, 20
  discarding, 33
  for clustering, 183
  name, 17
  occurrence counts, 118
  selection, 182
  sets, 248
  summaries, 196
  values, 16, 18, 35, 41, 96, 118

Autoencoders, 152
Axons, 122

B

Backpropagation, 127, 128
Backward heuristic approach, 165
Balanced Iterative Reducing and Clustering
    using Hierarchies (BIRCH), 195
Banking customer data, 28
Base classifier, 220, 224, 226, 229
  selection requirements, 224
Base learners, 217, 218, 220, 225, 226
  in classification, 220
  pseudotag, 218
Benchmarking data, 29
Biclustering, 204, 206
  algorithm, 206
  techniques, 207
Biclusters, 206–208
  quality, 208
  types, 206
Bidirectional approach, 165
Big Data
  analysis problems, 262
  analytics, 269
    architecture, 266
    platforms, 264
    techniques, 263
  characteristics, 260
Binary
  attribute, 16
  classification, 49, 63, 64, 111, 116, 126
    problem, 125, 126, 135
    tasks, 125
  classifier, 61, 64
  itemsets, 252
  linear classification, 108
  step function, 139

Bioinformatics tools, 269
Biological data cleaning, 33
Boolean attributes, 246
Bootstrapped dataset, 226
Bulk data generation, 31

C
Candidate
    feature set, 172
    generation, 240
    itemsets, 239, 241, 243
Categorical
    attribute, 16, 100
    features, 167, 231
CAtegorical data ClusTering Using Summaries
        (CACTUS), 196
Class
    boundary, 229
    distribution, 65, 98, 283
    examples, 62
    imbalances, 65
    label, 39, 49, 50, 53, 92, 93, 98, 170, 220–222,
        231
    negative, 61, 62
    probabilities, 220
Classification, 8, 181
    accuracy, 162, 163, 170
    algorithm, 12, 54, 55, 65, 93, 117
    binary, 49, 63, 64, 111, 116, 126
    CNN, 147
    efficiency, 163
    experiments, 65
    high-dimensional data, 172
    issues, 203
    model, 12, 51, 117
    neural network, 137
    problems, 61, 115, 116, 150
    problems linear, 116
    rate, 170
    results, 170
    task, 49, 55, 66, 161
    true positive, 62
Classifier
    binary, 61, 64
    ensemble, 216, 229

    linear, 108–110, 116
    models, 117, 225
    performance, 224
    random, 224
Clinical
    data, 25
    dataset, 24
Cluster
    analysis, 181
    data, 209
    database, 196
    quality, 212
Cluster-validity measures, 208
Clustering, 8, 93
    algorithms, 54, 55, 188, 189, 192, 201–203,
        208, 212, 220, 224, 227
    algorithms classification, 189
    approaches, 190, 194, 197, 208, 246
    ensemble, 220, 228
    functions quality, 182
    groups, 227
    methods, 195, 203
    models, 218, 219, 228, 231
    performance, 221
    problem, 182
    process, 196
    quality, 192, 204
    result, 191, 223, 224
    solutions, 210, 220, 223, 224
    strategy, 205
    techniques, 188, 190, 205, 209, 212
    unsupervised, 203
Clustering feature (CF), 195
Clustering Large Applications based on
        RANdomized Search (CLARANS), 193
Clustering LARge Applications (CLARA), 192
Clustering-feature tree, 195
Coclustering, 206
Combination learners, 220
Computer-vision tools, 271
Concept learning, 1
Consensus
    clustering, 220, 223, 224, 228
    clustering approach, 219
    voting decision, 221

Constant biclusters, 206
  quality, 206
Continuous data, 38
Convolutional layer, 145, 146, 148, 150, 152
Convolutional Neural Network (CNN), 12, 121,
        145, 146, 150, 151, 159, 161, 162, 217
Core
  distance, 202
  neighborhood, 202
  object, 197
Coregulated biclustering (CoBi), 208
Coronary artery disease (CAD), 280
Correlation mining, 250
Cosine similarity, 185
Count Distribution (CD), 254
Covertype dataset, 107
Crossentropy loss, 136
  function, 135
Crossvalidation, 57, 65, 83, 86
Curse-of-dimensionality, 202
Curve plotting, 289

D
Data
  acquisition, 23, 263
  analysis, 4, 6, 7, 22, 25, 161, 260
  analytics, 11, 23, 29, 278, 296
  attributes, 15, 20, 24
  augmentation, 24
  cleaning, 31, 32
  clusters, 209
  collection, 4, 38
  collection quality, 4
  consolidation, 44
  dependencies, 266, 274
  dictionary definitions, 2
  discovery, 9
  discrete, 38
  duplication, 32
  elements, 9, 15
  exploration, 3
  fake, 29
  features, 217
  field, 19
  flow, 269

generation, 18, 24, 25, 27, 29, 295
generation pipeline, 25
generator, 25, 27
heterogeneity, 23, 274
inconsistencies, 31
informative, 229
instances, 174, 181, 185–187, 192
integration, 31, 43, 44
items, 261
level, 43
massive amounts, 31, 259, 274
mining, 4, 9, 263
model, 19, 44
normalization, 12, 41
parallelism, 266
patterns, 5, 181
points, 70, 216, 295
preparation, 5, 12, 218
preprocessing, 31, 32
privacy, 25, 295
processing, 23, 262, 265, 295
quality, 29, 31, 32, 262
random, 29
randomization, 24
raw, 2, 3, 5, 6, 9, 21
reduction, 31, 36
redundancy, 35
regression, 80
repository, 18, 31
retrieval, 18
samples, 8, 24, 25, 27
scientist, 4, 5, 7, 27, 28, 173
segmentation, 181
selection, 44
sets, 4, 9, 37, 177
sheet, 21
sizes, 22
skewness, 43
sources, 2, 21, 23, 24, 31, 44, 295
storage, 6, 19, 28, 44, 262, 263
store, 44
structures, 5, 19, 43
transformation, 31, 37
transportation, 32
types, 15, 23, 166

unlabeled, 218, 229, 295
unseen, 58, 60, 61, 105, 215, 216, 219
visualization, 264
warehouse, 36, 260
warehousing, 273
Data Distribution (DD), 5, 7, 25, 27, 255, 266
Data Science, 1–4, 6, 7, 10, 28, 29, 47, 161, 277, 278, 295, 296
    applications, 10
    objectives, 9
    tasks, 7
    techniques, 279
Data-driven discovery, 4
Data-generation steps, 25
Data-storage formats, 19
Database
    clusters, 196
    designs, 2
    layout, 245
    management, 263
    queries, 1
    scanning, 241
    scans, 241–243
    systems, 4
    technologies, 6
Database-management system (DBMS), 19
Dataset
    examples, 91
    for supervised learning, 69
    header, 281
    label, 138
    loading, 280
    preprocessing, 281
    size, 174
Decimal-scaling normalization, 42
Decision trees, 97–99
Deep
    convolutional neural networks, 159
    learning, 87, 145, 157, 159, 231
    neural networks, 27, 141, 144, 152
Dendrites, 121
Dendrogram, 193
Density-based clustering, 197

Density-Based Spatial Clustering of Applications with Noise (DBSCAN), 199, 201
    algorithm, 199, 201
    approach, 201
Dependent metrics, 166
Descriptive
    analysis, 8
    features, 98, 100
Deterministic hot-deck prediction, 34
Diagnostic analysis, 9
Dimension subsets, 203
Dimensionality reduction, 37, 174, 190, 203, 204, 218, 231
Discarding data, 33
Discrete
    attributes, 17
    data, 38
    value attribute, 17
Discretization
    supervised, 39
    unsupervised, 39
Disjoint
    clusters, 210
    subsets, 102
Disjunctive Normal Form (DNF), 205
Dissimilarity, 93, 101, 186
Divergence measures, 186
Dot-Product similarity, 185
Dunn Index (DI), 210
Duplicate data
    detection, 35
Duplicated data, 37

E
Elastic Net Regression, 85
Elliptical clusters, 203
Embedded approach, 171
Encoder–decoder architectures, 150
EnDBSCAN, 201, 202
Ensemble
    approach, 173, 225, 226
    classifier, 216, 229
    clustering, 220
    feature selection, 166, 173

for supervised learning, 231
  generation, 218
  learners, 231
  learning, 12, 215–219, 224, 229, 231
  learning framework, 218
  methods, 220, 225, 227
Epoch
  learning, 134
  training, 134
Euclidean distance, 185
Evaluating linear regression, 74
Examples
  artificial, 58
  class, 62, 101
  dataset, 91
  negative, 61, 62
  unlabeled, 67
  unseen, 57, 92, 93, 104, 105
Exclusive clustering, 182, 190
  outcomes, 183
  techniques, 188
Exhaustive
  biclustering, 208
  search, 164
Expectation Maximization (EM), 203
  algorithm, 33
Exploding data, 273
Exploratory data analysis, 8, 52
Explosive data growth, 31
Extended MNIST (EMNIST), 23
Extensive database scans, 250
External evaluation, 208, 209
Extract, Load, Transform (ELT), 43
Extract, Transform, Load (ETL), 43
  method, 36

F
F-measure, 64
F-statistic, 75
F1-score, 64–66, 69
Fake data, 29
False negatives (FN), 65
False positives (FP), 65
Fault-tolerant graph architecture, 267

Feature
  categorical, 231
  data, 217
  extraction, 161, 162
  extractors, 161
  informative, 162
  level, 219
  numeric, 93, 165
  pairs, 170
  ranking method, 229
  reduction, 82, 174, 176
  selection, 85, 86, 161–166, 169–174, 190, 203,
    218, 229, 283, 295
    methods, 166
  set, 165, 166, 173, 174, 224
  similarity, 170
  space, 166, 224
  subset, 163–166, 170, 172, 173, 219
    evaluation, 166
    generation, 164
    quality, 174
  subspaces, 226, 228, 229
  values, 70, 91, 94, 107, 166, 168
  vectors, 167
Feedforward neural network, 126, 157, 158
Filter approach, 166
Forward heuristic approach,, 165
FP-growth, 242

G
General pipeline, 4
Generalization, 58
Generalized DBSCAN (GDBSCAN), 202
Generation methods, 26
Generative Adversarial Network (GAN), 27, 58
Generative learning, 159
Genetic Algorithm (GA), 172
Google dataset search, 23
Gradient descent, 130
Graphical Processing Units (GPU), 6
Graphics Processing Unit (GPU), 265
Greedy iterative search-based strategy, 208
Grouped data, 35
Grouping, 10
Guided learning paradigm, 48

H
Healthcare data, 28
Heterogeneous data, 2, 11, 270
    sources, 23, 44
Heuristic search, 164
Hidden
    knowledge discovery, 9
    patterns, 2, 10, 220
Hierarchical clustering, 193–195, 224
    techniques, 194, 195
High-dimensional data clustering, 202
Historical
    data, 263
    sales data, 10
Holdout dataset, 56
Homogeneous
    data sources, 23
    subsets, 103
Hot-deck and cold-deck methods, 34
Hybrid
    approach, 172
    ensemble learning, 218
Hyperparameters, 57
Hypothesis testing, 66

I
Inaccurate predictions, 215
Incorrect predictions, 59
Incremental
    classification, 117
    classifiers, 117
    decision trees, 117
    feature selection, 173
Independent metrics, 166
Inductive bias, 53, 54, 97, 107
Information-theoretic approach, 248
Informative
    data, 229
    features, 162
    features subsets, 172
    relationships, 248
Inherent
    patterns, 7, 47
    relationships, 52

Instances
    data, 174, 181, 185–187, 192
    unlabeled, 53, 229
    unseen, 61
Internal evaluation, 208, 210
Internet of Things (IoT), 11, 23, 31
    traffic data, 25
Interrelationships, 6
Intracluster similarity, 190
Intragroup similarity, 52, 182
Intrinsic clusters, 183
Intrusion detection system (IDS), 22
Iris dataset, 94, 116
Irrelevant
    attributes, 37
    features, 161
Itemset-mining algorithms, 238
Itemsets, 234, 236, 237
    binary, 252

J
Jaccard similarity, 209
Jensen–Shannon divergence, 187

K
K-means, 191, 192
    algorithm, 190, 290
k-Nearest-Neighbor (kNN) regression, 87
Kendall rank correlation coefficient, 186
Kernel similarity measures, 187
Knowledge Discovery in Databases (KDD), 4,
    15
Knowledge interpretation, 5
Knowledge-based methods, 35
Kullback–Liebler divergence, 187

L
Label
    class, 39, 49, 50, 92, 93, 170, 220–222, 231
    dataset, 138
    numeric, 165
Laplacian Kernel, 188
Las Vegas Filter (LVF), 174
Lasso regression, 82, 84–86
Last Observation Carried Forward (LOCF), 34

Latency, 44
Learned patterns, 91
Learning
    algorithm, 12, 49, 50, 55, 59, 66, 121, 170,
        174, 215, 217, 224, 225
    approaches, 48, 53, 217
    biases, 54
    deep, 87, 145, 157, 159, 231
    ensemble, 12, 215–219, 224, 229, 231
    framework, 153
    machines, 121
    model, 49, 59, 172, 215, 219, 221–223
    model construction, 5
    paradigms, 67
    perceptron, 124
    phase, 47
    predictive, 7
    procedure, 122
    process, 50, 52, 124
    rate, 125, 126, 132
    semisupervised, 12, 53, 67, 231
    supervised, 33, 48, 49, 52, 53, 67, 91–93, 144,
        152, 220
    supervision, 53
    technique, 215, 218, 224
    unsupervised, 12, 48, 52, 53, 67, 93, 144, 152,
        181, 212, 220, 231
    weights, 155
Lenses dataset, 98, 105, 106
Linear
    classifier, 108–110, 116
    kernel, 187
    least-squares regression, 70
    regression, 74–76, 81, 88
        fit, 78
        line, 75
        models, 76, 88
    separator, 108–110, 116, 117
Linearly
    related features, 86
    separable data, 125
Local
    database partition, 254
    receptive fields, 148
Localized clusters, 203

Logarithmic normalization, 43
Loose Coupling, 44
Loss, 128
    function, 82, 86, 121, 129–133, 135, 136, 152,
        155
    optimization, 130
    propagation, 128

M
Machine learning, 4, 6, 7, 12, 47, 48, 55, 58, 70,
        81, 87, 153, 161, 162, 171, 215, 216, 231,
        260, 263, 295
    supervised, 8, 49
    system, 7
    unsupervised, 67
Macroaveraging, 64
Majority voting, 96, 97, 220, 221, 225, 229
    decision, 221
Manhattan distance, 185
MapReduce architecture, 266, 267
Market-basket
    analysis, 234
    databases, 235
Massive amounts data, 31, 259, 274
Mean squared residue (MSR), 207, 208
Mean-squared
    error, 135
    loss, 135
Meaningful
    clusters, 204
    patterns, 52, 60
Measurement scale, 15
Medoids, 192, 193, 204
Memorizing model, 59
Merging Adaptive Finite Intervals And
        (MAFIA), 205
Message Passing Interface (MPI), 264
Metaclassifier, 226
Metadata, 2, 22
    standards, 23
Metaensemble learning, 215, 218
Metaheuristic-based strategies, 208
Microarray data, 269
Microaveraging, 65
Min–max normalization, 41

Minimum support, 237
Misclassification, 61
  rates, 97
Misclassified
  instances, 225
  samples, 225
Missing at Random (MAR), 33
Missing Completely at Random (MCAR), 33
Missing Not at Random (MNAR), 33
  missing data, 33
Missing values handling, 32
MNIST
  database, 23
  dataset, 23, 134
Multiclass classification, 49, 64, 65, 116, 147
Multidimensional
  data cubes, 43
  linear regression, 76
Multidisciplinary science, 6
Multilayer perceptron (MLP), 126, 127, 129, 141, 145
Multimodal data, 4
Multiple
  attributes, 182
  data sources, 44
  datasets, 23
  imputations, 34
Multisource data, 5, 43
Mutual Information (MI), 167, 169, 170, 210, 248

N
Natural language processing (NLP), 157, 272, 277
Nearest-neighbor classifiers, 93
Negative
  class, 61, 62
  examples, 61, 62
  feature set construction, 229
  patterns, 208
Negatively labeled samples, 229
Neighboring clusters, 211
Nerve impulse, 122
Nested clusters, 193
Network-security tools, 272

Neural network, 16, 27, 88, 122, 125, 127, 130, 133–136, 138, 141, 144–146, 148–156, 159, 218, 220
  classification, 137
  unsupervised, 152
  weights, 128
Nominal attribute, 16
Non-convex clusters, 197
Non-exchangeable database partitions, 254
Non-exclusive clustering, 183
Non-operational data, 2
Non-potential itemsets, 242
Non-redundant features, 162, 167
Non-relational database, 265
Normalized Mutual Information (NMI), 210
NoSQL database, 20, 263
Numeric
  features, 93, 165
  label, 165
  values, 86, 98
Numerical
  attribute, 17, 100, 106
  values, 169

O
Optimal
  clusters, 292
  subset, 203
Ordinal attributes, 16, 18
Outliers, 31, 36, 40, 182, 183, 191, 194, 197, 198, 216, 228
Overfitting, 60, 61, 215, 216
  in regression, 81
  in regression models, 88
Overlapping clusters, 183

P
Pairwise similarity, 182
Pandas data frame, 29
Parallel association mining, 254
Parallel-rule generation, 255
Parameter estimation, 33
Parametric data reduction, 36
Participating datasets, 43

Partitional clustering, 190, 192
Partitioning approach, 246
Partitioning Around Medoids (PAM), 192
Passive edges, 196
Patient data, 28
Patterns
    data, 5, 181
    negative, 208
    random, 81
Pearson's correlation coefficient, 167, 172, 186,
        250, 252
Perceptron, 123–126
    algorithm, 125, 126
    learning, 124
    model, 123
    rule, 124, 130
Perfect biclusters, 206
Performance measurement, 289
Polynomial
    kernel, 187
    regression, 78, 88
Popular data sources, 22
Prediction
    analysis, 8, 9
    capability, 57
    learning, 7
    model, 8, 10, 49, 53, 224
    performance, 51, 60
    purposes, 76
    task, 8
    techniques, 8
Predictor, 82
    variables, 82
Preprocessing data, 12
Prescriptive model, 9
Pretrained neural network, 88
Pretraining data amount, 88
Primary sources, 21
Principal component (PC), 175, 176
Principal-Component Analysis (PCA), 37, 174
Projected clustering, 203
Proteomic data, 11
Proximity measures, 52, 182–184, 212
Python libraries, 28

Q
Qualitative attributes, 15–17
Quality
    assurance, 270
    biclusters, 208
    cluster, 212
    clustering, 192, 204
    data, 29, 31, 32, 262
    data collection, 29
    function, 211
    raw data, 3
    score, 270
Quantile normalization, 42
Quantitative
    attributes, 17, 18, 38, 246–248
    databases, 246
Quantitative Association Rule (QAR), 245, 246
Query feature, 103

R
Radial Basis Function Kernel, 188
Rand Index (RI), 209
Random
    classifier, 224
    data, 29
    feature-subset generation, 165
    learners, 224, 225
    patterns, 81
Ranked features, 173, 174
Raw
    banking data, 28
    data, 2, 3, 5, 6, 9, 21
    feature set, 164
Reachability distance, 199, 201, 202
Receptive field, 145
Rectangle class, 101, 108
Rectified Linear Unit (ReLU), 141
    activation, 141
    activation function, 144
Recurrent neural networks, 145
Redundant data, 35
Regression, 70, 86
    analysis, 9, 12, 88
    coefficients, 82
    data, 80

Elastic Net Regression, 85
equation, 84–86
functions, 70, 87
hyperplane, 77
Lasso regression, 84
line, 70–72, 74, 77
linear, 74–76, 81, 88
Linear Least Squares
    Elastic Net Regression, 85
models, 12, 75, 77, 78, 80
prediction accuracy, 85
problems, 82, 88
regularization
    Lasso regression, 84
results, 88
techniques, 86
Regularization
Elastic Net Regression, 85
Relational
    data, 19, 261
    databases, 19, 202, 245, 261
Relationships informative, 248
Relative evaluation, 208, 211
Resampled instances, 228
Ridge regression, 82–86

S

Score quality, 270
Secondary
    data, 21
    data sources, 22
    sources, 21
Semistructured data, 20, 260, 262
    analysis, 20
Semisupervised
    ensemble learning, 218, 229
    learning, 12, 53, 67, 231
    linear classifier, 229
Sequential patterns, 8
Shifting patterns, 208
Shrinkage regression, 82
Sigmoid activation, 139
    function, 140
Signature features, 172
Signum function, 124

Similarity
    features, 170
    measures, 184, 187
Single-value imputation, 34
Singular Value Decomposition (SVD), 203, 204
Skewed data distributions, 37
Software tools, 28
Sparse data, 37
Spearman
    correlation coefficient, 167
    rank correlation coefficient, 186
Specialized data massive amounts, 272
Spike, 122
Standard
    error of regression, 75
    measures, 185
Standardized data, 177
Stanford Network Analysis Platform (SNAP), 22
Statistical measures, 185
Stochastic Gradient Descent (SGD), 131, 132
Stream data, 269
Streaming-graph architecture, 269
Structured data, 19
Subclusters, 194–196
Subsets features, 164, 172, 219
Subspace clustering, 190, 196, 204–206
    algorithm, 196
    approach, 204
Supervised
    classification problems, 203
    discretization, 39
    ensemble learning, 217, 224
    feature selection, 167
    learning, 33, 48, 49, 52, 53, 67, 91–93, 144, 152, 220
    machine learning, 8, 49
Support-Vector Machine (SVM), 108, 116, 117, 287
    classifier, 110, 116
Sustainable development goal (SDG), 278
Symmetric binary attribute, 16
Synthetic
    data, 24–28
        generation, 24
        quality, 26

datasets, 24, 26, 28
Synthetic Data Vault (SDV), 29

T
t-test, 66
Tabular data, 22
Testing, 48
Tight Coupling, 43
Traffic data, 1
Trained classifier, 61–65, 110
Trained model, 8, 26, 50, 51, 53, 56–58, 62, 93
Training, 48
Transaction
    data, 2
    database, 23, 233, 234, 236, 242, 250, 254
    dataset, 253
True
    classes, 50, 60
    clustering, 209
True negatives (TN), 65
True positives (TP), 65

U
UCI Machine Learning, 56
Unbalanced datasets, 37, 58
Uncorrelated features, 174
Underfitting, 59
Unlabeled
    data, 218, 229, 295
    data instances, 52
    datasets, 52, 229
    examples, 67
    instances, 53, 229
    samples, 229
    test instances, 49
    test instances class labels, 49
    training instances, 53
Unnecessary subsets, 239
Unseen
    data, 58, 60, 61, 105, 215, 216, 219
    data samples, 57

examples, 57, 92, 93, 104, 105
    instances, 61
    test instances, 49
Unstructured
    data, 3, 20, 260–262, 295
    data formats, 20
    heterogeneous data, 273
Unsupervised
    clustering, 203
    discretization, 39
    ensemble, 218
    ensemble learning, 218, 227, 231
    learning, 12, 48, 52, 53, 67, 93, 144, 152, 181, 212, 220, 231
    learning algorithms, 93
    learning groups instances, 53
    machine learning, 67
    neural network, 152
Unwanted features, 162
Unweighted Pair Group Method with Arithmetic mean (UPGMA), 194

V
Validation dataset, 57
Variational Autoencoder (VAE), 27, 58, 121, 153, 154
Video data, 262
Visible features, 154

W
Weak
    intergroup similarity, 182
    learners, 225
Weighted
    majority voting, 221
    sum rule, 222
Wrapper approach, 170

Z
Z-score normalization, 42

Printed in the United States
by Baker & Taylor Publisher Services